The Sun and the Wind

E.B Cohen

First edition, 2026

The Sun and the Wind: How Solar and Wind Power Are Remaking the Modern World

This book was produced with the assistance of artificial intelligence. It is a work of synthesis, not of original research: everything in it has been gathered, weighed, and retold from published scholarship — the reports of the International Energy Agency (IEA), the International Renewable Energy Agency (IRENA), the Global Wind Energy Council (GWEC), and the National Renewable Energy Laboratory (NREL); peer-reviewed papers; and the standard reference works and histories of energy.

First edition, 2026

Contents

THE SUN AND THE WIND

The Sun and the Wind

How Solar and Wind Power Are Remaking the Modern World

E.B Cohen

This book was produced with the assistance of artificial intelligence. It is a work of synthesis, not of original research: everything in it has been gathered, weighed, and retold from published scholarship — the reports of the International Energy Agency (IEA), the International Renewable Energy Agency (IRENA), the Global Wind Energy Council (GWEC), and the National Renewable Energy Laboratory (NREL); peer-reviewed papers; and the standard reference works and histories of energy. No new experiments were run and no new data were collected for these pages. Where the scholarship disagrees, or where a figure is moving too fast to pin down, the disagreement or the date is reported rather than hidden. Where a claim could not be sourced to something solid, it is not printed here at all. The aim throughout has been to be honest about what we know, how well we know it, and when we last checked.

A note on units, numbers, and money. Electricity is measured along a simple ladder of power, each rung a thousand times the last: a watt is the base unit; a kilowatt is a thousand watts; a megawatt is a million; a gigawatt is a billion; a terawatt is a trillion. Power is the rate at which energy flows, and energy is power multiplied by time — so a kilowatt is a rate, while a kilowatt-hour is an amount, the energy delivered by one kilowatt running for one hour (roughly what it takes to boil a large pot of water many times over, and the unit you will find on your electricity bill). Keeping the difference between the rate and the amount clear is half the battle in understanding energy, and this book returns to it often. The reader should also know that nearly every figure for how much capacity the world has installed, and for what it costs, carries a hidden expiration date: these numbers move quarter by quarter, and the ones quoted here are dated in the text so you can judge how fresh they are. Money is given in the currency of the day and, following the convention of this book, always written out in words rather than symbols.

First edition, 2026.

For everyone who has ever stood before a blue panel on a roof, or watched a white turbine turn on a distant hill, and wondered — honestly, and without an axe to grind — how the thing actually works, and whether machines like it could really run a world.

And for the long line of tinkerers who made the wondering possible: for Edmond Becquerel, a teenager in his father's laboratory, who noticed that light could stir a current; for the three at Bell Labs — Chapin, Fuller, and Pearson — who turned a curiosity into a cell that worked; and for the Danish schoolteachers and shipwrights and farmers who taught the wind to make electricity, and then taught the rest of us how they had done it.

An Author's Note

I want to begin with a fact that took me an embarrassingly long time to appreciate, and that reorders almost everything once you let it in: the wind is sunlight, one step removed.

We tend to file the sun and the wind in separate drawers. The sun is warmth and light, a thing you feel on your face; the wind is weather, a thing that pushes at you and rattles the windows. But the wind has no engine of its own. It blows because the sun heats the round, tilted, spinning Earth unevenly — the equator more than the poles, land faster than sea, morning differently than afternoon — and air, being a fluid, does what all fluids do when you heat one part more than another: it moves. Roughly one to two percent of the sunlight that reaches us is converted, by this vast and patient unfairness of heating, into moving air. So when a turbine blade turns on a ridge, it is harvesting a portion of the sun's energy that has simply changed its clothes on the way down. The two great renewable powers of our age are not cousins. They are the same story, told at two different removes from the same star.

That is the first idea this book is built around. Here is the second, and it is the one I most want you to carry away, because it is at once the single most important fact in modern energy and the one least understood by the public: the revolution now under way was not caused by the sun becoming more powerful, or the wind blowing harder, or some breakthrough that let us suddenly catch far more of either. The sun has poured down the same torrent of energy for as long as there have been people to stand in it. What changed — and changed with a speed that caught nearly every expert flat-footed — is that catching it became cheap.

How cheap, and how fast, is genuinely hard to believe, and I spent a good while checking the numbers because they read like a misprint. The price of a solar panel, measured per unit of power it can produce, has fallen by something on the order of ninety-nine percent over the span of a working lifetime. Not ninety. Not ninety-five. Close to ninety-nine, and by some accountings a shade past ninety-nine point nine. The machines that were once so absurdly expensive that the only customer able to afford them was a national space program are now so cheap that in the sunniest parts of the world they undercut coal and gas — the incumbents, the giants, the fuels that built the modern age. This did not happen because of one heroic invention. It happened because of a slow, compounding, decades-long grind of manufacturing improvement, the kind of thing that makes no headlines on any single day and yet, added up, quietly rewires the economics of civilization. Understanding that grind — why it happened, why so few saw it coming, and whether it can continue — is, I think, the most useful thing a curious person can take from the whole subject.

So let me tell you what this book is, and what it is not.

It is, first, an explanation — from first principles, and without a single equation you are required to solve — of how light and moving air are actually turned into electricity. I promise not to wave my hands at the hard parts. When we get to the photovoltaic effect, the small quantum miracle by which a particle of light knocks an electron loose in a wafer of silicon and sends it marching around a circuit, I will slow down and build it up from things you already know. When we get to why a turbine cannot, even in principle, capture more than about fifty-nine percent of the wind that passes through it — a real and beautiful limit, proved by a German physicist a century ago — I will show you where that oddly specific number comes from. And I will hand you a few numbers you can feel in your body: what a single square meter of sunshine is really worth; why doubling the wind's speed does not double a turbine's output but multiplies it eightfold; what a battery the size of a shipping container can and cannot do for a city after dark.

It is, second, a human story, and a much longer and stranger one than the "green energy" of recent headlines suggests. The wind has been doing our work for well over a thousand years, grinding grain in Persia and draining the sea out of Holland, long before anyone dreamed of making electricity from it. The dream of running an engine on captured sunlight is older than the light bulb: a Frenchman named Augustin Mouchot built solar-powered steam engines in the eighteen-seventies, terrified that Europe would burn through its coal. There were dead ends, spectacular failures, patents that changed the world and patents that changed nothing, and a cast of tinkerers, cranks, schoolteachers, and corporate laboratories whose blunders are often as instructive as their triumphs. I have tried to give them their due, because you cannot really understand a technology until you understand the people who were wrong about it, and why. The history of the sun and the wind is littered with premature funerals and premature victory laps in roughly equal measure, and both are worth studying: the machine that failed not because the idea was bad but because the moment was wrong, and the celebrated breakthrough that turned out to be a curiosity, teach you as much as the successes do. I have come to think that the most useful habit a reader of energy history can cultivate is a certain patience with time — an awareness that the distance between a laboratory marvel and a thing that quietly powers your neighborhood is often measured not in years but in decades, and that the decades are where the real work hides.

And it is, third — this matters to me more than anything — an honest book. The subject of renewable energy has been so thoroughly colonized by cheerleaders and doom-mongers that it can be hard to find anyone simply telling you what is true. One camp will assure you that the transition is all but finished, that only wickedness or stupidity stands between us and a clean and abundant future, and that every difficulty raised is a bad-faith delaying tactic. The other will tell you that the whole enterprise is a fragile bubble of subsidy and wishful thinking, that the machines are a blight and the physics a fraud, and that any day now reality will reassert itself. Both camps are loud, both are confident, and both are wrong in ways that a little patient arithmetic can expose. I have tried to be the person who does the arithmetic. When the enthusiasts overstate their case, I say so; when the critics land a real blow, I concede it. The intermittency problem — the plain fact that the sun sets and the wind sometimes does not blow — is real, and I will not pretend it away; it is one of the hardest engineering problems of the century. But it is an engineering problem, with the shape of a problem that gets solved, and not the brick wall that some insist it is. The myths get named and corrected, from both directions: no, a solar panel does not consume more energy in its making than it will ever produce; no, wind turbines are not, whatever you may have read, a leading slayer of the world's birds; but also no, we cannot simply flip a switch and run everything on sunshine tomorrow, and anyone who tells you the transition is easy, finished, or guaranteed is selling something. Neither doom nor boosterism. Just the thing itself, examined in good light.

Two images have stayed with me through the writing, and I offer them as a taste of what is coming. The first is a small blue-gray cell of silicon, built in the late nineteen-fifties, so ruinously expensive that there was quite literally no use for it on the surface of the Earth — no roof, no factory, no farm could justify the cost. Its one and only customer was the space program, which bolted it to a satellite because a satellite, alone among all the machines humans had made, had no other way to get power and no ability to complain about the bill. That is where solar electricity began its commercial life: as a technology so dear that only the vacuum of space could afford it. The second image is a desert auction, sixty-odd years later, in which companies competed for the right to build a solar farm and the winning bid came in at roughly one cent for a kilowatt-hour of electricity — a price that, a decade before, sober professionals in the field would have told you was flatly impossible. Between those two images lies the whole of this book: the distance a technology can travel when a great many people, over a long time, quietly make it a little better and a little cheaper, again and again and again.

This is a book for the curious, not the partisan. I do not much care, and this book does not ask, which political tribe you belong to; I care whether you want to understand one of the largest industrial transformations any of us will live through, and to understand it well enough to tell the real problems from the invented ones. If that is you, then we are going to get along.

So come and stand out in it with me — face to the warmth, back to the wind — and let us find out how these two ancient powers, older than agriculture, became the cheapest new electricity in the history of the world, and what it will take to build the rest of the way on them. We begin, as everything here must, with the star at the center of it all.

Prologue — The Cheapest Electricity in History

The number arrived, as these things now do, not with any drama but as a line in a spreadsheet and a figure read aloud in a conference room. It was a spring day in 2021, and a government tender in a sun-scorched corner of the Arabian Peninsula was being decided — the right to build a very large solar farm on a stretch of Saudi desert and sell the electricity it made, under a long contract, to the national grid. The developers had submitted their sealed bids. When the lowest of them was opened, it named a price of just over one cent for a kilowatt-hour of electricity — about one and four hundredths of a cent, to be exact, a figure precise enough to feel almost impudent.

Sit with that number for a moment, because everything in this book is, in one way or another, an attempt to explain it. A kilowatt-hour is a real and useful amount of energy: enough to run a modern refrigerator for the better part of a day, to wash and dry a load of clothes, to power a laptop for a week of evenings. To make that much electricity by burning coal or natural gas costs money for the fuel alone — money that must be paid every single hour the plant runs, forever, on top of the cost of building the plant in the first place. And here was a company volunteering, in a legally binding document, to deliver that same kilowatt-hour for roughly the price of a single stick of chewing gum, and to keep doing it for a quarter of a century. There is no fuel in that price, because the sun sends no invoice. There is only the cost of the glass and silicon and steel, the land and the labor and the borrowed money, spread thin across the torrent of free light that would fall on the panels every cloudless day for decades.

A decade earlier, that bid would not have been called aggressive or clever. It would have been called impossible, and the person who submitted it would have been quietly asked whether they had made an arithmetic error. In 2010, electricity from a new solar farm cost, on the global average, something north of thirty cents a kilowatt-hour — a boutique product, a green indulgence, three to five times the price of ordinary grid power and utterly dependent on subsidy to exist at all. The people who ran the world's energy system regarded solar with the fond condescension one reserves for a well-meaning hobby. The authoritative forecasts — the ones published by serious institutions and relied upon by governments and investors — projected that solar would get cheaper slowly, modestly, at the sedate pace of a mature industry. Year after year, those forecasts were not merely beaten but humiliated; the real price fell so much faster than the official projections that the forecasters had to redraw their charts almost every time they published, and still they lagged behind. The Saudi bid was not an outlier or a stunt. It was the leading edge of a collapse in price that ranks among the swiftest in the history of any manufactured good, and it was soon matched or bettered elsewhere — in the deserts of Chile and the sun-fields of the United Arab Emirates, in India's vast solar parks, wherever the sky is reliably clear and the land is cheap and the money is patient.

The same story, told at a slightly different pitch, belongs to the wind. On plains and ridgelines and out in the shallow seas, the descendants of the windmill have become so productive and so cheap that in many markets the least expensive electricity a new power plant can offer is not solar at all but onshore wind — a few cents a kilowatt-hour, cheaper than the cheapest fossil option, from a machine that runs on moving air and burns nothing. By the middle of the twenty-twenties, when the world's energy accountants totted up the cost of building new generation, the two ancient powers — the sun and the wind, the oldest sources humanity ever harnessed — sat together at the very bottom of the list, undercutting coal and gas and nuclear alike. The oldest had become the cheapest. That sentence is the hinge on which the modern energy world is turning, and it is worth stopping to feel how strange it is.

Because these are old powers — that is the part we forget. Long before the coal smoke and the oil derricks, before the atom was split, human beings ran their civilization on the sun and the wind and the falling water the sun lifted into the sky. Persian millwrights were grinding grain with the wind more than a thousand years ago. The windmills of Holland pulled a nation up out of the sea, spinning day and night to drain the polders and hold back the water. Sailors crossed oceans on nothing but canvas and a fair breeze. And the dream of capturing the sun's heat directly — of concentrating its light with mirrors until it could boil water and drive an engine — is older than the electric grid itself; there were working solar steam engines in the nineteenth century, built by men who feared, with what turned out to be premature accuracy, that the coal would one day run short. For a brief, roaring century and a half we set all of this aside, seduced by the astonishing convenience of fuels we could dig up and burn on demand, whenever and wherever we liked, with no regard for whether the wind blew or the sun shone. The fossil age was, among other things, a holiday from the weather. What we are living through now is not the arrival of something unprecedented but the return of the oldest powers there are — armed, this time, with a physics and a manufacturing base their ancient practitioners could not have imagined.

That physics is the first thing this book will give you, and it begins in two places at once. It begins at the center of the sun, where under crushing pressure and unimaginable heat the nuclei of hydrogen atoms fuse into helium and release, as a byproduct of that welding, a flood of energy so vast that the tiny fraction of it intercepted by our small planet, in a single hour, exceeds everything the whole of humanity consumes in a year. And it begins in the restless ocean of air above our heads, where that same solar energy, absorbed unevenly by a spinning and tilted world, sets the whole atmosphere in perpetual motion — the wind being, as I have said and will keep saying, sunlight wearing a different coat. Understanding how a particle of light becomes an electron in a wire, and how a parcel of moving air becomes a spinning shaft and a current, is not as hard as you may fear, and it is a good deal more wonderful than you may expect. There are two beautiful and unbreakable limits waiting for us there — one that caps how much of a beam of sunlight a single simple cell can ever convert, and one that caps how much of the wind's push any turbine can ever catch — and both, it turns out, can be understood by anyone willing to think carefully for a few pages.

Then comes the long human middle of the story, which is where the drama lives. It is a tale of tinkerers and obsessives and lucky accidents. A nineteen-year-old in a Paris laboratory in 1839 notices that shining light on certain chemicals nudges a needle, and cannot fully explain why. A man coats a sheet of selenium with gold so thin you can see through it and builds, in 1883, the first solid device to make electricity from light — at an efficiency of about one percent, a triumph and a near-uselessness at the same time. An inventor in Cleveland raises a windmill the size of a house in his backyard in 1888 and uses it to light his home for two decades. A Danish schoolteacher turns the windmill from a folk craft into a science. And then, in 1954, three men at a great American laboratory stumble — half by design and half by accident, following a crack in a piece of silicon — into the first solar cell efficient enough to be worth taking seriously, and the world's newspapers announce that limitless free power from the sun is surely just around the corner. It was not around the corner. It was decades and a near-total collapse in price away, and the road there ran through an oil shock that frightened the industrial world awake, through solar panels installed on a president's roof and then quietly torn down by his successor, through a wind boom in the California passes that inflated and burst, through a German law that bet a nation's money on machines almost nobody else believed in, and through the rise of a manufacturing colossus in China that learned to make solar panels by the hundreds of millions and, in doing so, drove their price through floor after floor after floor.

There is a particular pleasure in the middle of this history that I should warn you about in advance: it is full of people being magnificently wrong. The newspapers that announced limitless free power in 1954 were wrong about the timing by half a century. The forecasters who charted a slow, dignified decline in solar prices were wrong by an order of magnitude, and kept being wrong in the same direction for twenty years running. The utility executives who dismissed wind as a toy for cranks were wrong, and so were some of the enthusiasts who promised it would take over the grid by dates that came and went. If there is a single lesson in the human middle of this story, it is humility about prediction — a reminder that the future of technology is made not by the people who forecast it but by the people who, quietly and unglamorously, build the next slightly cheaper machine.

That price collapse is the heart of the book, and it deserves to be, because it is the least understood and most consequential fact in all of energy. It did not come from a single breakthrough. It came from a principle so reliable that it has the character of a law of nature applied to human effort: every time the cumulative number of solar panels ever manufactured has doubled, their price has fallen by roughly a fifth, with a regularity you could set a watch by, decade after decade after decade. Make twice as many of a thing and get systematically better at making it — that is the whole secret, and it is not much of a secret, and yet its compounding over forty years produced a fall in price of something close to ninety-nine percent, and almost no one in a position of authority took it seriously until it had already remade the world. We will spend real time on that idea, because if you understand it you understand not just where solar and wind came from but something deep about how technologies get cheap, and why the experts so often miss it.

And then, having earned the right to it, the book turns to the hard part — the genuinely unfinished part — and refuses to flinch. For all the wonder of cheap sunlight and cheap wind, there is a stubborn fact at the center of both: the sun sets, and the wind sometimes falls still, and they do so on their own schedule, indifferent to when we happen to want the lights on. A grid — the vast, humming, continent-spanning machine that keeps electricity in perfect instantaneous balance between supply and demand — was built over a century to run on power plants that burn fuel on command. Feeding it instead on the weather is one of the great engineering problems of our time. It is a problem of storage, of batteries and pumped lakes and stranger schemes for holding energy through a windless week or a dark winter; a problem of wires, of moving power across hundreds of miles from where the sun is shining to where the people are; a problem of markets and habits and the physical inertia of spinning machines. I want to be very clear, here at the outset, about the honest shape of this: it is a real problem, and it is not solved, and the people who wave it away are as untrustworthy as the people who insist it can never be solved at all. It has, so far, the shape of a problem that yields to money and engineering and time — but "so far" is doing real work in that sentence, and this book will not pretend otherwise.

For that is the tension I want to plant in your mind before we begin, and it is the tension the rest of these pages will hold. What is happening is a genuine revolution — not a public-relations campaign, not a subsidy mirage, but a true and measurable upheaval in the cost of the most fundamental commodity of modern life. The two oldest sources of power humanity ever tapped have become the cheapest new sources of electricity in human history, and that fact is already bending the trajectory of nations and industries and the atmosphere itself. But a revolution in the cost of making electricity is not the same as a finished transformation of the entire energy system, and it is certainly not a guaranteed one. Between the cheap electron at the solar farm's fence and a world actually run on sun and wind lies a mountain of work — physical, financial, political, and industrial — that has only begun to be climbed. This book is the story of how we reached the foot of that mountain, told as honestly as I know how: neither the triumphant press release nor the sneering dismissal, but the real thing, with its wonders and its unsolved problems given equal weight.

It all comes back, in the end, to that number read aloud in the desert — a penny for a kilowatt-hour, from a farm of silent glass drinking the light. To understand how it became possible, we have to go back to where all of this energy is made, in a furnace ninety-three million miles away that has been burning for four and a half billion years and will burn for billions more. We begin with the sun itself.

Chapter 1 — The Star at the Center

On a cloudless morning in the Mojave Desert, if you stand in the middle of a solar farm and hold out your open hand, palm up, something is landing on it. You cannot see it arriving and you cannot feel its weight, because it has none, but it is there all the same, pouring down in a steady, silent flood. In the time it takes you to read this sentence, the patch of skin on your palm has absorbed the tail end of a journey that began at the heart of a star ninety-three million miles away, in a furnace so hot and so violent that the word "fire" does not begin to describe it. Everything in this book — every spinning turbine blade on a ridge in Texas, every black rectangle bolted to a roof in Bavaria, every gust that fills a sail and every wheat field that ripens in the sun — traces back to that one source. Before we can understand how we harvest the sun and the wind, we have to understand what the sun actually is, and why it is so absurdly, extravagantly generous with its energy.

Start with a number that is almost impossible to hold in the mind. The sun radiates roughly 3.8 times ten to the twenty-sixth power watts. Written out, that is a 38 followed by twenty-five zeros, a figure so large it stops meaning anything. So let us shrink it to human scale. The total power used by all of human civilization at any given instant — every car engine, every power plant, every furnace and phone charger and jet engine on Earth combined — is around eighteen or nineteen terawatts, that is, eighteen trillion watts. The sun puts out about twenty trillion times more than that. Here is the fact that ought to reorganize your sense of the world: the amount of solar energy that strikes the Earth in a single hour is greater than the amount of energy the entire human species uses in a year. Not a close call, not a marginal comparison — the sun delivers a year's worth of humanity's appetite roughly every hour, and it has been doing so, without interruption or invoice, for four and a half billion years. We are not short of energy. We are standing in a downpour of it, holding a thimble.

So where does this torrent come from? The intuitive answer, the one most of us absorbed somewhere in childhood, is that the sun is on fire. It is a giant ball of burning gas. This is wrong, and the way it is wrong turns out to be the whole story. Fire — the ordinary combustion of a log in a fireplace or gasoline in an engine — is a chemical reaction. Atoms of carbon and hydrogen rearrange their partnerships with oxygen, and in the reshuffling, a little energy is released. But chemical reactions are feeble things. If the sun were made of coal and pure oxygen, burning as furiously as chemistry allows, it would exhaust itself in a few thousand years and go dark. The sun has been shining, at very nearly its present brightness, for billions of years. Chemistry cannot pay that bill. Something far more powerful is going on.

The story of how we learned this is itself a small drama of stubbornness and vindication. In 1920 the English astrophysicist Arthur Eddington stood before the British Association and proposed, against considerable skepticism, that the sun might be powered by the fusion of hydrogen into helium — that the sun ran on the transmutation of the elements. Many of his colleagues objected that the sun's core, hot as it was, was still not hot enough for atomic nuclei to overcome their mutual repulsion and fuse. Eddington's reply has become famous, and it captures the swagger of a man who knew he was right: he told the doubters that the helium they were quarreling about had to be made somewhere, and if they did not think the stars were hot enough, they should "go and find a hotter place." The full mechanism was not worked out until Hans Bethe and others detailed the actual chain of nuclear reactions in the late 1930s — work that later earned Bethe a Nobel Prize — but Eddington had seen the essential truth two decades early. The answer, worked out over the first half of the twentieth century by physicists including Eddington and Bethe, is that the sun does not burn — it fuses. At the sun's core, the pressure is unimaginable, the weight of six hundred thousand miles of overlying gas crushing inward, and the temperature reaches about fifteen million degrees Celsius. Under those conditions matter is stripped bare: atoms are torn apart into a soup of naked nuclei and free electrons. And the nuclei, ordinarily standoffish, are squeezed so hard and hurled at each other so fast that they occasionally slam together and stick. Four nuclei of hydrogen, the lightest and simplest element, are welded through a chain of steps into a single nucleus of helium, the next element up. This is nuclear fusion, and it is the engine at the center of every star in the sky.

Here is the beautiful, strange part, the part that reaches back to the most famous equation ever written. When you weigh the four hydrogen nuclei that go into the reaction, and then weigh the single helium nucleus that comes out, the helium is very slightly lighter — about seven-tenths of one percent of the mass has simply vanished. It has not gone anywhere in the ordinary sense. It has been converted directly into energy. Albert Einstein's equation, which we usually write with symbols but can say plainly in words, tells us the exchange rate: the energy you get equals the mass you lose multiplied by the speed of light, and then multiplied by the speed of light again. Because the speed of light is enormous — about three hundred million meters per second — and because you multiply by it twice, even a whisker of vanished mass yields a colossal amount of energy. A tiny bit of matter is worth a staggering amount of energy, and the sun trades in this currency on a scale we can barely imagine. Every second, the sun converts about six hundred million tons of hydrogen into helium, and in doing so it loses roughly four million tons of mass outright, four million tons of the sun's own substance flashing into pure energy every single second. It has been shedding weight this way since before the Earth existed, and it has enough hydrogen left to keep doing it for another five billion years.

If the mass-to-energy idea feels slippery, try this analogy. Imagine two lumps of modeling clay, and imagine that when you press them together very hard, they merge into one lump — but the merged lump weighs a fraction less than the two separate pieces did, and the missing weight comes flying out as a burst of heat and light. In the everyday world this never happens; press clay together and it just weighs the sum of its parts. But in the extreme conditions of a stellar core, the "glue" that binds a nucleus together is so powerful that snapping two nuclei into one releases part of their very mass as energy. Fusion is, in effect, the act of squeezing hydrogen so hard that it fuses and pays out a fraction of itself as light.

And what a fraction it is. Consider one of the sun's genuinely counterintuitive secrets, a fact that surprises nearly everyone who first hears it. Gram for gram, the core of the sun is not actually producing energy very fast. The rate of energy generation in the sun's core, averaged over its volume, is only a few watts per cubic meter — less heat, per unit of volume, than the metabolism of a compost heap or a reptile basking on a rock. The sun is blindingly powerful not because any small piece of it is especially fierce, but because it is so unspeakably vast; there are simply a tremendous number of cubic meters, all adding their modest contributions together. The sun's might is the might of the very large and very patient, not the very frantic. That is a lesson worth carrying forward, because it turns out to describe solar energy on Earth too: gentle everywhere, overwhelming in the aggregate.

There is a further surprise hiding in the sun's furnace, one that startles most people who meet it for the first time. Given the fifteen-million-degree heat and the crushing weight of the overlying star, you might expect fusion at the core to be a raging, near-instantaneous affair. In fact it is agonizingly slow and rare. The reason is that the very first step of the chain — two protons, two nuclei of ordinary hydrogen, sticking together — depends on a fantastically improbable quantum fluke that, for any given pair of protons, might not happen in billions of years of jostling. An individual proton at the center of the sun may wait, on average, something like a billion years before it finally fuses. The sun shines steadily not because fusion is fast but because it is slow and there is so unimaginably much hydrogen taking its turn. This is, in a roundabout way, extraordinarily lucky for us: a sun that burned its fuel quickly would have flared and died long before life had a chance to arise. The very sluggishness of solar fusion is what gives a star like ours a lifetime measured in billions of years, long enough for planets to cool, oceans to form, and creatures to evolve who would one day squint at the sky and work out what was happening up there.

Now let us follow that energy outward to us. By the time the sun's light has crossed the ninety-three million miles of empty space to reach the top of Earth's atmosphere, it has spread out and thinned, and what arrives is a well-measured quantity. On a surface held face-on to the sun, just above the atmosphere, the incoming sunlight delivers about 1,361 watts per square meter. Scientists call this the solar constant, though it is not perfectly constant — it wobbles by a fraction of a percent as the Earth's distance from the sun changes over the year and as the sun's own activity rises and falls. Still, 1,361 watts on every square meter facing the sun is the raw allowance, the budget nature hands us at the doorstep of our planet.

Some of that budget is spent before it reaches the ground. The atmosphere reflects a portion straight back to space off the tops of clouds, scatters some of it — the scattering of blue light is the very reason the sky is blue — and absorbs some in the ozone, water vapor, and gases overhead. So by the time sunlight reaches sea level, even at noon under a clear sky with the sun directly overhead, the peak has fallen to roughly a thousand watts per square meter. That round number, a kilowatt on a square meter, is so useful that the entire solar industry has enshrined it. When a manufacturer stamps a rating on a solar panel, it is telling you how the panel performs under a standard test that assumes exactly one thousand watts per square meter of sunshine falling on it. Engineers call this reference level "one sun." It is the yardstick against which every panel in the world is measured.

Let us make that tangible with a small worked example we will return to in later chapters. Picture a single square meter of ground at noon in a sunny place — a doormat-sized patch of desert, say. Falling on it is about one thousand watts of solar power, which is roughly the power drawn by a hair dryer or a small microwave oven, arriving free and silent on that one small square. Now imagine tiling an area the size of a modest parking space, say fifteen square meters, with that same sunshine. You are being handed something like fifteen thousand watts — fifteen kilowatts — of raw power, enough, if you could catch all of it, to run a whole street of homes. You cannot catch all of it; no technology can, and later chapters will explain exactly why not. But the sheer size of the gift is the point. The problem of solar energy has never been that there is too little. The problem has always been that it is spread thin, and catching it cheaply is hard.

This brings us to the two great misconceptions we need to clear away before going further. The first, which we have already met, is that the sun burns. It does not combust; there is no oxygen feeding a flame, no ash, no smoke. The sun fuses, converting matter into energy under crushing gravity, and the difference is not pedantic — it is the difference between a source that would last a few thousand years and one that lasts billions. The second misconception is subtler and more consequential for everything that follows. People often say that solar energy is "weak." You hear it in casual dismissals: the sun is too feeble, too unreliable, too gentle to run a modern industrial society. But solar energy is not weak. It is diffuse, which is an entirely different thing. A thousand watts on a square meter is not a small amount of power — it is genuinely substantial. What makes solar seem weak is that it is spread evenly across enormous areas rather than concentrated in a burning seam of coal or a spinning uranium core. A river delta and a fire hose can carry the same volume of water; the fire hose merely concentrates it. Sunlight is the delta: mild at any one point, staggering in total. All the engineering ingenuity of the solar age, from Becquerel's electrodes to today's mirror-finished silicon, is really the art of gathering the diffuse and making it useful.

There is a final misconception worth naming, because it recurs whenever people first meet these numbers. Some assume that because so much solar energy pours onto the Earth, harvesting it must be trivial, a matter of simply laying out a sheet and collecting. The abundance is real, but abundance is not the same as accessibility. The sun does not shine at night, it hides behind clouds, it slants low in winter, and turning its diffuse warmth into the concentrated, on-demand electricity that a hospital or a steel mill requires is a problem that took humanity the better part of two centuries to begin to solve. That long, human, error-strewn story — of teenage prodigies and stubborn Danes, of failed turbines and cracked silicon and a president's rooftop panels — is what the rest of this book will tell.

A quick recap before we move on. The sun is a fusion furnace, converting hydrogen to helium under fifteen-million-degree heat and crushing pressure, and in each fusion a sliver of mass becomes energy at the punishing exchange rate Einstein described. That process makes the sun radiate an almost inconceivable 3.8 times ten to the twenty-sixth watts, of which a well-defined trickle — about 1,361 watts per square meter at the top of the atmosphere, roughly a thousand at the sunlit ground — is our planet's share. That trickle still amounts to more energy in an hour than humanity uses in a year, not because sunlight is fierce but because it is vast and diffuse. But we have been speaking loosely of energy simply "arriving," as if it stepped across the void in an instant. It does not. Every photon that warms your palm made a long and astonishing journey to get there, and it is that journey — a hundred-thousand-year crawl followed by an eight-minute sprint — that we turn to next.

Chapter 2 — Eight Minutes to Earth

Let us do something a physicist would call a thought experiment and a novelist might call a small act of imagination. We are going to follow a single photon — one indivisible packet of light — from the moment it is born in the sun to the moment it strikes a solar panel on a roof in Arizona. The photon has no name and no memory, but its trip is one of the most extraordinary journeys in nature, and understanding it tells us almost everything we need to know about what sunlight is and what a square meter of it is actually worth.

The photon begins its life in the sun's core, in that furnace of fusing hydrogen we met in the last chapter. At the instant it is created, it is not the gentle golden light we associate with the sun. It is a gamma ray, a high-energy, invisible, ferociously powerful form of light, born from the mass that vanished when four hydrogen nuclei fused into helium. And now the surprises begin. You might assume that this newborn photon simply shoots outward and escapes. It does not. The sun is so dense, so packed with matter, that our photon travels only a tiny fraction of an inch before it slams into a particle, is absorbed, and is spat back out again in a random new direction. Then it happens again. And again. And again, uncountably many times, in a drunken, staggering, directionless zigzag that physicists call a random walk.

The photon lurches this way and that, sometimes outward, often sideways or backward, groping blindly through the crushing interior of the star.

How long does this take? Here is the first genuinely staggering fact of this chapter, one that stops most people cold. The journey from the sun's core to its surface — a distance of about four hundred and thirty thousand miles, which light in open space would cross in about two and a half seconds — takes, by many estimates, on the order of a hundred thousand years. Estimates vary widely, from thousands to hundreds of thousands of years depending on the assumptions, but the order of magnitude is the point. The sunlight warming your face today set out from the sun's heart around the time our species was learning to make its first tools, and it has been ricocheting through the sun's body ever since. The light is, in a real sense, ancient. By the time it finally reaches the sun's visible surface, it has been degraded and re-emitted so many times that it is no longer a savage gamma ray; it has softened into ordinary light — the mix of colors we call sunshine.

Then, at the surface, everything changes. Our photon breaks free into the vacuum of space, and now there is nothing in its way. It flies in a dead-straight line at the speed of light, three hundred thousand kilometers every second, and it crosses the ninety-three million miles to Earth in about eight minutes and twenty seconds. That is the number worth remembering: roughly eight minutes. The sunlight you see at any instant left the sun eight minutes ago. If the sun were somehow to blink out — it will not, but suppose — you would carry on enjoying its light for eight full minutes before the sky went dark. A hundred thousand years to escape the sun, eight minutes to reach us: the two halves of the journey could hardly be more different, and the contrast is one of the loveliest facts in astronomy.

Now, what exactly is this photon, this thing that has traveled so far? It is a quantum of electromagnetic radiation, which is a mouthful, so let us unpack it in plain terms. Light is a form of the same fundamental phenomenon that carries radio broadcasts, warms food in a microwave, and lets an X-ray machine see your bones. All of these are electromagnetic waves, and they differ only in their wavelength — the distance from one crest of the wave to the next. Stretch the wavelength out very long and you get radio waves; shrink it down very short and you get X-rays and gamma rays. In between lies a narrow band our eyes happen to be tuned to, which we call visible light. Within that band, the longest waves we can see look red, the shortest look violet, and the full sweep from red through orange, yellow, green, and blue to violet is simply wavelength changing smoothly. This is the electromagnetic spectrum, and it is best pictured as a piano keyboard stretching far beyond what the human ear can hear: our eyes play only a single octave near the middle, deaf to the vast ranges on either side.

Sunlight, as it happens, contains a broad slice of that keyboard. It arrives as a mixture: a little ultraviolet, the invisible short-wavelength light that tans and burns your skin; the full rainbow of visible colors; and a large helping of infrared, the invisible long-wavelength light we feel as radiant heat. When all the visible colors arrive together in roughly the proportions the sun sends them, our brains blend them into the sensation we call white. This is why sunlight looks white, or very faintly yellow, rather than any single color: it is all the colors at once. You can prove it to yourself with a prism, or with a garden sprinkler on a sunny afternoon, which splits the white light back into the rainbow it was always secretly made of. Isaac Newton did essentially this experiment with a prism in a darkened room in the 1660s and grasped that white light is not pure and elementary but a composite — a genuinely revolutionary idea at the time, and one people resisted for decades.

It is worth pausing on just how narrow our window onto this vast keyboard really is, because it is a genuinely humbling fact. The full electromagnetic spectrum spans an almost unimaginable range of wavelengths, from radio waves longer than a city block to gamma rays smaller than an atomic nucleus — a range covering something like twenty factors of ten. Out of all of that, human vision responds to a single, slim octave, a band so thin that if the whole spectrum were laid out as a keyboard stretching from one end of a country to the other, our eyes would see a strip of a few keys somewhere near the middle. Bees see a little further into the ultraviolet than we do, finding patterns on flowers invisible to us; pit vipers sense the infrared glow of a warm mouse in the dark. We are, in the grand scheme, nearly blind, and it is a happy accident of evolution that the narrow band we can see happens to sit right where the sun pours out most of its light. Our eyes are tuned to the sun's brightest notes — which is surely no coincidence, since they evolved under exactly that light.

There is a crucial detail buried in the spectrum, and it will matter enormously when we come to solar cells in a later chapter. Not all photons carry the same amount of energy. The shorter the wavelength — the bluer, the more toward the ultraviolet — the more energy the photon packs. A violet or ultraviolet photon is a heavyweight; a red or infrared photon is a lightweight. The energy of a photon depends only on its color, its wavelength, and nothing else. This is why ultraviolet light damages your skin while the far more intense warmth of infrared merely makes you feel cozy: each individual ultraviolet photon hits hard enough to break chemical bonds in living cells, while an infrared photon, however many of them arrive, simply cannot. Hold that thought. When we later ask why a solar cell responds to the color of light rather than merely its brightness, the answer will lie right here, in the fact that a photon's punch is set by its wavelength.

So much for what the photon is. Now for the question a solar engineer actually cares about: what is a square meter of sunshine worth? We established in the last chapter the peak figure — about a thousand watts on a square meter at noon under a clear sky, the "one sun" standard. But noon is not the whole day, and clear skies are not every day. To size a real solar installation, you cannot use the peak; you have to average over the relentless honesty of the calendar, over nights when nothing comes in at all, over dawns and dusks when the sun hangs low, over clouds and haze and the slow march of the seasons.

Here we can borrow the clear-eyed style of the physicist David MacKay, who insisted on doing this arithmetic without flinching. Averaged around the clock and around the year, a good sunny location — a desert in the American Southwest, the plains of North Africa, much of Australia — receives on the order of two hundred and fifty watts per square meter of horizontal ground. A cloudier, higher-latitude place — northern Germany, Britain, the Pacific Northwest — might average only a hundred to a hundred and fifty. Notice how far these numbers have fallen from the thousand-watt peak: by the time you account for night and cloud and the sun's low winter arc, the round-the-clock average is only a fifth or a tenth of the noontime maximum. This gap between the peak and the average is the single most important thing to understand about solar energy as an engineering problem. Sunlight is not just diffuse in space, as we saw in the last chapter; it is also intermittent in time. The sun keeps banker's hours, and worse ones in winter.

Why do noon and deserts win so decisively? The answer is the thickness of the air the light must punch through, a quantity engineers call air mass. When the sun is directly overhead at noon, its light takes the shortest possible path down through the atmosphere. When the sun hangs low in the morning or evening, its light slants in at a shallow angle and must plow through a far longer slug of air, which scatters and absorbs more of it along the way — which is exactly why a low sun is dim and red enough to look at directly, its blue light scattered away, while the noon sun is a blinding white you cannot face. Deserts win for a related reason: dry, cloudless air is the clearest window sunlight can pass through. A cloud is a curtain; a humid, hazy sky is a dirty pane. This is why the sunniest real estate on Earth is not the tropics with their afternoon thunderheads but the great subtropical deserts — the Sahara, the Mojave, the Atacama, the outback — where the air is thin of moisture and the clouds rarely gather.

Let us make it concrete with the worked example we promised, and carry through a real rooftop's worth of panels. Suppose you install two square meters of solar panels — a couple of standard modules, the kind you would see on a suburban roof — and suppose those panels convert about twenty percent of the sunlight that hits them into electricity, which is a fair figure for good modern panels as of the mid-2020s. At the noon peak of one thousand watts per square meter, your two square meters intercept two thousand watts of sunlight, and at twenty percent efficiency they turn four hundred of those watts into electric power. Four hundred watts at noon: enough to run a laptop, a couple of lightbulbs, and a phone charger with room to spare. But now apply the daily-and-yearly average. In a sunny location averaging, say, two hundred watts per square meter over the whole year, your two square meters intercept four hundred watts on average, and at twenty percent efficiency yield about eighty watts, averaged around the clock. Eighty watts sounds modest, but sustained day and night for a year it adds up to roughly seven hundred kilowatt-hours of electricity — a meaningful slice of a frugal household's needs from a patch of roof no bigger than a dining table. Multiply by ten or twenty such patches, and you begin to see how a roof can carry a home.

It is worth dwelling for a moment on the honesty this kind of arithmetic forces on us, because it is the antidote to a great deal of loose talk about solar power in both directions. David MacKay, the Cambridge physicist who insisted on it, liked to say that if you want to have an opinion about energy, you have to be willing to multiply and divide — to put real numbers on the back of an envelope rather than trading in vague enthusiasms or vague dismissals. Do the arithmetic and the truth sits calmly between the extremes. Solar is neither the effortless savior that will trivially power everything by next Tuesday, nor the feeble toy its detractors imagine. It is a real, substantial, but diffuse and intermittent resource that yields serious energy when you cover serious area with reasonably cheap hardware — no more and no less. The envelope, worked honestly, is worth more than a shelf of manifestos.

This exercise also lets us correct a pair of stubborn misconceptions. The first is that solar panels do not work on cloudy days or in cold countries. They work perfectly well; they simply produce less. A panel in cloudy Germany still generates a great deal of electricity over a year — Germany, not a sunny country by any stretch, has for years been among the world's leading solar generators — it just gathers less per panel than the same hardware would in Spain, and you compensate by installing more of it. Cold, in fact, is no enemy of solar at all: silicon cells are slightly more efficient when cool, and a bright, freezing winter day can outperform a hazy summer scorcher. The enemy of solar is cloud and darkness, not cold. The second misconception is that the efficiency number — that twenty percent — is a mark of failure, proof that the technology is primitive because it "wastes" eighty percent of the light. But efficiency is not the figure that matters to your wallet; cost per unit of energy is. Sunlight is free and effectively infinite, so throwing away four-fifths of a free resource is no scandal if the fifth you keep is cheap enough. A cheap panel that captures twenty percent beats an eye-wateringly expensive one that captures forty. The history of solar, as we will see, has been far less a race for efficiency than a relentless, extraordinary collapse in cost.

Where do you meet this photon's journey today? You meet it, quite literally, on the piece of paper or the web page that a solar installer hands you when you ask for a quote. Somewhere on that document, usually near the estimate of how much power the system will produce, sits a number describing the sunshine at your address — often given as "peak sun hours" per day, or as irradiance in kilowatt-hours per square meter per day. That unglamorous figure, which most homeowners skip right over, is the entire content of this chapter distilled into one line: it is the round-the-clock, cloud-and-season average of the eight-minute journey we have been tracing, measured for the exact patch of sky above your roof. When an installer in Phoenix quotes a bigger number than one in Hamburg, that difference — the air mass, the clouds, the latitude, the whole honest arithmetic of the average — is the difference between the deserts and the cloudy coasts made visible on an invoice.

To recap. A photon is born a violent gamma ray in the sun's core and spends something like a hundred thousand years staggering to the surface in a random walk, then crosses to Earth in a straight eight-minute sprint. It arrives as part of a broad spectrum — ultraviolet, visible, infrared — blended by our eyes into white, and the shorter its wavelength the more energy it carries, a fact that will matter greatly when we build a cell to catch it. A square meter of that light peaks near a thousand watts but averages only one to two hundred over a real year, because the sun keeps daytime hours and the air and clouds take their cut, which is why noon beats morning and deserts beat cloudy coasts. But the sun does more than shine on us. It also, indirectly, sets the whole atmosphere in motion — and that restless, moving air is the second great energy source of this book, the one we turn to now.

Chapter 3 — The Restless Air

Stand on a beach on a hot summer afternoon and you will feel it: a steady breeze coming in off the water, cool and reliable, pushing against your face as if the sea itself were exhaling. Come back to the same spot after dark, and the breeze has reversed — now the air is flowing gently out toward the water. Nothing has changed about the geography. The sea is where it always was, the sand where it always was. What has changed is the sun, and the way the land and the sea respond to it. That humble sea breeze, which sailors and children have known forever, contains in miniature the entire explanation of why the wind blows anywhere on Earth. Master the beach, and you have most of the physics of the planet's great wind belts, the trade winds that carried Columbus, the roaring westerlies that batter Patagonia, the jet streams that airliners ride across oceans. It is all, at bottom, the same trick.

Here is the trick, from first principles. Wind is air in motion, and to make air move you need a push — a difference in pressure, high on one side and low on the other, so the air slides from the crowded region to the emptier one, exactly as a crowd spills out of a stadium through an open gate. So the real question is not "why does air move?" but "what creates the pressure differences?" And the answer, almost always, is heat. Warm air is lighter than cold air. When you heat a parcel of air, its molecules move faster and spread apart, so the parcel becomes less dense and more buoyant, and it rises — the same reason a hot-air balloon climbs, the same reason smoke lifts from a chimney. When warm air rises off a patch of ground, it leaves behind a region of slightly lower pressure at the surface, and neighboring cooler, denser, higher-pressure air comes flowing sideways to fill the gap. That sideways flow of air, from the cool high-pressure zone toward the warm low-pressure zone, is wind. There is nothing more to it at the core. Every gust that ever spun a turbine is cooler air rushing in to replace warmer air that has floated away.

Now return to the beach and watch the trick play out. On a sunny afternoon, land heats up far faster than water. Sand and rock warm quickly under the sun; the sea, which can mix its heat down through a deep, sluggish body of water and which stubbornly resists changing temperature, stays comparatively cool. So the air over the land grows hot and buoyant and rises, leaving low pressure at the shore, and the cooler, denser air sitting over the sea slides in to take its place. That inflow is your afternoon sea breeze — cool because it comes off the water. At night the whole thing runs in reverse: the land loses its heat quickly and turns cold, the sea stays relatively warm, and now it is the air over the water that rises while cooler air drains off the land to replace it, giving the gentle nighttime land breeze. The same daily dance happens on a grander scale in the mountains, where slopes warmed by the sun send air rising uphill during the day, drawing a valley breeze up after it, and then, after dark, cold dense air pools and slides back down the slopes as a mountain breeze. These local winds are the sun writing its signature on the landscape, and they are the everyday proof that wind is nothing more exotic than uneven heating turned into motion.

Which brings us to the single most important sentence in this chapter: wind is sunlight one step removed. The sun does not merely shine on solar panels; it also, and quite separately, drives the entire circulation of the atmosphere. It heats the tropics, where its rays strike the Earth nearly head-on, far more intensely than it heats the poles, where the same sunlight arrives at a low, glancing angle and spreads thin across the ground — the very air-mass effect we met in the last chapter, now operating on a planetary scale. The equator is the beach and the poles are the sea, so to speak: a permanent, colossal imbalance in heating between the middle of the Earth and its ends. Warm air rises over the equator and cold air sinks over the poles, and the atmosphere spends its entire existence trying, and forever failing, to even out that imbalance. Roughly one to two percent of all the solar energy that reaches the Earth gets converted into the kinetic energy of moving air. It sounds like a small fraction, and it is, but recall from the first chapter just how enormous the sun's total delivery is: one or two percent of an almost unimaginable number is still an almost unimaginable number. This is why wind and solar are not rival, unrelated technologies, as people often assume. They are two harvests from the same field. The wind is simply the sun's energy after it has been laundered through the atmosphere, converted from raw light into moving air.

If the Earth stood still, this would be a simple picture: warm air rising at the equator would flow toward the poles up high, sink at the poles, and return along the surface toward the equator, one great overturning loop in each hemisphere. But the Earth does not stand still. It spins, once a day, and that spin bends every moving thing on its surface — a subtle, relentless deflection that atmospheric scientists call the Coriolis effect. The way to feel it in your imagination is to picture a merry-go-round. Stand at the center of a spinning carousel and try to roll a ball straight out to a friend at the rim. To you, standing on the moving platform, the ball seems to curve away to the side, even though it traveled in a perfectly straight line as seen from above. The ground turned beneath it. The whole Earth is that carousel, and every wind, every current, every long-range missile is deflected — to the right in the Northern Hemisphere, to the left in the Southern. Air trying to flow straight from the equator to the pole gets bent sideways instead, and this bending is what breaks the simple loop into the banded pattern of prevailing winds that has organized human trade and travel for millennia.

The result is the great wind belts. Near the equator, air converges and rises in a hot, calm, rainy zone that old sailors dreaded and called the doldrums, where ships could sit becalmed for weeks. On either side, from roughly the subtropics back toward the equator, blow the trade winds — remarkably steady easterlies, bent by the Coriolis effect, that fill sails with dependable force and that carried European ships to the Americas and back for four centuries. Farther from the equator, in the temperate zones where most of the industrialized world lives, blow the westerlies, the prevailing winds that sweep from west to east across the middle latitudes, driving weather systems across North America and Europe and howling almost without interruption around the bottom of the Southern Hemisphere, where little land stands in their way — the "roaring forties" and "furious fifties" of maritime legend. And high above, near the top of the weather, run the jet streams: narrow ribbons of ferociously fast wind, often well over a hundred miles per hour, snaking around the globe at the boundaries between warm and cold air masses. Airline pilots know them intimately, riding them eastbound to save fuel and time and dodging them westbound, which is why a flight from New York to London is reliably shorter than the return.

This planetary map is why some places are wind-rich and others are not, and here is a surprising fact worth pausing on: the world's best wind is astonishingly concentrated. A relative handful of regions catch a wildly disproportionate share of the usable wind. The Great Plains of North America, a corridor running from Texas up through Oklahoma, Kansas, and the Dakotas, sit squarely in the path of the westerlies with no mountains or forests to slow the flow across their flat expanse — which is why that corridor is sometimes called the Saudi Arabia of wind. The North Sea, shallow and open and lashed by the North Atlantic westerlies, has become the great proving ground of offshore wind. And Patagonia, at the wind-scoured southern tip of South America, sits directly in the roaring forties and endures some of the most relentless surface winds on the planet, so constant that trees there grow permanently bent. These are not random lucky spots. They are the places where the geometry of a spinning, unevenly heated planet happens to funnel moving air across accessible ground.

The Coriolis effect has a human backstory worth telling, because it shows how long it took people to grasp something that governs their daily weather. It is named for Gaspard-Gustave de Coriolis, a French mathematician who worked out the underlying mechanics in the 1830s while studying, of all things, the behavior of rotating machines like waterwheels and turbines — not the atmosphere at all. The connection to winds and ocean currents came later, and even then it was resisted and misunderstood, because the effect is so counterintuitive: nothing actually pushes the moving air sideways; it merely appears to curve because the solid Earth is turning underneath it while we, standing on that turning ground, forget that we ourselves are spinning. This is also the source of one of the most persistent pieces of folklore in all of science — the claim that the Coriolis effect makes water spiral one way down a drain in the Northern Hemisphere and the other way in the Southern. It does not. The effect is real and powerful over the vast scales of hurricanes and trade winds, where air travels hundreds of miles and the Earth's rotation has room to bend it, but it is utterly negligible in a sink or a toilet, where the swirl is set entirely by the shape of the basin and the way the water happened to be moving when it started to drain. The next time someone tells you the drain proves the Coriolis effect, you may gently correct them: the effect is grand and planetary, not domestic and plumbed.

Now for the numbers, because they contain the second great surprise. The total kinetic energy stored in the world's moving air, and the rate at which the sun keeps replenishing it, utterly dwarfs everything humanity uses. Estimates vary, and the exercise is inherently rough, but the power continuously present in the Earth's winds runs to many tens of terawatts at the very least — recall that all of human civilization draws around eighteen or nineteen terawatts at any moment. There is, in the crudest accounting, vastly more wind energy blowing past us than we could ever need. And yet — here is the catch that keeps the enthusiasts honest — only a thin sliver of it is anywhere we can reach. The overwhelming majority of the atmosphere's kinetic energy is locked up miles overhead in the jet streams and the deep circulation, far beyond the reach of any tower we can build. The wind we can actually harvest is the wind in the lowest hundred meters or so of the atmosphere, skimming the surface where a turbine's blades can catch it, and that near-ground layer is only a small fraction of the whole. So the honest picture is neither the boosters' fantasy of limitless free power nor the skeptics' dismissal of a feeble breeze. It is this: the resource is genuinely enormous, far larger than our needs, but reaching it is the whole engineering problem, and only a modest, accessible slice will ever turn a rotor.

Let us clear away two misconceptions that cloud almost every conversation about wind. The first is that wind is random — that it blows when it feels like it, chaotically and unpredictably, so no serious grid could ever depend on it. This is wrong in a specific and important way. Wind is variable, yes: it rises and falls, it gusts and lulls, and no one can promise a particular breeze at a particular minute a month from now. But variable is not the same as random. Wind is deeply patterned. The trade winds have blown in the same direction for all of recorded history. The westerlies are as dependable as the calendar. Sea breezes come every sunny afternoon. And because the atmosphere obeys physical laws, its behavior can be forecast — not perfectly, but remarkably well, hours and days ahead, with an accuracy that has improved dramatically in recent decades. Modern grid operators lean on wind forecasts the way farmers lean on weather forecasts: not as certainties, but as good, actionable predictions that get better every year. The wind is not a coin flip. It is a system with structure, memory, and rhythm.

The second misconception is the one we have already begun to dismantle but which deserves to be stated flatly: the idea that wind power and solar power are separate, unrelated technologies that happen to both be fashionable. They are not separate at all. They are siblings, children of the same star. Solar power catches the sun's energy as light, directly, the instant it arrives. Wind power catches that same solar energy after the atmosphere has transformed it into motion — a second-hand harvest of the identical source. This kinship is not a poetic flourish; it has practical consequences that run through the rest of this book. The sun and the wind often trade shifts: solar peaks at midday and vanishes at night, while wind frequently strengthens in the evening and blows through the dark, and across the seasons a sunny summer and a windy winter can complement each other on the grid. Two harvests from one field, gathered at different hours — that complementarity is one of the quiet reasons a grid built on both can be steadier than a grid built on either alone.

Let me close with one small anecdote that captures how old this understanding really is, even if the physics is young. Long before anyone spoke of Coriolis or pressure gradients, the Dutch had read the wind so well that they bet their country on it. From the twelfth century onward they built windmills by the thousands, not merely to grind grain but to pump water — to lift the sea back off the land they were reclaiming, turning marsh and shallow bay into dry polder fields behind their dikes. They did not know that the westerlies sweeping in off the North Sea were solar energy laundered through a spinning atmosphere. They only knew that the wind came, reliably, off the water, and that a well-built mill could turn it into work. They were, without the equations, harvesting the sun. And the deep question they never had to ask — but that any engineer building a modern turbine must answer precisely — is exactly how much energy is hiding inside that moving air, and how much of it a machine is allowed to take.

So, to gather the threads: wind is the sun's energy at one remove, born from the fierce difference between a sun-baked equator and a frozen pole, driven by warm air rising and cool air rushing in to replace it, and bent by the spinning Earth into the trade winds, the westerlies, and the jet streams that organize the planet's weather. The total resource dwarfs human demand many times over, yet only the thin near-ground layer is ours to catch, and that wind — patterned, forecastable, and kin to the sunlight on a panel — is what our turbines must convert into power. To build those turbines wisely, we need to know precisely how much muscle is packed into a moving river of air, and that is a question with a startling, cubic answer.

Chapter 4 — The Energy in a Moving River of Air

There is a moment familiar to anyone who has ever ridden a bicycle into a stiff headwind. At a gentle pace, the breeze in your face is a mild nuisance, easily ignored. Push a little harder, pedal a little faster, and the wind seems to stiffen out of all proportion — suddenly it is shoving against your chest, roaring in your ears, fighting you for every yard. It feels as though the air has grown teeth. It has not, of course; the air is the same. What has changed is your speed through it, and the resistance you feel does not climb in step with your speed. It climbs far faster. That disproportion, that sense of the wind's force exploding as speed rises, is not an illusion of tired legs. It is the single most important fact in all of wind engineering, and it is the subject of this chapter. Get it into your bones and everything about turbines — why they are so tall, why their blades keep growing, why a modest hill can be worth a fortune — falls into place.

Let us build the idea from first principles, gently. Moving air carries energy simply by virtue of being in motion; physicists call this energy of motion kinetic energy, and everything that moves has it, from a thrown baseball to a river in flood to a gust across a field. The faster a thing moves and the more of it there is, the more kinetic energy it carries. Now picture the wind as exactly what the chapter title calls it: a river of air, flowing invisibly past you. Ask how much power that river delivers to a wheel you place in its path — a turbine rotor — and three things turn out to matter.

The first is how thick the air is, its density. Thick, heavy, cold air at sea level carries more punch than the thin, light air of a mountaintop, for the same reason a river of water shoves harder than a river of steam. The second is how big a bite your rotor takes out of the passing stream — the swept area, the circle traced by the spinning blade tips. A wider circle intercepts more of the river. And the third, by far the most consequential, is the speed of the wind. Here is the whole of it, stated in plain words: the power available in the wind is proportional to the density of the air, times the swept area of the rotor, times the wind speed multiplied by itself three times over. That last phrase is the crux. Not the wind speed, not the wind speed squared, but the wind speed cubed. Engineers call it the cube law, and it is as close to a sacred text as wind power possesses.

Why cubed, and not simply the speed? It is worth seeing where the three factors of speed come from, because the logic is beautifully simple once you slow it down. Two separate effects each pull in a factor of wind speed, and a third comes along for the ride. First, faster-moving air carries more energy in each parcel — the kinetic energy of a moving mass grows with the square of its speed, so doubling the speed of a given parcel of air quadruples the energy it holds. That is two factors of speed right there. Then comes the third: faster wind also delivers its parcels more quickly. In a given second, a wind blowing twice as fast sweeps twice as much air through your rotor. So you have not only quadrupled the energy in each parcel of air, you have also doubled the number of parcels arriving per second. Multiply the two together — four times the energy per parcel, twice as many parcels — and you get eight. Double the wind speed and you get eight times the power. That is the cube law made visible: two from the energy, one from the delivery rate, three in all.

Sit with that number for a moment, because it is genuinely astonishing and it violates almost everyone's intuition. We naturally think in straight lines. If a wind of ten miles an hour gives some amount of power, surely twenty miles an hour gives twice as much? It does not. It gives eight times as much. And thirty miles an hour, three times the speed, does not give three times the power — it gives twenty-seven times the power, because three cubed is twenty-seven. The wind does not add its strength; it multiplies it, explosively. A merely breezy day is not a little better than a calm one for a turbine. It can be enormously, transformationally better. This is the merciless heart of the whole business, and every decision a wind engineer makes bows to it.

Let us make it concrete with the worked example the cube law demands. Take a single spot of wind and imagine the same turbine standing in it on three different days. On the first day the wind blows at five meters per second — a light breeze, enough to flutter a flag, roughly eleven miles an hour. Call the power the turbine can draw from that wind some baseline amount; let us simply say one unit, to keep the arithmetic clean. On the second day the wind blows at ten meters per second — a fresh breeze, about twenty-two miles an hour, the kind of day when you'd hold your hat. The speed has doubled, so by the cube law the available power is two times two times two, or eight units. Eight times the power, from a wind that a casual observer would call merely "a bit windier." Now the third day: fifteen meters per second, about thirty-three miles an hour, a proper gale that bends the trees. That is three times the original speed, so the power is three times three times three — twenty-seven units. From the calm day to the gale, the wind's speed tripled, but the power it can yield rose twenty-seven-fold. The very same machine, standing in the very same field, harvests twenty-seven times more on the windy day than on the light one. No other common energy source behaves with such violent leverage on a single variable.

This cube law is why the wind industry is, above all else, obsessed with location, and it explains a fact that surprises newcomers: two wind sites that look almost identical on paper can differ wildly in value. Suppose one ridge averages a wind speed just twenty-six percent higher than a neighboring one — the difference between, say, a decent site and a good one, barely noticeable to a person standing on either. By the cube law, that twenty-six percent speed advantage becomes very nearly a doubling of the power, because 1.26 cubed is almost exactly two. A quarter more wind, twice the energy. This is why wind developers spend years and considerable sums erecting measurement masts, hoisting anemometers, and studying a candidate site season by season before committing. A small error in estimating the average wind speed becomes a large error in estimating the electricity, and therefore the money. It is also why the best wind sites — the exposed ridgelines, the open plains, the windswept coasts and the seas beyond them, the places we met in the last chapter — are worth so disproportionately much. A little more wind is not a little more power. It is a great deal more.

The cube law also governs how a turbine is designed to behave across the range of winds it will meet, and this is where three key speeds enter the story, terms you will encounter on any turbine's specification sheet. Below a certain gentle threshold, the cut-in speed — typically around three to four meters per second, a light breeze — there is simply too little energy in the air to bother turning the rotor against its own friction, so the machine sits idle. Above cut-in, the turbine begins to generate, and as the wind strengthens its output climbs steeply along that cubic curve, doubling and redoubling with astonishing speed. But it cannot climb forever, because the generator and gearbox and blades can only handle so much force before they risk tearing themselves apart. So at the rated speed — often around twelve to fifteen meters per second — the turbine reaches its maximum, its nameplate power, and from there on the control system deliberately spills the excess, pitching the blades to shed the surplus wind and holding the output flat no matter how much harder it blows. And finally, if the wind grows truly violent — above the cut-out speed, around twenty-five meters per second, a storm — the turbine shuts itself down entirely and turns its blades edge-on to the gale to protect itself from destruction. This whole operating strategy, idle then climbing then capped then parked, is a direct response to the cube law's ferocity: a machine that must gently coax power from light winds yet survive the twenty-seven-fold blows of a gale needs a gearbox of behaviors, not a single setting.

There is a second scaling law braided together with the first, and it explains the most visible trend in the whole industry: the relentless growth of the blades. Recall that the power in the wind depends not only on speed but on the swept area of the rotor — the size of the circle the blades trace. And the area of a circle grows with the square of its radius. So if you double the length of a turbine's blades, you do not double the area they sweep; you quadruple it, and therefore roughly quadruple the power the machine can catch. This is why turbine blades have grown from the length of a bus to the length of a football pitch within a few decades, and why the largest offshore machines now spin rotors well over three hundred meters across, wider than the tallest skyscrapers are tall. Every extra meter of blade buys area faster than it costs length. Between the cube law rewarding height — since wind blows faster and steadier the higher you climb, away from the drag of the ground — and the square law rewarding breadth, the entire evolutionary pressure of the industry points one way: taller towers, longer blades, bigger circles cut from the moving river of air.

Let me offer one more everyday analogy to lock the cube law in place, and then a real-world echo. Think of the wind not as a breeze but as a river current pushing against a paddle. In a lazy, slow-moving stream, the paddle barely tugs. But anyone who has stood in fast-flowing water knows how suddenly and how brutally the force builds as the current quickens — a knee-deep flow at walking pace is a mild push, but the same depth of water moving at a run can knock a grown adult off their feet. The river's grab on you does not rise gently with its speed; it surges. The wind grabs a turbine blade in exactly that surging, disproportionate way, and the cube law is simply the precise accounting of the surge.

Now for the misconceptions, because this chapter exists largely to demolish one very common and very costly error. The everyday assumption is that a windy day is "twice as good" as a calm one, or that a site with a bit more wind is worth a bit more money — that the relationship is a gentle, proportional one. We have seen that it is not. A windy day can be eight times as good, or more; a modestly windier site can be worth double. People consistently, dramatically underestimate the value of stronger wind, because human intuition adds where nature multiplies. A related misconception runs the other way: that a calm, low-wind region can be made worthwhile simply by installing a great many turbines, brute-forcing the numbers. But the cube law is merciless here too. If your average wind speed is low, each turbine produces so little — because the power falls off cubically as the wind slackens — that no amount of quantity easily rescues the economics. The wind resource is stubbornly local and stubbornly nonlinear, and there is no getting around it: you must go where the wind actually blows hard, and a little harder is a lot more. A final, subtler misconception is that the rated power stamped on a turbine — say, a "three-megawatt" machine — is what it typically produces. It is not; it is the capped maximum the turbine hits only when the wind reaches rated speed and holds there, which in most places is a minority of the time. What a turbine actually delivers, averaged over a real year of light breezes and gales and dead calms, is a much lower figure, and reconciling that gap between the nameplate and the reality is a theme we will return to when we discuss how much of the year a turbine truly earns its keep.

There is a further consequence of the cube law that engineers must live with daily, and it is worth naming because it explains why wind power feels so different from a coal plant. Because power rises with the cube of speed, a wind farm's output is not steady but wildly peaked: it does relatively little in light and moderate winds, then a great deal in the strong ones, so that a modest number of genuinely windy hours can dominate a whole year's production. A site might generate a large share of its annual electricity during a minority of its hours, when the wind is really blowing. This is the cube law casting a long shadow over the grid, and it is why wind, however patterned and forecastable, delivers its energy in an uneven, lumpy rhythm that the rest of the power system must learn to accommodate — a challenge we will take up in full when we come to storage and the grid.

Where do you meet the cube law today? You meet it, if you know to look, in the very shape and placement of every wind farm you have ever driven past. The reason the turbines stand on the highest available ground, spaced far apart, on towers that seem almost comically tall for the blades they carry, is the cube law and its partner the square law working in concert — reaching up for faster wind, spreading wide for more area, each turbine positioned so as not to steal the wind from its neighbors and rob it of that precious, cubically valuable speed. You meet it too in the news whenever a manufacturer unveils a new record-breaking machine with a longer blade than the last: the headline is really the square law made visible, another few meters of radius bought to quadruple the catch. And you meet it, invisibly, every time a wind forecast is folded into a grid operator's plans, because a forecast that is off by a couple of miles an hour in the wind speed is off by far more than that in the power — a small meteorological error magnified, by the cube, into a large one on the grid.

To recap the essential truth of this chapter: the power carried in moving air rises with the density of the air, with the swept area of the rotor — which grows as the square of blade length — and, above all, with the cube of the wind speed, so that doubling the wind gives eight times the power and tripling it gives twenty-seven. That single fact explains the towering height of turbines, the ever-lengthening blades, the obsessive hunt for windy ground, and the strict choreography of cut-in, rated, and cut-out speeds. We now understand how much energy the wind offers and why location rules everything. But we have been quietly assuming a turbine can simply take that energy — and it cannot take all of it, not even close, for reasons we will come to. First, though, we must return to the sun and to a nineteen-year-old in his father's Paris laboratory, who stumbled onto the strange effect that lets light become electricity at all.

Chapter 5 — The Photovoltaic Effect

In a laboratory in Paris in 1839, a nineteen-year-old was tinkering with jars of liquid and metal plates when he noticed something that had no business happening. The young man was Edmond Becquerel, working in the laboratory of his father Antoine-César Becquerel, a distinguished physicist, in a family that would go on to produce four generations of scientists, one of whom — Edmond's own son Henri — would later win a Nobel Prize for discovering radioactivity. But that was decades away. On this day the teenager was experimenting with what we would now call an electrochemical cell: two electrodes, plates of metal, dipped into an acidic liquid, wired together so that any current could be measured. Cells like this were the cutting edge of the age; the battery was young, electricity itself was barely understood, and much of physics still consisted of poking at nature to see what jumped. What Edmond noticed was this. When he let sunlight fall on one of the electrodes in the liquid, the current flowing in his little circuit changed. Shade the plate, and the current dropped. Light it, and the current rose. The light itself was doing something electrical.

This was the discovery of the photovoltaic effect — the direct generation of electricity from light — and it is worth lingering on how quietly and accidentally it arrived. There was no thunderclap, no eureka in the bath. A teenager in his father's lab shone light on a wet electrode and watched a needle twitch. He wrote it up, the scientific world took modest note, and then, for the most part, everyone moved on, because nobody, least of all Becquerel himself, had the faintest idea why it happened. The effect was real, reproducible, and completely mysterious. It would remain a puzzle at the frontier of physics for the better part of the next seventy years, defeating some of the finest minds of the nineteenth century, until it took a young Albert Einstein to crack it open — but we are getting ahead of the story.

Let us do what Becquerel could not, and explain from first principles what was actually going on, in plain language. Everything is made of atoms, and every atom has electrons — tiny negatively charged particles — bound to it. In most materials these electrons are held firmly in place, and that is why most materials do not spontaneously conduct electricity; the electrons cannot move freely to carry a current. Now here is the crucial idea, the one that makes solar cells possible at all: light carries energy, and if a particle of light — a photon, as we called it in earlier chapters — strikes an electron and hands over enough energy, it can knock that electron loose from its atom, freeing it to move. A freed electron that can wander is the raw material of an electric current. So at the most basic level, a solar cell does one thing: it uses the energy of incoming light to shake electrons loose, and then it arranges matters so that those loosened electrons all flow the same way, around a circuit, doing useful work. Light in, loosened electrons out, current flowing. That is the whole trick.

But there is a catch, and it is the catch that turns a curiosity into a technology. Knocking an electron loose is not enough. In a plain lump of metal, sunlight might jostle electrons free, but they would simply rattle about aimlessly and settle back down, producing no net flow — a crowd milling in a plaza, going nowhere. To get a current, you need the loosened electrons to march in one direction, and for that you need some built-in asymmetry, some hidden slope in the material that rolls the freed electrons consistently toward one side. Think of it as the difference between a flat parking lot, where a spilled bucket of marbles just scatters and stops, and a tilted parking lot, where every marble rolls the same way downhill and out the gate. The genius of a working solar cell, which took more than a century to perfect, lies almost entirely in engineering that tilt — a one-way street for electrons, so that light striking the material produces not a pointless jostle but a steady, directional flow of current. In Becquerel's wet cell, the boundary between the metal electrode and the liquid provided a crude version of that asymmetry. In modern cells it is provided by a cunning arrangement inside solid silicon, which later chapters will explore in detail. For now, hold the essential picture: light frees electrons, and a built-in slope makes them all flow one way.

The next chapter in the human story belongs to an American inventor named Charles Fritts, and it moves the effect out of the liquid and into a solid. In 1883, working in New York, Fritts took the element selenium — a curious substance whose electrical behavior changes in light — pressed it into thin wafers, and coated one side with a film of gold so thin it was semi-transparent, thin enough for light to pass through and reach the selenium beneath. When he exposed this little sandwich to light, it produced a current, all on its own, with no liquid, no acid, no external battery. Fritts had built the world's first solid-state solar cell — the direct ancestor of every silicon panel on every roof today. It was, by any modern standard, hopeless: it converted perhaps one percent of the light that struck it into electricity, meaning ninety-nine percent was wasted, and gold is not the material of cheap mass production. But Fritts grasped, with startling clarity, what he was holding. He shipped one of his selenium cells to the great German industrialist and physicist Werner von Siemens, who examined it and pronounced the result "scientifically of the most far-reaching importance." And Fritts himself made a prediction that reads today as either lunacy or prophecy, depending on when you read it: he foresaw that his solar cells might one day compete with, and even displace, the coal-fired power stations that Thomas Edison was just then beginning to build. In 1883, with a one-percent gold-and-selenium wafer, a man looked at the burning heart of the industrial age and calmly suggested the sun might replace it. He was about a century and a quarter early, but he was not wrong.

Here we arrive at the deep mystery that haunted the effect for two generations, and it is one of the most illuminating puzzles in the history of physics, so it is worth setting out carefully. As scientists studied light knocking electrons loose from metals — a closely related phenomenon called the photoelectric effect — they found something that made no sense under the physics of the day. Everyone assumed that light was a wave, like a wave on water, and that a brighter light was simply a bigger wave carrying more energy. So a brighter light, they reasoned, should knock electrons loose more forcefully, giving them more energy, the way a bigger ocean wave hits the shore harder. But that is not what happened. It turned out that the color of the light, not its brightness, determined whether electrons came loose and how much energy they carried. Shine a dim blue or ultraviolet light on the metal and electrons flew off energetically. Shine an intensely bright red light — far brighter, far more total energy — and nothing happened at all; no electrons came loose, no matter how brilliant the beam. This was baffling. It was as though a swimming pool full of gentle ripples could shatter a rock on the shore while a single towering tsunami left it untouched, purely because of some quality of the ripples the tsunami lacked. Classical wave physics had no answer. The brightness should have been what mattered, and it stubbornly, mockingly did not.

The resolution came in 1905, from a twenty-six-year-old patent clerk in Bern named Albert Einstein, in one of the papers of his miraculous year. Einstein proposed something radical: that light is not purely a smooth wave but comes in discrete packets, individual grains of energy — the photons we have been speaking of all along. And crucially, the energy of each individual photon is set by its color, its wavelength, exactly as we saw in the second chapter: a blue or ultraviolet photon is a heavyweight, a red one a lightweight. Knocking an electron loose, Einstein reasoned, is a one-on-one collision between a single photon and a single electron, and it takes one photon with enough individual energy to do the job. A blue photon, packing plenty of punch, frees an electron in a single hit. A red photon, no matter how many of them arrive in a bright beam, simply lacks the individual energy to dislodge anything — a million gentle taps will never do what one firm knock can, if each tap falls below the threshold. Brightness only counts the number of photons; it is color that sets each photon's strength. That single insight dissolved the whole paradox, and it was so important, so foundational to the new quantum physics, that it was for this work — not for relativity — that Einstein was awarded the Nobel Prize in 1921. The mystery Becquerel had stumbled into as a teenager, that had defied physics for sixty-six years, was finally understood: light is grainy, and each grain's power is its color.

Let me offer a small worked example to make the photon-by-photon picture tangible, because it explains something you can observe. Imagine two flashlights of exactly equal total brightness pointed at a solar cell — say, the same number of watts of light coming out of each. One is a deep red light, the other a blue. Even though they deliver identical total energy, they will not produce identical electricity, because what a cell cares about is not the total wattage but whether each individual photon carries enough energy to free an electron, and the blue photons individually carry more. A stream of feeble red photons may waste much of its energy or fail to free electrons efficiently, while the blue stream frees them cleanly. This is why the response of a solar cell depends on the spectrum of light falling on it — on its mix of colors — and not merely on how bright it seems to the eye. It is Einstein's 1905 insight, sitting quietly inside every panel, sorting photons one at a time by their color.

Now to clear away two misconceptions that trip up nearly everyone new to the subject. The first is the belief that solar cells store energy — that a panel is a kind of battery, holding sunlight for later. It is not, and the distinction matters enormously. A solar cell is a converter, not a container. It turns light into electricity only in the very instant the light is falling on it, and the moment the sun goes behind a cloud or sets for the night, the cell's output stops, as instantly and completely as a tap turning off. There is no reservoir inside it, no charge banked from the bright hours. This is precisely why solar power, on its own, cannot run through the night, and why the whole enormous and separate business of batteries and other storage — which we will come to later in this book — exists at all. The panel makes; something else must store. Conflating the two has confused public debate about solar for decades.

The second misconception is subtler and just as widespread: that photovoltaics and solar-thermal power are the same thing, or two versions of one idea. They are fundamentally different, and the difference runs right back to Becquerel versus the inventors who tried to boil water with sunshine. Solar-thermal technology — the mirrors and troughs of inventors like Augustin Mouchot and Frank Shuman, whose stories belong to another chapter — uses sunlight as heat: it concentrates the sun's warmth to boil a fluid, make steam, and drive a conventional engine or turbine, exactly as a coal plant does, only with the sun playing the role of the furnace. It is, at heart, an old idea — sunlight as fire. Photovoltaics does something entirely different and, in a sense, far stranger. It skips heat altogether. It converts light directly into electricity through the quantum shuffle of photons and electrons that Becquerel discovered and Einstein explained — no boiling, no steam, no moving parts, no engine at all. A photovoltaic panel sitting silently in the sun, generating current with nothing whatsoever moving inside it, is doing something with no precedent in the entire history of machinery, which had always, until then, meant heat and motion. That silence is the signature of the photovoltaic effect, and it is worth pausing to appreciate how deeply odd it is that a slab of specially prepared rock, lying still in the light, simply produces electricity.

It is worth adding one more thread to the human story, because it corrects a common impression that the photovoltaic effect leapt straight from Becquerel to Einstein. It did not; there was a quiet, crucial middle chapter in England. In 1876, three years before Becquerel's own son would be born, two British scientists, William Grylls Adams and his student Richard Evans Day, were experimenting with selenium — the same peculiar element Fritts would soon use — and found that it produced an electric current when exposed to light, with no heat and no external battery involved. This was the first demonstration that a solid material, not a wet chemical cell, could turn light directly into electricity, and it planted the seed Fritts would harvest seven years later. What is striking about this whole lineage — Becquerel in his liquid, Adams and Day with their selenium, Fritts with his gold-coated wafers — is that every one of these discoveries was made before anyone had the faintest theoretical idea why it should work. For nearly seven decades, the photovoltaic effect was a purely empirical fact, a thing that happened, catalogued and reproduced and even patented, with no explanation underneath it. It is a striking reminder that in science, the ability to do something often runs decades ahead of the ability to understand it, and that some of the most important technologies begin as unexplained tricks that stubborn tinkerers refuse to ignore.

Where do you meet Becquerel's teenage discovery today? You meet it, most likely, several times before breakfast. The little cell that keeps a pocket calculator alive without a battery; the panel on a parking meter or a highway sign or a garden lamp; the roof of a suburban house feeding power to the grid; the vast fields of silicon in the deserts we visited in earlier chapters; the wings of spacecraft and the roofs of electric cars — every one of them is a direct descendant of a nineteen-year-old's twitching needle in an 1839 Paris laboratory. The effect he found by accident, that no one could explain for two generations, now generates a substantial and fast-growing share of the world's electricity, silently, everywhere, one photon and one electron at a time. It is a rare thing in the history of technology: a pure scientific curiosity, useless and inexplicable for decades, that grew up to help reshape the energy supply of an entire planet.

To recap. In 1839 Edmond Becquerel discovered, by accident and at nineteen, that light shining on an electrode produces electricity — the photovoltaic effect. In 1883 Charles Fritts built the first solid-state cell from selenium and gold, a hopeless one percent efficient, yet he correctly foresaw solar rivaling coal. The deep puzzle of why color, not brightness, governed the effect defeated classical physics until Einstein's 1905 insight that light comes in photons whose energy is set by their color — the work that won him the Nobel Prize and laid the quantum foundation beneath every solar cell. Light frees electrons; a built-in slope makes them flow; and that, not storage and not heat, is what a panel does. But the sun has no monopoly on hard limits set by physics. Just as a solar cell can never catch every photon, a wind turbine can never catch every gust — and in 1919 a German physicist named Albert Betz proved exactly how much the wind will let a machine take.

Chapter 6 — Betz's Limit

In the German university town of Göttingen, in the years just after the First World War, there gathered one of the most remarkable concentrations of scientific talent the world has ever seen. At its center sat Ludwig Prandtl, the quiet, methodical man who had essentially invented the modern science of how air flows around solid objects — the science of aerodynamics — and around him clustered a generation of brilliant students and colleagues who would go on to shape aviation, rocketry, and fluid physics for the rest of the century. Among them was a young physicist named Albert Betz. In 1919, working within this hothouse of ideas, Betz asked and answered a question so fundamental that it has governed every wind turbine built from that day to this, and every one that ever will be built. The question was disarmingly simple: of all the energy carried by the wind blowing through a turbine, what is the very most that any machine, however cleverly designed, could ever possibly extract? Not a particular turbine — any turbine, ideal or real, present or future. Betz proved there is a hard ceiling, a law of nature no engineering can break, and he calculated exactly where it sits: at roughly 59.3 percent. No turbine, ever, can capture more than about three-fifths of the energy in the wind that passes through it. That number, 59.3 percent, is called the Betz limit, and understanding why it exists tells you something profound about the nature of harvesting a flowing resource.

To see why there must be such a limit, begin with a tempting but wrong idea — the idea that a perfect turbine would extract one hundred percent of the wind's energy. It sounds like the obvious goal. But follow the logic and it collapses on itself. The energy in the wind is its energy of motion, its kinetic energy, which it carries by virtue of moving. For a turbine to extract all of that energy, it would have to bring the air to a complete and total stop — every last bit of motion converted to spin in the rotor, the wind emerging on the far side utterly still, dead calm, not a breath. But now ask the fatal question: if the air behind the turbine is perfectly motionless, where does the next gust of incoming wind go? A wall of stationary air sits directly in its path, blocking the way. The oncoming wind cannot pass through, because there is nowhere for it to flow — the spent air has piled up behind the rotor like traffic behind a stalled car, with no room to move on. And if the wind cannot pass through the turbine, it cannot deliver any energy to it at all. So the very act of extracting one hundred percent of the energy would stop the flow, and stopping the flow would extract nothing. The two goals destroy each other. A turbine that took everything would, in the same instant, get nothing.

There is a homely analogy that captures the whole paradox, and it is worth keeping in mind because it makes the abstract concrete. You cannot drink a river dry and still expect it to keep flowing past you. If you want to take energy from a moving stream — of water, of air, of anything — the stream has to keep moving after it has passed you, which means you must leave some of its motion behind. Take too little and you have wasted an opportunity; take too much and you choke off the flow that brings you more. Somewhere in between lies a sweet spot, a best possible compromise where the air is slowed enough to give up a good share of its energy but still moving fast enough to get out of the way and let the next parcel through. Betz's achievement was to find that sweet spot precisely, using nothing more than the two great conservation laws of physics — that mass cannot be created or destroyed, and that momentum cannot either — applied to an idealized picture of a rotor. He modeled the turbine not as blades but as a simple, perfect disk that slows the air passing through it, a device engineers call an actuator disk, and he asked how much you slow the wind to extract the most power. The answer that fell out of the mathematics is that the ideal turbine slows the wind to two-thirds of its original speed at the rotor, and to one-third of its speed downstream, and at that optimal slowing it captures exactly sixteen twenty-sevenths of the wind's energy. Sixteen divided by twenty-seven is 0.593 — the famous 59.3 percent. It is a pure number, born of geometry and conservation laws, owing nothing to the materials or the cleverness of any particular machine.

Here is the first genuinely surprising thing to absorb: the Betz limit is not a temporary engineering shortfall to be overcome by better technology. It is a law, as absolute and unbudgeable as a speed limit written into the fabric of physics itself. We are accustomed, in the modern world, to the idea that today's technological limits are tomorrow's solved problems — that a clever enough engineer, given enough time and money, can push past almost any ceiling. Betz's limit is not that kind of ceiling. No breakthrough in blade design, no exotic new material, no artificial intelligence, no amount of funding will ever produce a wind turbine that captures seventy percent of the wind's energy, or eighty, let alone a hundred. The ceiling at 59.3 percent is as permanent as the fact that you cannot travel faster than light or build a perpetual-motion machine. In an age that instinctively distrusts the word "impossible," it is bracing to meet a genuine, proven, forever impossibility — and to realize that the entire wind industry has made its peace with it and built a thriving enterprise underneath it.

Which brings us to the second surprise, and it cuts the other way. Real turbines, out in the world, do not reach the Betz limit either — but they come remarkably, satisfyingly close to the practical maximum. A real turbine loses some energy that the idealized actuator disk did not: friction in the bearings and gearbox, the drag of air over the blades, the swirl the spinning rotor imparts to the air behind it, the electrical losses in the generator. Engineers bundle a turbine's real-world catch into a single figure called the power coefficient, which measures the fraction of the wind's energy the machine actually captures. The best modern utility turbines achieve a power coefficient somewhere around 0.35 to 0.50 under good conditions — that is, they capture roughly thirty-five to fifty percent of the wind's energy. Set that against the Betz ceiling of 0.593 and you find that the finest turbines are running at something like seventy-five to eighty-five percent of the theoretical maximum. For a machine wrestling a chaotic, gusting, invisible fluid, that is an astonishingly good performance — as if a runner, told the fastest a human could conceivably move was some fixed speed, were routinely clocking eighty percent of it. The wind engineer's job is not to beat Betz, which is forbidden, but to close as much as possible of the gap between a real machine and Betz's perfect disk, and a century of refinement has closed most of it.

Now, why do modern turbines almost all have exactly three blades? This is one of the most common questions people ask when they first really look at a wind farm, and the answer weaves together physics, economics, and a surprising amount of aesthetics — and it is emphatically not the intuitive answer. The intuition is that more blades should catch more wind: a rotor solid with blades, like a child's pinwheel or an old American farm windmill with its two dozen metal vanes, ought to grab more of the moving air than a spare three-bladed one. But it does not work that way, and here we can reach back to the last chapter's Danish pioneers. Poul la Cour, running the world's first wind-electricity laboratory in Denmark in the 1890s, discovered a counterintuitive truth that still governs turbine design: fewer, faster-moving blades are more efficient than many slow ones. A dense rotor of many blades presents the wind with something close to a solid wall; the air, meeting that near-wall, tends to pile up and spill around the edges rather than flow smoothly through, and much of its energy slips past unharvested. A small number of slender blades, spinning fast, sweeps the whole disk of air just as effectively — each blade passing quickly through the space its neighbor just vacated — while offering far less obstruction and far less drag. So beyond a certain point, adding blades does not add power; it adds cost, weight, and drag, and can actually reduce efficiency.

But if fewer blades are better, why not one blade, or two? Why did the world converge on three? Here the reasons become mechanical and economic. A single-bladed rotor, though it has been built, is violently unbalanced and must carry an awkward counterweight, and it spins so fast it looks unnerving and stresses its own structure. A two-bladed rotor is better but suffers a subtle and punishing problem: as the pair of blades swings from vertical to horizontal, the rotor's resistance to being turned by the wind changes sharply, and this sets up heavy vibrations and stresses every rotation, especially as the machine yaws to follow shifting winds. A three-bladed rotor, by a happy quirk of geometry, is dynamically smooth: with three blades evenly spaced, the rotor's balance stays essentially constant as it turns, so it runs steadier, quieter, and with far less fatigue on its bearings and tower. Three blades captures nearly as much energy as any sensible design can, spins slowly enough to look calm and to spare the structure, and balances beautifully — and adding a fourth or fifth blade buys almost no extra energy while adding real cost and weight. Three is the sweet spot where efficiency, cost, mechanical smoothness, and even the way the machine looks against the sky all come into alignment. It is one of those rare engineering answers that is right for half a dozen independent reasons at once, which is exactly why virtually every large turbine on Earth now has three blades. There is even a small anecdote in this convergence: the "Danish concept" three-blade upwind design, refined by la Cour's intellectual heirs like Johannes Juul, whose 200-

kilowatt Gedser turbine of 1957 became the direct ancestor of the modern machine, won out over decades of experiments with one, two, four, and even many-bladed rotors, until the whole world quietly agreed the Danes had been right.

Let me carry a small worked number through, to make the Betz limit tangible in the way earlier chapters made the cube law tangible. Suppose the wind blowing through a particular turbine's rotor carries, at a given moment, one thousand kilowatts of raw power — a nice round figure for the energy in the moving air passing through that circle. Betz tells us that no machine, however perfect, could extract more than 59.3 percent of it, so the absolute ceiling is about five hundred and ninety-three kilowatts; the remaining four hundred-odd kilowatts must, by the iron logic of the flow, be left in the departing wind to keep it moving out of the way. Now bring in reality: a good modern turbine, running at a power coefficient of, say, 0.45, will actually capture about four hundred and fifty kilowatts of that original thousand — a little over three-quarters of what Betz allows, and just under half of the raw total. So of the thousand kilowatts nature offered, physics itself forbade us more than five hundred and ninety-three, and good engineering delivered four hundred and fifty. That is the honest accounting of a wind turbine: not a machine that squanders most of the wind through incompetence, but one that takes very nearly all that the laws of nature permit it to take.

This lets us correct two stubborn misconceptions in one stroke. The first is the belief that a bigger, or faster-spinning, or more powerful turbine could somehow catch all the wind, or nearly all — that the 59.3 percent ceiling is just a limitation of current designs waiting to be smashed. It is not. Betz forbids it, permanently and for every conceivable machine. You can build a turbine with a bigger rotor to intercept more wind — recall the swept-area square law from the last chapter — but for any given wind passing through any given rotor, you will never extract more than about three-fifths of its energy, and no scale or speed or ingenuity changes that. Bigger turbines catch more wind because they reach into more of it, not because they beat Betz. The second misconception is the pinwheel intuition we have already dismantled: that more blades mean more power. They do not, beyond a point; a many-bladed rotor is often less efficient than a lean three-bladed one, because it obstructs the flow it is trying to harvest. The old multi-bladed farm windmills of the American West had many vanes not for efficiency but for high starting torque, to lift water against a load from a standing start — a different job entirely from generating electricity, which rewards fast, spare, high-efficiency rotors. Blades are not like buckets, where more is more. They are like the teeth of a comb through flowing hair: a few, well-shaped and fast, do the work better than a dense thicket.

Where do you meet Betz's limit today? You meet it every single time you see a wind turbine turning and notice, perhaps for the first time, that the air on the far side is still moving — that the turbine has not stopped the wind, only slowed it. That un-stopped, departing breeze is Betz's law made visible: the machine is deliberately, necessarily letting most of the wind's mass flow on through, taking its permitted share and no more, because to take more would be to choke itself. You meet the limit, too, in the very three-bladedness of nearly every large turbine on the horizon — a silhouette so universal we no longer see it as a choice, though it is the frozen conclusion of a century-long argument that la Cour started and the Danes won. And you meet it, quietly, in every honest energy calculation, every wind-farm feasibility study, every engineer's spreadsheet, where the number 0.593 sits as a permanent boundary, the wall past which no amount of money or genius can push, marking the edge of what the moving air will ever surrender.

To recap: in 1919, out of the great aerodynamics school at Göttingen, Albert Betz proved that no wind turbine can ever capture more than about 59.3 percent — sixteen twenty-sevenths — of the energy in the wind passing through it, because taking all of it would stop the air and block the flow, exactly as you cannot drink a river dry and expect it to keep running. Real turbines reach thirty-five to fifty percent, remarkably close to that hard ceiling, and the world settled on three blades because fewer fast blades beat many slow ones while three balance most smoothly of all. We have now assembled the physics of both harvests: the sun's light freeing electrons, the wind's motion turning rotors, and the firm natural limits on each. But machines to catch the wind are not new — humans were harvesting the restless air for grain and water and reclaimed land many centuries before anyone had heard of Betz or Coriolis, and it is to that long, inventive prehistory that we turn next.

Chapter 7 — The First Machines to Catch the Wind

On the plain of Sistan, where the modern border smudges between eastern Iran and southwestern Afghanistan, the wind has a name and a season. The locals call it the bad-e sad-o-bist-ruz, the Wind of a Hundred and Twenty Days, and it does exactly what the name promises: from late spring into the heart of summer it pours out of the north, day and night, week after week, so steady and so hard that it drives fine sand into every hinge and hem and mouth. Travelers have hated it for as long as anyone has written the place down. But somewhere around the eighth century of our era, in a settlement on that gritty, sun-blasted plain, someone did the opposite of hating it. Someone built a wall of mud brick, cut a slot in it to aim the gale, and set behind the slot a vertical spindle hung with bundles of reeds. When the wind struck the reeds they swung, the spindle turned, and above it — or below it — a millstone began to grind. For the first time in history, a machine was doing the work of a wind. No ox leaned into a yoke. No slave bent over a quern. The Wind of a Hundred and Twenty Days, cursed by every caravan that ever crossed the desert, had been put to work grinding somebody's flour.

We do not know that person's name. We almost never do, for the deep inventions. But we know the machine, because it survived — not just in old Arabic and Persian geographies but in the flesh, standing and turning in a handful of Iranian villages such as Nashtifan into the twentieth century, and lovingly restored there today. These are the asbad, the windmills of Sistan, and they are the oldest windmills on Earth that we can actually describe. To stand under one is to understand something the textbooks rarely convey: that the story of catching the wind does not begin with the tidy four-sailed silhouette of a Dutch postcard. It begins here, in the dust, with a wall and a bundle of reeds, roughly five hundred years before Europe built its first mill.

Let me explain how such a thing works, because the physics is beautifully simple and the trick at the heart of it is the whole secret. A windmill turns because moving air pushes on something. In the Sistan design, that something is a set of flat sails or reed bundles fixed to a vertical shaft — imagine a revolving door, or the little paddle wheel of a river steamboat stood on its end. Now here is the problem such a design has to solve. If the wind blows straight across a revolving door, it pushes just as hard on the panel coming toward it as on the panel going away. The two pushes cancel, and the door does not spin; it just shudders and stalls. A child on a merry-go-round learns this instinctively: to make the thing go, you have to push on one side only. So the Sistan builders pushed on one side only. They wrapped the rotor in a thick wall — really a small building of mud brick — with a single opening cut to face the prevailing wind. The wall shields the half of the rotor that would otherwise be shoved backwards, and it funnels the whole force of the gale onto the exposed half. The wind can only reach the sails that are traveling with it. The other sails hide behind the masonry, and the rotor spins, patient and one-directional, for as long as the season blows.

That shielding wall is the entire invention. It is why these are called, in the technical literature, "panemone" mills —

from Greek roots meaning "all winds," a slightly optimistic label, since in practice the fixed wall means the mill only works when the wind comes from the one direction the slot is built to catch. But on the Sistan plain that is not a limitation at all; it is a perfect fit. The Wind of a Hundred and Twenty Days comes reliably from the north, so you build your slot to face north and you are in business for a third of the year. This is the first great lesson of the prehistory of renewable energy, and it will echo through every chapter to come: the machine is always married to its place. A device that would be useless in the fickle breezes of an English valley was, on this particular plain with this particular wind, a small revolution.

Here is a fact to keep in your pocket. The oldest of these mills used a vertical axis — the spinning shaft stands straight up like a spinning top — whereas every wind turbine you have ever photographed on a hillside uses a horizontal axis, the shaft lying flat like a car's axle, the blades spinning like a propeller. The two families are as different as a paddle steamer and an airplane, and, as we will see, the horizontal design eventually won almost everywhere because it is far more efficient. But it did not come first. The wind's first machines spun sideways, wall-shielded, in the deserts of central Asia. And a second fact, to unsettle the tidy timeline further: the very first wind-driven machine we can name may not have ground anything at all. It made music.

Around the first century of our era, in Alexandria — the great Mediterranean city of scholars, its famous library still standing — an engineer named Hero (or Heron) wrote up a device in his treatise on pneumatics. Hero was the age's great tinkerer, the man who described a spinning ball driven by steam, coin-operated holy water dispensers, and self-opening temple doors — theatrical marvels meant to astonish worshippers. Among them was an organ, an instrument of pipes fed by a bellows, and Hero's twist was to drive the bellows not by a laboring human arm but by a small wind-wheel: a set of vanes that the breeze turned, working a piston that pumped the air that sounded the pipes. If his description means what it appears to mean, this is the oldest wind-powered machine on record — older than the Sistan mills by half a millennium. It did no useful labor; it was a toy, a curiosity for a temple courtyard. But it proves a point that matters enormously for how we think about our ancestors: the idea of harnessing the wind to a machine was in human heads long, long before anyone made it pay. The gap between imagining a thing and making it economically worthwhile is often the longest gap in the whole history of technology. We will meet that gap again and again — most poignantly with the solar engineers of the nineteenth century, whose brilliant machines arrived a century too early.

The Sistan mill and Hero's organ were not the only ancestors. Far to the east, in China, vertical-axis windmills of a wholly different and rather ingenious kind came into use — the earliest firm records date from around the thirteenth century, though some historians argue for earlier origins and possible influence flowing along the Silk Road in one direction or the other. The Chinese design dispensed with the fixed shielding wall and solved the revolving-door problem another way. Its sails were mounted on masts around the rim of a great horizontal turntable, rigged rather like the sails of a Chinese junk, and hinged so that they swung open to catch the wind on one side of the circle and feathered — swiveled edge-on — to slip through it with little resistance on the other. Picture a carousel of little boats, each one hoisting its sail as it runs downwind and dropping it as it beats back upwind. No wall required; the sails did their own shielding, automatically, every rotation. These mills turned the great brine pumps of the Chinese salt fields and lifted irrigation water for the paddies, and some worked into living memory. That two civilizations, at opposite ends of Asia, independently cracked the same problem in two different ways tells you the problem was ripe — that the pressure to escape the tyranny of muscle was being felt wherever grain had to be ground and water had to be raised.

And that pressure is the deepest theme of this chapter, so let me put it plainly. For almost the whole of human history, the only sources of power on Earth were living muscles — your own, your family's, your slaves', your oxen's and donkeys' and horses'. Everything that got done, got done because something alive got tired doing it. Grinding a day's grain for a household by hand, shoving a stone back and forth across a saddle quern, was hours of brutal, monotonous labor, and in most societies it fell on women and on the enslaved. To grind grain with the wind, or with falling water, was therefore not a minor convenience. It was the first time our species reached past its own body for power, the first withdrawal from an energy account we had never before been able to touch. Water got there slightly earlier and, as we will see in a later chapter, on a far larger scale — the watermill was already spreading through the Roman world. But wind was the other great prize, and it was available exactly where water often was not: in dry, flat, windy country where no stream fell fast enough to turn a wheel. The Sistan plain is the perfect illustration. It has almost no reliable falling water. What it has, four months a year, is a howling, inexhaustible river of air. The genius of the asbad was to tap the one abundant resource the desert actually offered.

Let me make the scale of that gift tangible with a small worked example, the kind we will return to throughout this book. The power carried by moving air rises with the cube of its speed — a relationship so important that I will spend a whole later chapter on it — which means fast, steady wind is worth wildly more than slow, gusty wind. The Wind of a Hundred and Twenty Days can blow at well over one hundred kilometers an hour in its fury, and even a modest working breeze on that plain carries far more energy per second than the puffs of a sheltered European lowland. A single one of these mud-walled mills, historians estimate, could grind on the order of a few hundred kilograms of grain in a day when the wind was up. Set that against a human being turning a hand quern, who might manage only a handful of kilograms in a day of aching, blistering effort, and you see the multiplication: one mill was doing the flour-grinding of dozens of exhausted people, and it never asked to rest, never fell ill, never had to be fed. It ate only wind. In a world where every calorie of work had a human or animal price, a machine that worked for free was close to miraculous — and it is no accident that these mills clustered where the wind was most reliable and the human alternative most miserable.

Now let me clear away some misconceptions, because this corner of history is thick with them. The first and most stubborn is that windmills were a European invention — that they are somehow essentially Dutch, or English, a product of the medieval West. They are not. The horizontal-axis mill of the European postcard is indeed a European development, and a magnificent one that we will honor in the next chapter. But the windmill as such — a machine to turn the wind's push into useful rotary work — was up and grinding in the Islamic world of Persia and central Asia centuries before it appeared in Europe, and it very likely traveled westward, along the trade and pilgrimage routes, as an idea whose time had come. The medieval Muslim geographers wrote about the Sistan mills with pride and precision; the technology was theirs first. To call the windmill European is a little like calling paper or the number zero European: it flatters the wrong ancestors.

The second misconception is broader and more insidious — the quiet assumption that ancient and medieval people had no real machines, that "technology" is a modern thing that begins with the steam engine, and that everyone before then simply hauled and heaved by hand in a fog of ignorance. Hero's wind-organ should be enough to dispel this on its own, and if it is not, the Sistan mills finish the job. These were sophisticated devices: they involved gearing to transmit rotation from the shaft to the stone, an understanding of how to shield and aim a rotor, careful choice of site, and generations of refinement passed down by builders who could not have written a single equation about any of it. People have always been ingenious. What changed in the modern era was not the arrival of cleverness but the arrival of cheap, concentrated, portable fuel — a change so total that it made us forget how much our ancestors had wrung from wind and water and muscle alone.

A third, gentler misconception is worth naming: that these old wind machines were crude, feeble things, quaint but negligible. In truth they were, for their setting, superbly matched to the job, and some of them worked continuously for centuries — the same mill grinding for the great-grandchildren of the people who built it. The reed sails wore out and were replaced; the mud walls were patched; the millstones were re-cut. But the design endured because it was right. When engineers went to study the surviving Nashtifan mills in the twentieth century, they found a machine that a modern person could understand in five minutes and could not obviously improve for its purpose. That is not crudeness. That is elegance under constraint.

Where do you meet all this today? More places than you might think. If you have ever seen the towering, sail-decked profile of a modern vertical-axis wind turbine — the eggbeater-shaped Darrieus rotor, or the S-curved Savonius drum that spins on top of some ventilation ducts and highway signs — you are looking at the direct conceptual descendants of the Sistan panemone: rotors that spin about an upright shaft and do not care which way the wind blows. Engineers keep returning to the vertical-axis idea for small-scale and urban settings precisely because of the old virtue — it accepts wind from any direction without needing to turn to face it. And in Iran, the restored windmills of Nashtifan still stand and, on windy days, still turn, tended now not for their flour but as one of the oldest working machines on the planet, a UNESCO-recognized monument to the first people who ever made the wind do a day's work. Stand there in season and you can hear it: the groan of ancient timber, the whir of reeds, the low grinding note that is, quite literally, the oldest sound of the renewable age.

So this is where the story of catching the wind truly opens — not in the Low Countries but on a desert plain, with a wall, a slot, and a bundle of reeds; not with a single genius but with anonymous builders in Persia and independent inventors in China, all of them wrestling the same revolving-door problem to the ground; and, before any of them, with a Greek engineer in Alexandria who wired a breeze to a set of organ pipes just to hear it play. The wind's first machines were vertical, wall-shielded, married to their windy homes, and quietly revolutionary, because they were among the very first devices that let human beings reach past the limits of living muscle. They ground grain and lifted water and never grew tired. What they could not yet do was face the wind and follow it wherever it came from — and it was in the valleys and coastlines of medieval Europe, with a wholly new idea about which way a windmill's shaft should point, that the next great leap would come.

Chapter 8 — The Windmills of Europe

In the flat green heart of the Netherlands, a little south of Rotterdam, there is a place called Kinderdijk where nineteen windmills stand in a row along the water like a congregation waiting for a service. They were built in the eighteenth century, most of them around 1738 and 1740, and they were not built to grind anything. They were built to fight the sea. This whole landscape sits below the level of the rivers and the ocean that surround it — some of it several meters below — and without ceaseless pumping it would simply fill up and drown, as it has drowned catastrophically more than once in Dutch history. The Kinderdijk mills were the pumps. Each one caught the wind on four great sail-clad arms, turned a wheel of wooden scoops or a spiraling screw, and lifted water from the low fields up into higher channels, from those channels into higher ones still, and finally out into the river and away. They worked in relays, mill lifting to mill, a bucket brigade made of wind. On a blustery day you can stand on the dike and watch the whole line turning at once, and understand something that no diagram teaches: that a substantial piece of one of the world's richest countries is dry land only because, for centuries, the wind agreed to bail it out.

To reach these mills we have to leave the deserts of the last chapter and cross both a continent and a conceptual divide. The windmills of Sistan spun about a vertical shaft and hid half their rotor behind a wall. The windmills of Europe did something completely different and, it turned out, far more powerful: they turned the shaft on its side. The European mill is a horizontal-axis machine. Its sails spin like a great four-armed propeller in a vertical plane, facing the wind head-on, the axle running roughly level into the body of the mill. This is the silhouette that has become, for most of the world, the very picture of a windmill — and it appeared, so far as the records show, only around the twelfth century, first in the windy lowlands of northwestern Europe: England, northern France, the Low Countries. Roughly five hundred years after Persia, Europe arrived at the windmill by a different door, and the door it came through changed everything.

Why does turning the axle sideways matter so much? Here we need a little first-principles physics, and I will give you two everyday pictures to hold onto. The Sistan mill worked by simple push — the wind shoved flat sails the way it shoves an open umbrella you are foolishly holding into a gale. Push works, but it is wasteful, because the sails moving away from the wind fight you and have to be walled off. The European horizontal mill could exploit a subtler and stronger effect: it could set its sails at an angle, like the blades of a fan or the slats of a venetian blind, so that the wind did not merely shove them but was deflected off them, giving the sail a sideways force that swept it around. Think of how a sailing boat can sail across the wind, even somewhat into it, because its angled sail redirects the airflow rather than just catching it like a parachute. The angled windmill sail is doing a cousin of the same trick. Because every sail on the wheel is angled the same way, the whole rotor is driven around continuously and evenly, and — crucially — there is no half of the rotor that needs to be shielded, because all four sails are working all the time. A propeller-style rotor sweeps a big circle and takes power from the whole of it. This is fundamentally the same reason that, eight centuries later, every giant turbine on every modern wind farm is a horizontal-axis machine with slender angled blades, and not an upright wall-shielded drum. The Europeans, without any theory of aerodynamics whatever, had stumbled onto the more efficient family of design.

But a horizontal rotor buys its efficiency at a price, and the price is a problem the desert mills never had to solve. A wall-shielded vertical mill is happy as long as the wind blows from its one appointed direction. A horizontal rotor, by contrast, must face directly into the wind to work at all — and the wind in Europe is a notorious weathercock, swinging around the compass from hour to hour. So the whole history of the European windmill becomes, in large part, the history of clever ways to keep the sails pointed into a shifting breeze. The first solution was gloriously brute-force: the post mill. In a post mill, the entire body of the windmill — the sails, the machinery, the grinding stones, the miller's whole workplace, tons of oak and stone — is balanced on a single massive vertical post and can be physically rotated as one piece. When the wind shifted, the miller went outside, put his shoulder to a long tail-beam that stuck out the back like a tiller, and shoved the entire building around until the sails caught the wind square. Imagine turning a small house by leaning on a broom handle. That is the post mill, and thousands of them dotted medieval Europe, each one a building that the wind could make dance.

The post mill was heroic but heavy, and in time a smarter idea prevailed: leave the building still and turn only the top. This is the tower mill, built of stone, and its wooden cousin the smock mill, so called because its sloping weatherboarded sides reminded people of a countryman's smock. In these, the sails are mounted on a rotating cap at the very top of a fixed tower, and only that cap — carrying the sails and the top of the drive shaft — swings to face the wind, while the miller and his machinery stay put on solid ground floors below. Turning just the cap is far easier than turning the whole mill, and it can be done from inside. The

Dutch and the English devised an automatic gadget to do it without human muscle at all: a small secondary wind-wheel, called a fantail, mounted at right angles on the back of the cap. When the main sails were correctly facing the wind, this little fantail sat edge-on to the breeze and did not turn. But the moment the wind shifted, it struck the fantail at an angle, spun it, and through a train of gears the fantail crept the cap around until the main sails once more faced the wind square and the fantail once more stalled. It is a self-correcting machine — a wind-powered device for aiming a wind-powered device — and it anticipates, by centuries, the yaw motors that swing a modern turbine's nacelle to face the wind today. Here is a surprising truth worth pausing on: the medieval windmill already contained, in wood and gearing, the same three functional problems every modern turbine still solves — how to catch the wind efficiently, how to keep facing it, and how to avoid tearing yourself apart when it blows too hard. The miller reefed his sails in a storm the way a sailor reefs his; the modern turbine pitches or feathers its blades. Different hardware, identical logic.

Inside, the mill was a machine of real sophistication, and to call it simple — as people often do — is the first misconception I want to correct. The great sails turned a horizontal windshaft; that motion had to be turned through a right angle to drive vertical millstones, and it had to be geared up, because the sails turned slowly and grand and the stones needed to spin briskly. So inside every mill sat a big wooden gearwheel, the brake wheel, meshing with a smaller gear, the wallower, on an upright shaft, which drove another great wheel, which drove the stones. This is a gear train, the same class of device as the transmission in your car, built entirely of shaped oak teeth that a millwright carved, fitted, and replaced as they wore. The miller had to regulate the gap between the millstones to a fraction of a millimeter to grind fine flour without scorching it; he had to reef and set his sails as the wind rose and fell; he had to keep the whole thing from running away and destroying itself in a gale, which could and did happen — a mill whose sails were caught unbraked in a sudden storm could spin until the friction set the timber alight and burned itself to the ground, a fate common enough to have its own dread among millers. This was not a quaint gadget. It was, for its time, a genuinely dangerous and demanding piece of heavy machinery.

Which brings me to the humans, and to the single most useful way of understanding what these mills meant: the European windmill was the factory of the pre-industrial world. We think of factories as belching brick sheds full of steam and iron, a nineteenth-century thing. But a factory is really just a place where power is concentrated and applied to make things at a scale no household could match, and by that definition the windmill — and its cousin the watermill — was the first factory Europe had. Yes, mills ground grain into flour, the essential daily work of feeding a town. But wind power was quickly harnessed to far more. Mills sawed timber into planks, their reciprocating saws driven by the sails through a crank, turning out lumber for the Dutch shipbuilding boom that made a small wet country into a global naval and trading power. Mills fulled cloth, pounding woolen fabric with great wooden hammers to thicken and clean it. Mills crushed seeds for oil, ground pigments for paint, made paper, and pumped water. The town of Zaandam near Amsterdam became a roaring industrial district powered entirely by wind, with hundreds of mills packed together, sawing and grinding and pounding — a genuine industrial cluster three centuries before the word "industrial" was coined, its power coming not from coal but from the sea breeze.

The scale of it is genuinely startling, and here are the facts to prove it. At the peak of the windmill age, before steam swept them aside, Europe had, by careful estimates, on the order of two hundred thousand windmills turning — some counts run higher still. The Netherlands alone had perhaps ten thousand of them at the height of the polder-draining era; England had thousands; the great plain of northern Europe, from Brittany to Poland, was stitched with them, a sail turning on nearly every rise. And they made land. This is the third surprising fact, and it is close to miraculous: a very large fraction of the Netherlands — the polders, the neat rectangular fields you fly over on the way into Amsterdam, home to millions of people and some of the most productive farmland on Earth — exists as dry land only because generations of windmills pumped the water out and kept pumping it, year after year, century after century. The Dutch have a proud saying, "God made the world, but the Dutch made the Netherlands," and the tool they made it with, for four hundred years, was the windmill. It is one of the largest and longest engineering projects in human history, and its engine was the wind.

Let me anchor the scale of the work with the running kind of example this book likes to keep. A good drainage mill of the Kinderdijk sort might have been rated, in modern terms, at only a few tens of kilowatts on a strong day — a fraction of what a single electric water pump you could buy today produces. It could lift its scoop wheel of water perhaps a meter or a meter and a half, no more, which is why the mills had to work in relays, each passing the water up to the next like a chain of hands passing buckets up a ladder. One mill could not lift the water all the way from the sunken field to the high river; three or four in series could. Multiply one modest, tireless mill by ten thousand of them, each nudging its portion of water uphill through every windy hour of every year, and you have enough continuous pumping to hold back the ocean from a nation. That is the arithmetic of the polders: not one enormous machine but a vast, distributed fleet of small ones, each doing a little, all doing it constantly. It is, I would gently point out, exactly the arithmetic of a modern renewable grid — thousands of modest, weather-driven machines summing to a mighty total — and the Dutch were running it, in oak and canvas, three hundred years ago.

Now to the misconception everyone arrives with, the one

wearing armor and carrying a lance. In Cervantes's novel of 1605, the deluded knight Don Quixote spies a row of windmills on the plain of La Mancha, mistakes their turning sails for the flailing arms of monstrous giants, and charges — only to be flung from his horse when a sail catches his lance. It is one of the most famous scenes in all of literature, and it has quietly taught generations of readers to file windmills under "charming, obsolete, faintly ridiculous." But look again at what the scene actually shows. Quixote is mad precisely because he cannot see the windmills for what they are: modern machinery, the cutting-edge technology of his own day, brand-new arrivals on the Spanish plain. The joke is on the man who romanticizes the past and misses the powerful device right in front of him. There is a lesson there for our own age, which is prone to filing solar panels and wind turbines under "nice idea, not serious." The windmills of La Mancha were not medieval relics; in 1600 they were the future, and Cervantes knew it. To "tilt at windmills" ought to mean fighting the real and formidable, not the imaginary and quaint.

A second misconception worth naming is the one I have already been chipping at: that windmills were simple, primitive things, all creaking charm and no engineering. I hope the gear trains and fantails and the fine flour and the burning brakes have laid that to rest. A working windmill demanded a skilled resident operator, the miller, who was a person of real standing and real suspicion in his community. He controlled the one machine everyone needed — you could not eat until your grain was ground — and he was paid in a portion of the flour, the "multure," which gave him ample opportunity, real or imagined, to skim a little extra. The miller with his thumb on the scale is a stock villain of medieval tales from Chaucer onward, and the resentment was structural: here was a man who owned the town's power plant and charged rent on daily bread. A third, quieter misconception is that windmills were purely rural and pastoral. Tell that to the roaring wind-powered sawmill district of the Zaan, where the air was full of sawdust and the pounding of a hundred mills — an industrial soundscape as harsh as any coal-fired one to come.

Where do you meet the European windmill today? Everywhere the word "mill" hides inside another. The family names Miller, Müller, Molnar, Molina, and their cousins across Europe all descend from the man who ran the mill — one of the commonest surnames on the continent, a genealogical fossil of how central these machines were. You meet it at Kinderdijk itself, now a World Heritage Site, where those nineteen mills still turn for visitors and, on occasion, still pump. You meet it in the American farm windmill, that skeletal multi-bladed steel fan on a lattice tower that pumped water for cattle and homesteads across the drying Great Plains in the nineteenth century — a direct descendant, re-engineered for a new frontier, and still spinning on ranches today. And you meet it, transformed almost past recognition but unmistakably related, every time you pass a modern wind farm: the horizontal axis, the rotor swung to face the wind, the blades angled to deflect rather than merely catch it, the machinery to spill power in a storm. The turbine on the hill is the great-great-grandchild of the mill on the dike.

So the wind, which we last left spinning a wall-shielded drum in the Persian desert, arrived in medieval Europe wearing a new and better shape: the horizontal-axis rotor, propeller-faced and angled-sailed, mounted first on a whole revolving building and then on a nimble turning cap steered by a self-correcting fantail. Tens of thousands of these machines became the factories of a world that had no coal — grinding and sawing and fulling and, most astonishingly, pumping a nation's worth of land up out of the sea and holding it there for four centuries. They were not simple, they were not quaint, and they were not, whatever the postcards imply, the beginning of the windmill's story or the whole of it. But powerful as they were, they and their watery cousins shared one hard limit that no amount of oak and canvas could break: they could only give as much as the wind and the falling water happened to offer, when and where they offered it. To understand why that limit ruled human life so completely — and why breaking it would change everything — we have to step back and look at the entire energy budget of a world that ran on nothing but muscle, water, and wind.

Chapter 9 — The Muscle of the World

Open the Domesday Book, the great survey of England that William the Conqueror's clerks completed in 1086, and among the counts of villages and ploughs and pigs you will find, recorded with the same bureaucratic patience, the mills. There were roughly six thousand of them turning in England at the time — water mills, almost every one, their wheels dipping into the country's streams and rivers to grind the kingdom's grain. Six thousand mills for a population of perhaps two million people. That is one mill for every few hundred souls, which tells you two things at once: that England had already mechanized its grain-grinding to a remarkable degree eight hundred years before the Industrial Revolution, and that this was very nearly the ceiling of what the country could do, because there were only so many streams, and every good mill site was already taken. The Domesday survey is, among other things, an inventory of a civilization pressing hard against the limits of the only power it had. To understand where solar and wind are taking us, we first have to feel, in the body, how tight those old limits were — and how the entire shape of human society was pressed into being by them.

Let me lay out the whole of the old power economy, because it is short. Before fossil fuel, there were exactly four sources of usable power on the planet, and you have now met all of them. There was human muscle. There was animal muscle. There was falling water. And there was wind. That is the complete list. Everything human beings built, moved, ground, forged, hauled, and made — every cathedral, every galley, every acre ploughed, every ton of flour — was accomplished by drawing on one or more of those four, and nothing else. There was no fifth option. Fire, of course, gave heat and light, and heat was indispensable for cooking and smelting and warmth; but fire before the steam engine could not be turned into motion, into mechanical work, into the pushing and lifting and turning that gets things done. Heat and work were two separate kingdoms with no bridge between them. Building that bridge — learning to make heat do work — is the single event that ends this prehistory and begins the modern world, and we are almost to it. But first we must sit inside the world without it.

The defining feature of that world was scarcity of power so severe that it is genuinely hard for a modern person to imagine. Here is the most useful way I know to frame it: think of a whole civilization living on a strict energy budget, the way a household on a tight income lives on a strict money budget. Every task costs power, and the total power available each day is fixed and small, because it all has to come, ultimately, from muscles that must be fed and rested, or from the wind and water that blow and flow when they please and not otherwise. You cannot borrow against tomorrow's wind. You cannot stockpile a river. If you want more ploughing done, you need more oxen, and more oxen need more pasture, which is land you cannot then use to grow human food, so the ox competes with the family for the fields. Every increase in power comes out of the same fixed pie of land and daylight and calories. This is what economists studying the period call the "organic economy," or the "photosynthetic constraint," and the phrase is exact: essentially all the energy in that world, whether it moved through a man's arm or a horse's shoulder or the grain that fed them both, had arrived recently as sunlight, been caught by growing plants, and could be spent only as fast as the sun and the soil could replace it. It was a solar economy already — a real and total one — but a solar economy running with no storage and a tiny battery.

Let me put some numbers on the muscle, because they are humbling. A healthy human laborer can sustain something like seventy or eighty watts of useful mechanical work over a full day — call it a tenth of one horsepower — which is to say that a person pedaling flat out to light a single old-fashioned hundred-watt bulb could not quite keep it burning. Everything that a pre-industrial peasant accomplished in a lifetime of labor was delivered at roughly the power of a dim light bulb. A strong draft horse does far better, sustaining perhaps five hundred to seven hundred watts — which is precisely why the horse was so precious, and why we still measure our engines in "horsepower," a unit James Watt coined to sell his steam engines by comparing them to the animal they replaced. But even the horse is a feeble engine by our standards. The car idling in your driveway develops, at a modest cruise, the power of a hundred or more horses working at once, and a jetliner at takeoff the power of hundreds of thousands. We are so drowned in power today that we cannot feel how astonishing it would look to any human being who lived before about 1750, for whom a single horse was a serious capital asset and a team of them the difference between prosperity and hunger.

Falling water did better than any muscle, which is why watermills, not windmills, were the true workhorses of the pre-industrial world — a fact often forgotten by people who picture the age as a landscape of turning sails. A watermill could develop several times the steady power of the best horse, and it did so day and night without eating, which is why a good mill site on a reliable stream was among the most valuable pieces of property a medieval lord could own. There were two main ways to take a stream's power, and the difference between them is worth understanding, because it is the same difference that governs a hydroelectric dam today. The undershot wheel simply dipped its lower paddles into a flowing stream and let the current push them, like a hand trailed in a river — cheap to build, but weak, capturing only the water's speed. The overshot wheel was cleverer and far stronger: water was led along a channel to the top of the wheel and poured into buckets on its rim, so that the sheer weight of the falling water, not merely its motion, turned the wheel down and around. An overshot wheel harvests the full drop of the water, and a well-built one could be several times as powerful as an undershot on the same stream. This is why the great medieval monasteries, the most sophisticated engineers of their day, went to enormous trouble to build weirs and leats and races that gathered water and dropped it onto overshot wheels — they were maximizing their power the way a modern utility maximizes the head of water behind a dam. The Cistercian monks in particular ran what were, in effect, water-powered industrial estates, with mills grinding, fulling, and hammering iron, all from the fall of a carefully managed stream.

And here is where the tightness of the budget carved itself directly into the shape of society, which is the deepest point of this chapter. Because water power could only be had where water fell, and because there were only so many places it fell, geography became destiny. Towns and industries clustered where the power was, which meant on rivers — and not just any stretch of river, but the stretches where it dropped fast enough to turn a wheel, at fords and falls and rapids. A vast number of the world's older cities sit exactly where they sit because that was where a river could be made to work: at a fall line, a bridging point, a place where the water gave power and the land gave a crossing. The map of pre-industrial civilization is, to a striking degree, a map of where the four powers could be gathered. Windy coasts and lowlands got windmills; falling rivers got watermills; and everywhere else, where neither wind nor water served, the work fell back onto muscle — which meant, with grim frequency, onto the muscles of people who were not free.

I want to state this plainly, because it is the darkest fact of the old energy economy and it is too often left out of the pretty picture of a world of turning sails. When power is scarce and can only come from muscle, human muscle becomes a resource to be owned, and societies again and again solved their energy shortage by forcing other human beings to supply it. The rowers chained to the benches of a war galley were an energy system. The enslaved people who ground grain, hauled stone, and worked the fields of the ancient and medieval and early-modern worlds were an energy system. Serfdom, which bound the medieval peasant to his lord's land and labor, was in large part a way of guaranteeing a supply of muscle power to those who commanded it. I do not mean that cheap energy alone explains slavery and serfdom — human cruelty needs no single cause, and these institutions had many roots. But it is no coincidence that unfree labor was near-universal in the age of the muscle economy and that it collapsed, across much of the world, in the same century that fossil-fueled machines began to supply mechanical power at a price no gang of enslaved workers could match. When a steam engine could do the work of five hundred men for the cost of some coal, the economic ground beneath human bondage shifted. This is not a comforting story of moral progress alone; it is also, uncomfortably, a story about energy. The abolitionists supplied the conscience. The coal supplied, in part, the opportunity.

Now the rupture. For all its ingenuity, the organic economy had a hard roof over it, and no amount of cleverness could raise that roof, because it was set by the sun's yearly gift of growth and by the fixed geography of wind and falling water. A society could get more efficient — better wheels, better gears, better breeds of horse, more mills packed onto every stream — but it could not fundamentally get more, because the total was capped by nature's annual budget. Then, over the eighteenth and nineteenth centuries, human beings learned to do the one thing that broke the cap: they learned to burn coal to make heat, and to make that heat do mechanical work, through the steam engine. Coal changed everything because it was not this year's sunlight or even this century's — it was hundreds of millions of years of ancient sunlight, caught by long-dead forests and pressed into the ground and stored, a colossal battery charged over geological time that humanity suddenly learned to discharge. Where the old economy could spend only as fast as the sun refilled it, coal let us spend a vast inheritance all at once. The energy ceiling did not rise. It vanished. And with it vanished the world of the six thousand watermills pressing against the limits of England's streams — for the new engines could be built anywhere, needed no river and no wind, ran day and night through every calm and every drought, and delivered power in quantities that would have seemed to any medieval miller like the work of gods. This is the great rupture that separates all the chapters before it from all the chapters after, and every triumph and catastrophe of the modern age, the renewable revolution included, flows from it.

Let me now correct two misconceptions that this history invariably breeds, because both are common and both matter for how we think about our own moment. The first is that pre-industrial life was static — a flat, changeless millennium of peasants grinding grain the same way forever until steam woke everyone up. It was nothing of the kind. The spread of the watermill and then the windmill, the shift from undershot to overshot wheels, the harnessing of water and wind to sawing and fulling and forging, the slow perfecting of gear trains and the horse collar and crop rotation — these were real and consequential advances that steadily raised what a given amount of land and labor could produce. The organic economy was not asleep; it was innovating hard, right up against a ceiling it could not see how to break. Recognizing this matters, because it shows that human ingenuity was never the missing ingredient. What was missing was a new source of power, and when it came, the ingenuity that had been straining against the old ceiling for centuries was released all at once.

The second misconception is the one most relevant to everything that follows in this book, and I want to name it carefully. It is the idea that turning to solar and wind power now means "going backwards" — returning to some pre-modern, pre-industrial past, trading the mighty engines of the coal age for the feeble sails and paddle wheels of the medieval one. This is a natural thought, given that the sun and the wind are precisely the ancient powers this chapter has been describing, and it deserves a fair and full answer rather than a dismissal. The answer is that a modern solar panel or wind turbine has almost nothing in common with the old organic economy except the raw resource it draws on. The medieval mill was limited by the ceiling of the organic economy because it could not store or concentrate or move its power, could not work when the wind fell, could not be built where the resource was absent, and captured only a small and crude fraction of what came its way. A modern turbine is a precision aerodynamic machine that harvests a large fraction of the wind's energy through a deep understanding of physics the medieval world entirely lacked; a modern solar cell converts sunlight straight into electricity through quantum effects no one had even imagined; and both feed a grid that can move their power across continents at the speed of light and, increasingly, store it in batteries against the calm and the dark. The old sun-and-wind economy failed at its ceiling for lack of storage, transmission, and conversion. The new one is being built precisely out of storage, transmission, and conversion — the very things the medieval miller never had. To call it "going back" is like calling a jet aircraft a return to the age of birds because both use wings. The resource is ancient. The machine is not.

Where do you meet the muscle economy today? You meet it every time you say "horsepower," a unit that still rates the engine in your car by the number of animals it replaced. You meet it in the water mills that survive as picturesque restaurants and museums on old millstreams across Europe and New England, their great wheels turning for tourists where they once turned for flour. You meet it, most viscerally, at the gym, where affluent people now pay to generate, on a stationary bicycle, the same tenth of a horsepower that their ancestors spent every waking hour trying to escape — a species that fought for ten thousand years to stop pedaling now pedals for pleasure, because our machines have made muscle power so worthless that we must manufacture reasons to use our bodies at all. And you meet the whole story, quietly, on your electricity bill, where the kilowatt-hours you buy for a few cents each represent an amount of work that would have taken a strong laborer weeks to deliver by hand. The average home in a wealthy country now commands, at the flick of a switch, the equivalent of dozens of tireless invisible servants — the "energy slaves" that some writers use to make the comparison vivid — none of them fed, none of them chained, all of them descended from the rupture that ended the world of muscle.

So this is the whole of the old power economy, and its hard ceiling: four sources only — human muscle, animal muscle, falling water, and wind — every one of them a form of recent sunlight, all of them together capping what a civilization could do, pressing cities onto rivers and, at its cruelest, pressing human beings into bondage to supply the muscle that machines could not. That ceiling shaped everything until coal shattered it, unleashing the modern world and, with it, both the prosperity and the atmospheric burden that our own century must now reckon with. Wind, as we have seen, was one of the two great ancient powers that strained against that ceiling. The other, the deeper and more abundant one — the very sunlight from which wind and water and muscle all secretly descended — had scarcely been touched directly at all. To capture the sun's warmth head-on, not through the slow patience of growing plants but through the focused fire of a mirror or a lens, was a dream that reached back to antiquity, to a Greek besieged in his city with the enemy fleet in the harbor and, so the legend runs, nothing but the sun and a set of polished shields to defend him.

Chapter 10 — The Burning Mirror

In the year 212 before our era, the Roman fleet of Marcus Claudius Marcellus lay in the harbor of Syracuse, on the eastern coast of Sicily, besieging the greatest city of the Greek west. Inside the walls was an old man of about seventy-five, the finest mathematician and engineer of the ancient world, Archimedes. He had already tormented the Romans with cranes that reached over the walls to pluck ships from the water, with catapults ranged to every distance, with claws of iron that dropped on the galleys and shook them to pieces. And then, according to a tradition that would grow and glow across two thousand years, he reached for the sun itself. He is said to have arranged mirrors — polished bronze shields, in some tellings, angled and aimed by his soldiers — so as to gather the Sicilian sunlight into a single burning point and throw it onto the Roman ships until their timbers smoked and burst into flame. The invaders, it was said, fled in terror from a weapon they could not see and could not fight: sunlight sharpened into fire.

It is a magnificent story, and I have to tell you at once that it is almost certainly not true. But the reason it is not true, and the reason it has nonetheless refused to die for two millennia, together open the door onto one of the oldest and most seductive dreams in the history of energy: the dream of taking the sun's diffuse, gentle warmth and concentrating it, focusing it, sharpening it into something fierce enough to do real work. This chapter is about that dream in its long infancy — the burning mirrors and the sacred flames and the great lenses — and about the quieter, humbler eighteenth-century experiment, a simple insulated box, that turned out to matter far more than any death ray and that leads in a straight line to the solar collector on a modern roof.

Let me first say why the Archimedes legend should be treated with fond skepticism, because doing so fairly is more instructive than either swallowing it whole or sneering at it. The earliest sources for the siege of Syracuse — the historians closest to the events, writing within a couple of centuries — describe Archimedes's cranes and catapults in loving detail and say nothing whatsoever about a burning mirror. The blazing-sun story appears only much later, growing more vivid with each retelling across the Byzantine and medieval centuries, which is exactly the signature of a legend accreting rather than a fact recorded. And then there is the physics, which is stubborn. To set a ship's timber alight with focused sunlight, you must gather the light falling on a large area and concentrate it onto a very small spot, and hold that spot precisely on the same patch of wood, without wavering, for a considerable time — long enough to raise damp, salt-soaked, tarred marine timber to its ignition temperature, which is a good deal harder than igniting a dry scrap of paper. The ships were bobbing on the swell. The sun was moving across the sky. Each mirror-bearing soldier would have had to track a rolling target and keep his own contribution of light pinned to the exact spot every other soldier was aiming at, all at once, all the while under fire. In the modern era, people have actually tried to test this. A group at the Massachusetts Institute of Technology famously attempted it with a hundred and something mirrors aimed at a mock wooden ship and, under ideal, motionless, cloudless conditions, managed after a long while to get the wood to char and briefly flame — and concluded that against a moving fleet in the chaos of battle it would have been hopeless. The popular myth-testing television show that tried the same thing reached the same verdict: busted, at least as a practical weapon of war.

And yet — here is the fact that keeps the legend honest — the underlying physics is entirely real. You genuinely can burn things with concentrated sunlight; any child who has ever aimed a magnifying glass at a dry leaf and watched it smoke and catch has performed the experiment. The reason the myth is so durable is that it is not physically impossible, only militarily implausible; the sun really does carry enough energy to start a fire if you gather and focus it well enough. So the deep idea at the heart of the Archimedes story is sound, and it is the idea this whole chapter turns on. Let me explain it from first principles, because it governs everything from the child's magnifying glass to the vast mirror fields of a modern solar power tower.

Sunlight arrives at the ground spread thin. On a clear day with the sun high, the flux of solar energy striking a surface aimed straight at the sun is roughly one kilowatt — a thousand watts — on every square meter. That is a real amount of power, about the draw of a small electric heater, but it is spread over a whole square meter, and spread thin it feels only pleasantly warm on your skin; it will never start a fire, because at any one point the energy is arriving too slowly to raise the temperature faster than the surface can shed it. The trick of concentration is to gather the sunlight falling on a large area and funnel it all onto a small one. A magnifying glass the width of your palm collects the sunlight falling across its whole face and bends it to a point the size of a pinhead; all the energy that was spread across the lens is now crammed into that tiny spot, and there the arrival of energy overwhelms the surface's ability to shed it, the temperature races upward, and the leaf ignites. Here are two everyday pictures for the same principle. The first: think of rain falling on a broad, funnel-shaped roof, gentle everywhere, that all drains to a single downspout — stand under the open sky in that rain and you are lightly wetted, but stand under the downspout and you are knocked flat by the same water, now gathered. Concentration does to sunlight what the funnel does to the rain. The second picture: a crowd leaving a stadium moves slowly and harmlessly across the wide concourse, but funnel that same crowd through one narrow gate and the press becomes dangerous — same number of people, concentrated. A curved mirror or a lens is a funnel for light, and what it concentrates, at its focus, is heat.

The ancients knew this in practice, if not in theory, and used it for purposes closer to reverence than to war. In Rome, the sacred fire of the goddess Vesta, tended perpetually by the Vestal Virgins in the heart of the city, was never supposed to go out; but if through some failure it did, ritual forbade relighting it from any common flame. One permitted method of rekindling the sacred fire was to draw it fresh from the sun, using a concave vessel or mirror to gather the sunlight to a focus and catch new flame in a scrap of dry tinder — fire born, purely and without human contamination, from the heavens. The same practice appears in the ancient world for lighting ceremonial flames, and it echoes still: the flame for the modern Olympic Games is kindled, to this day, at Olympia in Greece by a priestess who holds a scrap of tinder at the focus of a parabolic mirror and takes fire directly from the sun, exactly as the ancients did. If you have ever watched that ceremony on television before an Olympic Games and seen the little curl of smoke catch in the polished bowl, you have watched a genuinely unbroken thread of solar-concentrating technology reaching back past Archimedes into the mists. That is our first "where you meet it today," and it is older than almost any other living technology on Earth.

The burning mirror and the burning lens passed down through the centuries as objects of philosophy, magic, and increasingly of serious science. By the seventeenth and eighteenth centuries, natural philosophers were building "burning lenses" and "burning mirrors" of real size and power to probe the nature of heat and matter, and here we meet a genuine scientific triumph and a wonderful anecdote. In 1774, in Paris, the great chemist Antoine Lavoisier — the man who would name oxygen, establish that combustion is a chemical union with that gas, and lay the foundations of modern chemistry before losing his head to the guillotine in the Revolution — needed the highest temperatures he could reach, hotter than any furnace of the day, to see how substances behaved under extreme heat without contaminating them with the smoke and ash of a fire. So he turned to the sun. He had built for the French Academy of Sciences an enormous burning lens, actually a compound of two great lenses filled with wine spirits or fluid, mounted on a massive wheeled carriage so it could be turned to track the sun, the whole apparatus large enough to require a team to operate. With it he focused sunlight to a spot fierce enough to melt platinum, one of the most stubbornly heat-resistant metals known, and to burn diamond — proving, in the cleanest possible way because there was no flame and no fuel to muddy the result, that a diamond is made of carbon and can be made to combine with oxygen just like a lump of coal. It is a marvelous image: the founder of modern chemistry, in the last years of the old regime, wielding a wagon-sized lens to pour the concentrated sun onto a diamond and watch it burn. The sun, for Lavoisier, was the cleanest furnace in the world.

But the deepest and quietest breakthrough of the whole prehistory of solar heat came not from concentration at all, and not from a famous name, but from a Swiss gentleman-scientist named Horace-Bénédict de Saussure, better known to history as a pioneer of mountaineering and Alpine science, who in the 1760s built something so simple it is easy to underrate: a hot box. De Saussure took a small wooden box, insulated it, painted its interior black, and closed its top with panes of glass — sometimes several panes with air gaps between them. Then he set it in the sun and measured the temperature inside. What he found astonished him and should astonish you: the inside of this little glazed box climbed to well over a hundred degrees Celsius, past the boiling point of water, hot enough to cook. He carried versions of it up into the cold, thin air of the Alps and found that even there, in the chill of altitude where the surrounding air was near freezing, the interior of his glass-topped box still baked. There was no lens, no mirror, no concentration whatever — just ordinary, unfocused sunlight, a black surface, glass, and insulation. De Saussure had built the world's first solar oven, and in doing so he had stumbled onto and demonstrated one of the most consequential principles in all of energy and, as it happens, in all of climate science.

Here is why the hot box works, and it is a different trick from the burning mirror, so I will explain it carefully with its own analogy. The concentrating mirror gathers a lot of sunlight onto a small spot. The hot box does not gather anything; instead, it lets sunlight in and then traps the heat so it cannot get back out. Sunlight, which is mostly visible light, passes easily through the glass, strikes the black interior, and is absorbed, warming the surface. That warm surface then tries to radiate its heat back away — but a warm object radiates not in visible light but in longer-wavelength infrared, the kind of heat you feel from a radiator in a dark room, and glass, which is transparent to incoming visible light, is nearly opaque to that outgoing infrared. So the light gets in but the heat cannot get out; it is admitted as light and imprisoned as warmth. The everyday version of this you already know intimately: it is why a parked car with the windows up becomes an oven on a sunny day even when the outside air is merely warm — the same one-way trap, sunlight in through the glass, heat unable to escape back through it. This is, in miniature, the very mechanism that scientists would later name the greenhouse effect, the same physics by which certain gases in our atmosphere let sunlight down to the surface but slow the escape of the Earth's outgoing heat and so keep the planet warm — a warming that is entirely natural and necessary in moderation, and that becomes the central problem of our age only when we thicken those gases too far. De Saussure's little box was thus a double ancestor: it is the direct forerunner of every flat-plate solar collector that heats water on a rooftop today, and it is one of the first clean demonstrations of the trapping effect that governs the temperature of the whole Earth. It is worth pausing on the strangeness of that: the same simple principle that warms your parked car and heats a solar hot-water panel is the principle at the center of the climate question that the rest of this book keeps circling.

Let me draw the crucial distinction cleanly, because these are the two great families of solar heat and the reader should carry them forward. There is concentrating solar, the line of Archimedes and Lavoisier — mirrors and lenses that gather sunlight from a wide area onto a small target to reach ferocious temperatures, good for melting metals, driving engines, and, in the modern world, boiling the fluid in a solar power tower. And there is flat-plate solar, the line of de Saussure — no concentration at all, just a dark absorber under glass that traps ordinary sunlight to reach modest but genuinely useful temperatures, good for warming water and air. Concentration reaches high heat over a small area; flat-plate reaches gentle heat over a large one. Both catch the sun; they simply do it in opposite ways, one by funneling, the other by trapping. Nearly every solar-thermal device ever built descends from one line or the other, and the two next great characters of our story, Augustin Mouchot and Frank Shuman, will build their sun-engines squarely in the concentrating tradition of the burning mirror.

Now to correct the misconceptions, of which this chapter has more than most because legend has grown so thick over it. The first, which I have already worked to dismantle fairly, is the confident belief that Archimedes burned the Roman fleet at Syracuse — a story you will find repeated as fact in a hundred places. The honest position is that the underlying physics of igniting wood with concentrated sunlight is real, but that the historical event is almost certainly a legend: unattested by the earliest sources, wildly impractical against a moving fleet in battle, and repeatedly shown by careful modern experiment to fail under anything but perfect, motionless, laboratory conditions. Believe in the burning mirror as a principle; doubt it as a weapon of war. The second misconception is the assumption that using the sun's heat is a modern idea, born of the 1970s oil crisis or of our own climate anxiety. It is nothing of the kind. Human beings were kindling sacred fire from the sun before Archimedes; Lavoisier was burning diamonds with a solar furnace in 1774; and de Saussure was cooking food in a glass box on an Alpine slope in the 1760s. The dream of catching the sun is one of the oldest technical dreams we have. A third and subtler misconception is worth naming: that solar heat and the "greenhouse effect" are separate, even opposed things — one benign and useful, the other a modern menace. In fact they are, at bottom, the same physics. The trapping of the sun's warmth by glass that makes a solar water heater work is a cousin of the trapping of the Earth's warmth by atmospheric gases; the difference is one of place and degree, not of kind. Understanding that the two are one physics is the beginning of thinking clearly about both solar energy and the climate.

So the burning mirror of Syracuse, whether or not it ever scorched a single Roman plank, lit a fire in the human imagination that has never gone out. The ancients drew sacred flame from the sun with polished bowls; Lavoisier melted platinum and burned diamond beneath a wagon-sized lens; and a Swiss mountaineer with a black box under glass discovered, without fanfare, the trapping principle that heats both a rooftop panel and the whole living Earth. Two lineages were now in the world — concentration and flat-plate, the funnel and the trap — and both stood ready to be pushed past mere heat toward the thing the age most wanted from any new power: not just warmth, but work, the turning of wheels and the driving of machines. And in the coal-shadowed France of the 1860s, a schoolteacher who feared his country was about to run out of fuel would take the burning mirror and try, in earnest, to make it run an engine.

Chapter 11 — Mouchot's Sun Engines

At the Universal Exhibition in Paris in the summer of 1878 — the great world's fair that filled the city with the machines and marvels of the industrial age, spread across the Champ de Mars and the heights of the Trocadéro — visitors came upon a strange apparatus glinting in the sun. It looked like an enormous truncated cone of polished metal, a great shining funnel open to the sky, tilted to face the sun and lined inside with silvered mirror. At the narrow throat of the cone sat a blackened boiler. And from that boiler, fed by nothing but the concentrated Paris sunlight pouring down the mirrored funnel, ran a small steam engine — and the engine, in one of the demonstrations that made the crowd gasp, drove a refrigerating machine that produced, out of the blazing August heat, a block of ice. Sunlight was being turned into ice before the visitors' eyes. The man responsible was a quiet, sickly, brilliant mathematics teacher from the provinces named Augustin Bernard Mouchot, and for that machine the exhibition awarded him a gold medal. It was the high-water mark of a life's obsession, and very nearly the last good day that solar power would have for the better part of a century.

Mouchot is one of the great might-have-been figures of energy history, and to understand him you must understand what frightened him. He was born in 1825, the son of a locksmith, and rose by intelligence and effort to become a teacher of mathematics in French provincial schools — Alençon, then Tours. He was working in the France of the 1860s, at the roaring height of the coal-fired Industrial Revolution, when steam was remaking the country and coal seemed the very blood of progress. And Mouchot, looking at all this, worried about something almost nobody else was worrying about: that the coal would run out. He reasoned, correctly in principle, that the coal beds were finite, laid down once over unimaginable ages and now being burned through at a furious and accelerating rate, and that a civilization which had staked everything on a fuel it could only dig up and never replace was building its prosperity on a foundation that must, someday, give way. "The time will come," he wrote, in words that ought to be carved somewhere prominent, "when the industry of Europe will cease to find those natural resources, so necessary for it. Coal will undoubtedly be used up. What will industry do then?" This was in the early 1870s. He was, in other words, warning about the exhaustion of fossil fuels roughly a century and a half ago — long before "running out" of fossil energy became a modern anxiety. That is our first surprising fact, and it is worth letting it land: the fear that we cannot burn buried carbon forever is not a child of the 1970s oil crisis or of the modern climate movement. It was already keeping a French schoolteacher awake in the age of Napoleon III.

Mouchot's answer to his own question was the sun. France, and more to the point France's sun-drenched colony of Algeria, had in his view an inexhaustible alternative fuel pouring down from the sky for free, and the task of the engineer was simply to catch it and put it to work. So from about 1860 he set himself to build a machine that would do with sunlight what the coal-fired engines did with coal: make steam, and make steam do work. Here we can pick up the running thread of first principles that has carried through the last chapter, because Mouchot's machine is the burning mirror of Archimedes and Lavoisier, finally harnessed to a wheel. The chain of conversions is short and elegant. Sunlight, spread thin across a wide mirror, is concentrated by that mirror's curve onto a small black boiler. The concentrated sunlight heats the boiler and boils the water inside it into steam. The steam, expanding, pushes a piston, and the piston turns a wheel or a pump — mechanical work, at last, drawn straight from the sky. Sun to heat, heat to steam, steam to work. Every step of that chain was already well understood for coal; Mouchot's contribution was to replace the firebox with a mirror and prove that the rest of the chain did not care whether its heat came from burning carbon or from focused sunshine. Heat is heat. That was the whole audacious wager, and it worked.

Let me put a tangible number on why this was so much harder than it sounds, using the worked example this book keeps returning to. Recall that sunlight arrives at the ground spread thin — on the order of one kilowatt, a thousand watts, on each square meter of surface aimed at a high sun on a clear day. A kilowatt is real power, but to boil enough water to run even a small engine you need far more heat than one square meter can supply, and you need it delivered onto a boiler small enough to get genuinely hot rather than merely warm. So Mouchot had to gather the sunlight from a large collecting area and pour it onto a small one — exactly the funnel trick of the burning mirror. His great conical reflectors were sometimes several meters across, gathering the sunlight from many square meters of open sky and concentrating it down onto a boiler you could put your arms around. Only by multiplying the thin free sunshine over a wide mouth and cramming it onto a narrow throat could he reach the temperatures steam demands. This is why every solar engine, then and now, is mostly mirror: the working part is small, but it must be fed by a large and costly apparatus of reflectors, and that apparatus is where the money goes. Hold that thought about cost; it is the reason for everything sad that follows.

The 1878 Paris ice machine was the triumph, but the demonstration I love best belongs to Mouchot's gifted assistant, an engineer named Abel Pifre, and it came a few years later, in 1882. Pifre set up one of the concentrating solar engines in the gardens of the Tuileries in the middle of Paris, its mirrored dish drinking the sun, and connected its steam engine not to a pump or an ice maker but to a printing press. And there, in the open air, powered by nothing but focused sunlight, the press ran off copies of a newspaper — printed, appropriately and rather wonderfully, with the title Le Journal du Soleil, the Journal of the Sun. A crowd watched sunlight set type and turn out newspapers at a rate of some hundreds of copies an hour. It is one of the most charming demonstrations in the whole history of energy, and it made the same point the ice machine made, from the opposite direction: the sun could be made to do the concrete, commercial, mechanical work of the industrial age, whether that meant freezing water or printing the news. If you wanted a symbol for the solar dream in its first youth, you could hardly do better than a printing press in a Paris garden, running the Journal of the Sun off the power of the sun itself.

Mouchot was not alone in this. Across the Atlantic, the Swedish-American engineer John Ericsson — already famous as the designer of the ironclad warship Monitor, whose revolving turret helped remake naval warfare in the American Civil War — spent the last decade or so of his life, through the 1870s and 1880s, building his own solar engines. Ericsson favored a different mirror shape, a trough-like parabolic reflector focusing sunlight onto a line rather than a point, and he built a series of working machines that drove pumps and small engines from the sun. He was a more practical and hard-headed engineer than the dreamy Mouchot, and he grasped clearly the central difficulty that would doom them both, writing bluntly about the trouble that the sun's energy, though colossal in total, arrives so diffuse and so intermittent — gone every night, hidden behind every cloud — that gathering enough of it to compete with a lump of coal you could hold in your hand was fearsomely expensive. Ericsson, like Mouchot, built machines that unquestionably worked. Neither could make them pay.

Why did the sun engines die? The answer is a hard economic lesson that echoes through this entire book, and it is worth stating without sentiment. First and most simply, coal was cheap and got cheaper. The very resource whose exhaustion Mouchot feared was, in his own lifetime, abundant and falling in price as mining and railways improved, and a coal-fired steam engine was far smaller, far cheaper to build, worked day and night in any weather, and could be placed anywhere. Mouchot's solar engine, by contrast, needed a large and expensive apparatus of precision mirrors, worked only in bright sun, stopped every night and every cloudy day, and even in sunny Algeria could not match the coal engine's cost per unit of work. The economics were merciless: why buy an costly machine that runs only when the sun shines when a cheaper machine runs on cheap fuel around the clock? Second, the politics turned against him. Much of Mouchot's support had come from a French state interested in powering its North African colonies, where coal had to be shipped in at great expense and local sun was free; but when a cheaper route to Algerian coal opened and the strategic calculation shifted, the government's interest cooled, Mouchot's funding was cut, and the sickly inventor, his health broken, returned to obscurity and teaching. He died in 1912, largely forgotten, in the year that — as the next part of our story will tell — another inventor, Frank Shuman, was building a far larger solar plant in Egypt that would run headlong into the very same wall.

Here is the second great surprising fact of this chapter, and it is the tragic one: Mouchot and Ericsson were not wrong. Everything they built worked. Their machines really did turn sunlight into ice and print and pumped water and mechanical work. They failed not because the science was mistaken but because they were roughly a century too early — because the sun's energy, though free, was too diffuse and too intermittent to compete with the astonishing cheapness of fossil fuel in an age that had no reason yet to count the hidden costs of burning it, and no technology yet to store solar power through the night or manufacture mirrors at low cost. This is the recurring heartbreak of the solar story, and we have seen its shape before, with Hero's wind-organ that stayed a toy: the gap between a machine that works and a machine that pays can be a hundred years wide. Mouchot proved the principle in 1878. The world would not be ready to act on it until the principle could be made cheap, and that would take the collapse in solar costs of our own century — the very collapse that has, at last and in our own time, made the sun the cheapest source of new electricity in much of the world, vindicating a forgotten French schoolteacher a century and a half after he sounded the alarm.

Let me correct the misconceptions, because this chapter overturns some widely held ones. The first, which Mouchot's whole life refutes, is the belief that worrying about running out of fossil fuel — or looking to the sun as an alternative — is a distinctly modern concern, a product of the 1973 oil shock or of late-twentieth-century environmentalism. It is not. Augustin Mouchot published explicit warnings that coal would be used up and that industry must therefore turn to the sun in the early 1870s, a full century before the oil embargo, and he backed the warning with working machines. The anxiety about fossil finitude, and the solar answer to it, are Victorian, not modern. The second misconception is the assumption that early solar power was a failure because it did not work — because the machines were crude fantasies that could not really do anything. The opposite is true, and it is the whole point: the machines worked beautifully. They made ice and printed newspapers and pumped water. They failed on cost and timing, not on physics, and confusing "uneconomic in 1880" with "impossible" is precisely the error that has caused each generation to underrate solar power right up until the decade it suddenly became the cheapest power there is. A third, gentler misconception is that these Victorian solar pioneers were lone eccentrics, cranks tinkering in isolation. In fact they worked with government backing, won gold medals at the greatest exhibitions of the age, published serious technical treatises, and were taken seriously by the engineering establishment of their day. Mouchot's book on solar heat and its industrial applications was a substantial scientific work. They were not cranks; they were pioneers whose market had not yet been born.

Where do you meet Mouchot's sun engines today? You meet them, at giant scale, in the concentrating solar power plants that dot the world's sunbelts — the fields of parabolic troughs in the Mojave Desert, the mirror arrays of Morocco's Noor Ouarzazate complex at the edge of the Sahara, the towering "power towers" surrounded by acres of sun-tracking mirrors that focus light onto a single boiler atop a mast. Every one of these is Mouchot's cone and Ericsson's trough, grown vast and joined to a modern turbine: the identical chain of sun to heat to steam to work that ran the ice machine in 1878, now producing electricity for the grid. Some of these modern plants have even solved the problem that most tormented the pioneers — the sun's disappearance at night — by storing the collected heat in tanks of molten salt hot enough to keep the steam turbine running for hours after sunset, so that the plant can deliver power into the evening from sunlight it caught at noon. That is the storage Mouchot never had, and it is precisely what turns his intermittent curiosity into a dispatchable power station. You also meet him, more humbly, in the solar cooker used in sun-rich, fuel-poor regions of the world — a curved reflector concentrating sunlight onto a black pot, a direct descendant of the burning mirror, cooking a family's meal with no fuel at all, exactly the kind of application Mouchot first imagined for the sunny colonies.

So the sun engines of Augustin Mouchot and Abel Pifre and John Ericsson were not failures of science but casualties of timing and cheap coal — machines that genuinely made ice from sunlight and printed the Journal of the Sun by the sun's own power, built by men who warned, a century and a half early, that the fossil age could not last forever, and who were proved right in principle and simply too early in practice. They stand at the end of the long line of solar heat that began with the burning mirror: the line of concentration, of catching the sun's warmth and turning it into work. But all of them, from Archimedes to Mouchot, were chasing the same thing — heat. None of them had made the far stranger and more powerful leap of turning sunlight directly into electricity, with no boiler, no steam, and no moving parts at all. That leap had, in fact, quietly been made in a New York rooftop laboratory in 1883, in the same decade Pifre was printing his solar newspaper, by an inventor coating a rare element with a whisper-thin film of gold — and it is to that first true solar cell, and the electric mystery at the heart of it, that we turn next.

Chapter 12 — Charles Fritts and the First Solar Cell

On a rooftop in lower Manhattan, sometime around 1884, a handful of flat panels sat tilted toward the sky, doing something that no machine on Earth had ever done in a useful, self-contained way: turning daylight directly into electricity, with no fire, no steam, no moving part of any kind. To a pedestrian on the street below they would have looked like nothing much — dark plates in wooden frames, silent, giving off no smoke. But wired to a meter, they twitched a needle whenever the Sun came out from behind a cloud. The man who put them there was a New York inventor named Charles Fritts, and he believed, with a confidence that would take another century to vindicate, that arrays like his would one day "compete with" the coal-fired dynamos of Thomas Edison, whose Pearl Street station had begun lighting the neighborhood only two years earlier. Fritts was wrong about the timing by roughly a hundred years. He was right about almost everything else.

To understand what Fritts actually built, we need to go back to a French teenager and a jar of liquid. As we saw earlier in this book, the photovoltaic effect — light producing a voltage — was first observed by Edmond Becquerel in 1839, when he was nineteen years old, working in his father's laboratory. Becquerel's apparatus was a "wet" cell: two metal electrodes dipped in an acidic liquid, an electrolyte. When he shone light on one of the electrodes, a small current flowed between them. It was a genuine, reproducible, astonishing result. It was also, from an engineering point of view, nearly useless. A device full of sloshing acid, that must be kept upright and sealed and refilled, is not something you bolt to a roof and forget. Becquerel had proved that light and matter could be coaxed into making electricity. He had not built anything you could sell.

The leap Fritts made — and it is the leap that defines the entire modern industry — was from wet to dry, from liquid to solid. His cell had no fluid in it at all. It was a sandwich: a base plate of metal, a thin layer of the element selenium pressed onto it, and, on top of the selenium, a film of gold so unimaginably thin that light could pass straight through it. This is what engineers mean by a solid-state device — one in which the action happens entirely inside solid materials, with no liquid, no gas, no vacuum tube, nothing that flows or burns or wears against another part. It is the same word we now use for the "solid-state drive" in a laptop, and it means the same thing: no moving parts. Fritts's little selenium sandwich was, as far as anyone can tell, the first solid-state device that generated its own electric power from light. Everything from the silicon cell on a calculator to the vast solar farms of the Mojave descends from that idea.

Why gold, and why so thin? Here the first-principles story gets lovely and strange. For light to knock electrons loose inside the selenium, the light has to reach the selenium. But to collect the electrons once they are freed, you need a conductor — a metal — touching the selenium's surface. Those two demands fight each other: metal is exactly the stuff that light cannot pass through, which is why a mirror is a polished sheet of metal. Fritts's solution was to make the top electrode so thin that it stopped being an ordinary mirror and became a veil. Beat gold thin enough — and gold can be beaten astonishingly thin, into "gold leaf" a fraction of a wavelength of light across, the same material gilders press onto picture frames and cathedral domes — and it turns semi-transparent. Hold a sheet of gold leaf up to a window and you can see the Sun through it, glowing an eerie green. Fritts laid that translucent film over his selenium so that light could sneak through the gold to do its work below, while the gold still formed a continuous electrical skin to carry the freed electrons away. It is a trick of having your cake and eating it, and every modern solar panel plays a version of it, using a transparent conducting oxide or a fine metal grid instead of gold leaf.

Think of it two ways. First, imagine a greenhouse with a wire-mesh roof: you want sunlight to pour in on the plants, but you also want the mesh to hold together as a single sheet. Make the wires too fat and you shade the plants; make them too thin and gaps open up. Fritts found the sweet spot in gold. Second, picture a fishing net dragged through a school of fish. The water — the light — must pass through freely, but the net must still be continuous enough to catch what it is after. Fritts's gold film was a net for electrons that let the light swim past.

Now, the humbling number. Fritts's cell converted roughly 1 percent of the light energy falling on it into electrical energy. Put starkly: for every hundred units of sunlight that struck the panel, ninety-nine were wasted and one became electricity. A modern silicon panel of the kind you can buy today converts something like 20 to 23 percent, and laboratory record cells have passed 34 percent. By that yardstick Fritts's device was a near-total failure. It could not compete with a coal dynamo; it could not even compete with a good oil lamp on a cost basis. Selenium, we now understand, is simply a poor material for the job — its internal properties waste most of the energy the light delivers, for reasons that would not be properly explained until the quantum physics of the twentieth century, which is a story for a later chapter. But here is the first thing worth being surprised by: a 1 percent solar cell, sitting on a New York roof in the 1880s, was a real, working demonstration of a principle that the world's largest companies would eventually pour hundreds of billions of dollars into perfecting. The efficiency was pathetic. The proof of concept was complete.

Fritts had not, in truth, invented the underlying phenomenon in selenium from nothing. A few years earlier, in 1876, two Englishmen — William Grylls Adams, a professor at King's College London, and his student Richard Evans Day — had shown that the element selenium, when illuminated, produced an electric current without any heat, wire, or moving part: light alone, striking a solid, making electricity flow. It was a startling laboratory result, and it established that the solid-state photovoltaic effect was real. But Adams and Day had demonstrated a principle, not built a device. Fritts's originality lay in engineering that principle into an actual usable cell — the base plate, the pressed selenium, the gold-leaf window — something you could wire up, mount, and array on a roof. The distinction between showing that nature permits a thing and building a thing that exploits it runs all through the history of technology, and Fritts stands squarely on the building side of it.

Fritts himself grasped the wider meaning of his work. He sent samples of his cells across the Atlantic, and one reached the workshop of Werner von Siemens, the German industrialist and physicist who was one of the towering electrical engineers of the age — the founder of the firm that, more than a century later, would build some of the largest wind turbines on the planet. Siemens might have dismissed the feeble American gadget. He did the opposite. He examined the selenium cells, confirmed that they genuinely produced current from light, and pronounced the finding "scientifically of the most far-reaching importance." Coming from Siemens, that was a coronation. He is said to have presented the results to the Berlin Academy and argued that here was a phenomenon that deserved serious study, not because it was practical yet, but because it revealed something new about the relationship between light and electricity. It is one of the quieter examples in the history of technology of a great man recognizing a small thing for what it would someday become.

And what did Fritts's technology become, in the meantime? Not a power source — that door stayed shut for seventy years — but something you have almost certainly held in your hands without knowing it. The selenium photovoltaic cell, refined over the following decades, turned out to be an excellent light detector. It did not need to be efficient; it only needed to respond reliably and instantly to how bright the light was, producing a little current proportional to the illumination. That is exactly what you want in a light meter. From the 1930s onward, the exposure meters built into cameras — the small honeycomb-textured window on the front of a classic film camera that told the photographer how bright the scene was — were very often selenium photovoltaic cells, direct descendants of Fritts's sandwich. They needed no battery, because like Fritts's rooftop panel they made their own trickle of electricity from the light they measured. Generations of photographers framed their shots using a sliver of the same physics that Fritts had bolted to a Manhattan roof. Here is the second surprise: the first solar cell technology found its long career not in powering the world but in helping people take photographs, and it held that job for the better part of fifty years.

The selenium light meter deserves one more moment, because its long reign hides a second lesson about how technologies survive. A camera exposure meter did not need to be efficient; it needed to be honest and consistent. Feed it twice the light and it should give you twice the current, cleanly and quickly, so the photographer could trust the reading. Selenium did this beautifully. It was also self-powered, needing no battery at a time when small batteries were bulky, expensive, and unreliable — a real virtue in a handheld device. So a material that was hopeless as a power source turned out to be excellent as a sensor, and it earned its keep in that role for half a century, until light meters shifted to more sensitive materials and, eventually, to the silicon sensors in digital cameras. It is a reminder that a technology rejected for one purpose can quietly thrive at another, and that "too inefficient to matter" is always a judgment about a particular job, not a verdict on the physics.

There is a worked example hiding in Fritts's rooftop that is worth carrying through, because it explains why his enthusiasm and his failure were both entirely rational. Consider one square meter of the Earth's surface on a clear day, with the Sun high. Physicists have measured the power arriving there at roughly 1,000 watts per square meter — the so-called "one sun" figure, about the same as a small electric space heater running full blast, spread over a card table. That is a lot of free energy, raining down silently everywhere the Sun shines. Now run it through Fritts's cell at 1 percent. One square meter of his selenium would have produced about 10 watts in bright sun — barely enough to trouble a single dim bulb of the era. To light a modest room you would have needed many square meters of hand-made gold-and-selenium panel, in an age when gold was gold and selenium was a laboratory curiosity, and even then only while the Sun shone. You can see why the coal men slept soundly. But now run the same square meter through a modern panel at, say, 20 percent: about 200 watts, twenty times more, from the identical sunlight, using cheap, mass-produced materials. Nothing about the Sun changed between Fritts's day and ours. Everything about our ability to catch it changed. That gap — between the generous flood of energy the Sun offers and the thin trickle we could once capture — is the whole drama of this book, and Fritts stood at the very start of it.

This is the right moment to correct some misconceptions that cling to the solar story like barnacles. The first, and most common, is that solar cells are a modern invention — a product of the space age, or the 1950s, or Silicon Valley. They are not. The photovoltaic effect was discovered in

1839, the first solid-state cell was built in 1883, and rooftop solar existed, in a real if feeble form, by around 1884. When people say solar power is "unproven" or "new," they are off by more than a century. What is new is not the idea but the price and the efficiency. The second misconception is that Fritts's cell "didn't work." It worked exactly as it was supposed to; it simply converted very little of the incoming light, because selenium is a poor photovoltaic material. Working and efficient are different things, and confusing them has misled people about renewable energy for generations. The third misconception is subtler and worth naming: that early solar was a wet, fragile, chemical affair like a battery. That was Becquerel's cell, not Fritts's. Fritts's great contribution was precisely that his device was dry, solid, and stable — the true ancestor of the thing on your neighbor's roof, not a laboratory beaker of acid. And a fourth, quieter myth: that no one important took early solar seriously. Werner von Siemens, one of the most respected electrical minds alive, took it seriously enough to tell the Berlin Academy it mattered.

Where do you meet Charles Fritts today? More places than you would guess. Every time you see a rooftop covered in dark panels — on a suburban house in California, a barn in Bavaria, a slum roof in Bangladesh charging a phone — you are looking at the great-great-grandchild of the array Fritts installed on that New York rooftop, and you can fairly say that rooftop solar, as a concept, is around a hundred and forty years old. If you have ever picked up an old film camera at a flea market and seen a textured little window beside the lens, that is very likely a selenium cell, Fritts's technology still keeping time with the Sun. And in the deeper sense, you meet Fritts every time someone claims solar is too new, too untested, too futuristic to rely on — because the honest answer is that the principle was demonstrated on a Manhattan roof before the first practical American power grid was fully built, and it was declared scientifically important by one of the founders of modern electrical engineering. The technology took a century to grow up. It was never an infant idea.

Fritts died in 1903, having lit no cities, undercut no coal barons, and won no fortune from his cells. On the face of it he was one of history's noble failures — a man a hundred years ahead of his materials. But that framing sells him short. He had answered the question that Becquerel's wet cell left open: could you make light into electricity in a solid, portable, durable device, with no fluids and no fuss? Yes, you could. Everything after Fritts is a matter of finding better materials and cheaper ways to make them. He had drawn the map; others would build the road.

The tools he lacked were, above all, two: a material far better than selenium, and a theory that explained why some materials drink light greedily while others let it slip away. Both would arrive in the twentieth century, and when they did, the thin trickle would become a flood. But while Fritts was gilding selenium on the solar side of our story, another American inventor, four hundred miles to the west in Cleveland, was doing something just as audacious with the other free force in the sky — not the Sun's light, but its restless child, the wind.

Chapter 13 — Charles Brush's Automatic Dynamo

In the winter of 1888, if you had walked past a certain mansion on Euclid Avenue in Cleveland, Ohio — the street they called "Millionaires' Row," lined with the palaces of oil and steel and railroad fortunes — you would have seen something in the back garden that belonged to no other house in the world. Rising out of the lawn on a stout wooden tower stood a wheel of cedar slats the size of a small building: seventeen meters across, more than fifty feet, studded with a hundred and forty-four blades arranged like the vanes of an enormous pinwheel. When the wind blew off Lake Erie the whole colossus turned, slowly and with tremendous force, and a tail like a barn-door swung it to face the gusts. Below ground, in the cellar of the house, that turning wheel spun a dynamo that fed a room full of batteries — twelve of them, in glass jars — and those batteries, in turn, lit the lamps of the mansion above. The man who built it, Charles Francis Brush, had done something no one had ever managed before: he had made a machine that generated electricity from the wind entirely on its own, day and night, tended by no one, for years on end. It was the world's first automatically operating wind turbine, and it powered a home before the electric grid reached most of America.

Brush was not a crank tinkering in a shed. He was one of the richest and most accomplished inventors in the United States, a man whose name meant to the 1880s roughly what Edison's did. His great achievement was the arc light — a brilliant, sputtering electric lamp bright enough to illuminate streets and public squares — and the dynamo, or electrical generator, robust enough to run it. Brush arc lights had turned night into a passable day in cities across America and Europe; the fortune they earned him built the Euclid Avenue mansion. So when Brush turned his attention to the wind, he brought to it the finest electrical engineering of the era. This was not a windmill grinding grain, of the sort that had dotted Europe for seven centuries. This was a windmill wired to a generator, a machine for making the newest and most magical commodity of the age — electricity — out of moving air. To grasp how strange and new that was, remember that Edison's first power station had opened in New York only in 1882. Brush's backyard turbine, running by 1888, made household electricity in an age when almost no household had any.

How did it work? Start with the wind itself, and a piece of physics we will return to again and again in this book. Moving air carries energy — the same energy that fills a sail, topples a tree, or stings your face on a cold platform. A wind wheel's job is to intercept some of that moving air and let it push the blades around, converting the air's motion into the spinning of a shaft. Brush's design belonged to an ancient and intuitive family: the more blades, the better. His rotor had a hundred and forty-four of them, packed so densely that the wheel was almost a solid disc to the oncoming wind. A wind wheel built this way turns slowly but with enormous twisting force — high torque, in the engineer's word — like a low gear on a bicycle that lets you grind up a steep hill without ever pedaling fast. That was exactly what an old-fashioned water-pumping windmill wanted: slow, strong, relentless turning to work a pump. Brush inherited that logic and scaled it up to monstrous size.

But a dynamo does not want slow, strong turning. A dynamo wants speed — it makes more electricity the faster you spin it. So Brush faced a mismatch, and he solved it with gears. Inside the machinery he stepped up the leisurely turn of the great wheel, spinning at perhaps ten turns a minute, into the fast whirl the dynamo needed, a step-up of about fifty to one. It is the same trick as the chain and sprockets on a bicycle, trading the slow powerful push of your legs for the fast spin of the back wheel. The dynamo then charged the bank of batteries in the cellar, and this was the second stroke of Brush's genius. The wind is fickle; it gusts and dies and comes again. Lamps need steady power. By pouring the turbine's output into batteries and drawing the household's electricity back out of them, Brush smoothed the wind's moods into a constant supply. When the air was still, the house ran on stored charge. He had built, in his own cellar in 1888, the essential architecture of a modern off-grid solar-and-battery home — generation, storage, steady load — more than a century before it became a suburban fashion.

There was one more piece, and it is the piece that earns the machine its title of the first "automatic" wind turbine: a governor. A windmill left to itself is at the mercy of the weather, and a strong enough gale can spin a wheel to destruction, tearing off blades or shaking the tower apart. Brush built in a regulating mechanism that sensed when the wind grew too strong and turned the great wheel partly out of the wind, spilling the excess so the machine held a safe, steady speed instead of racing itself to pieces. This is why the turbine could be left alone. It started itself when the wind rose, protected itself when the wind raged, and stopped feeding the batteries when they were full — no operator required. For twenty years, by most accounts, it ran in Brush's garden with only modest attention, a silent servant turning Lake Erie's weather into light. Twenty years of reliable, unattended service from an experimental machine built before the automobile is the first thing here that ought to surprise you: this was no fragile stunt, but a genuinely dependable power plant.

The scale of the thing is hard to overstate, and a second surprising number drives it home. The great wheel and its tower together weighed on the order of forty tons — a mass of timber and iron the equal of a locomotive, balanced so cleverly on its pivot that the mere pressure of the wind could swing the whole assembly around to face a shifting breeze. Brush had to solve problems that no one had solved before at that scale: how to keep the tower from shaking itself apart, how to lubricate bearings that would turn for years unattended, how to route the electricity down from a rotating machine at the top of a tower into a fixed cellar below. He was, in effect, inventing wind engineering as he went, with no textbook and no precedent, drawing only on his mastery of dynamos and his considerable fortune. That a machine assembled from these first principles ran for two decades is a measure of how carefully he built.

It did not go unnoticed. In December 1890, Scientific American — then, as now, the great popular chronicle of invention — devoted a detailed, illustrated feature to the Brush machine, complete with engravings of the vast wheel and the cellar full of batteries and instruments. The magazine called it, in effect, the first successful example of a windmill generating electric light, and marveled at the scale and the ingenuity. For its readers it must have seemed a glimpse of a plausible future: every prosperous house with its own private wind-electric plant, independent of any central station, drawing its light from the sky. That future did not arrive on schedule, for a reason we will come to, but the vision was printed and admired. It is worth pausing on the image of Gilded Age Americans reading, in 1890, an enthusiastic account of clean, self-sufficient, backyard wind power — because we tend to imagine such ideas were born in the 1970s, or last decade. They are older than the light bulb in most homes.

Here, though, we reach the paradox at the heart of Brush's

magnificent machine, and it is a paradox that will govern the whole future of wind engineering. That enormous wheel, with its hundred and forty-four blades, was for all its size and reliability deeply inefficient. It produced only about twelve kilowatts at its best — enough for a large house, but a feeble return for a rotor seventeen meters across. The very feature that made it so intuitive — pack in as many blades as possible, present a nearly solid face to the wind — was the feature that limited it. This runs so hard against common sense that it deserves a careful explanation, because "more blades must catch more wind" is one of the most stubborn misconceptions about wind power, and Brush's own machine is the great cautionary tale.

Picture the wind as a crowd of people trying to hurry down a corridor. A wind wheel extracts energy by making the air push on the blades as it passes. Now, if you pack the corridor with too many turnstiles — too many blades, too close together — the crowd simply piles up in front and much of it never gets through at all. The air, meeting a nearly solid disc, mostly deflects around the whole wheel rather than passing through and spending its energy on the blades. A dense, slow, many-bladed rotor stalls the wind before it can do useful work. Now picture instead just two or three slender blades sweeping quickly through the air. Each blade, moving fast, slices through a fresh column of wind before the next blade arrives, so that over one rotation the small number of fast blades actually "work" nearly all the air passing through the circle — and crucially, they let the air keep flowing so that more of it can come behind. It is the difference between mopping a floor with a wide heavy pad you can barely drag, and a few quick strokes of a fast, light mop that covers the same ground with far less effort. Fewer, faster blades, properly shaped, beat many slow ones. Brush had built the mightiest example of exactly the wrong instinct.

He can hardly be blamed. The physics that explains why — the aerodynamics of lift, the ceiling on how much energy any wheel can pull from the wind — would not be worked out for another generation. The German physicist Albert Betz would not publish his famous proof, that no wind wheel can capture more than about 59 percent of the energy in the air passing through it, until 1919, thirty years after Brush's turbine first turned. And the practical demonstration that a lean, fast rotor trounces a fat slow one was being made in these very years, quietly, by a Danish schoolteacher named Poul la Cour, whose work we will meet in the next chapter and who deserves the credit for turning wind power from folk craft into science. Brush was an electrical genius applying an old mechanical tradition; he optimized the dynamo and the batteries and the governor brilliantly, and simply inherited the wrong shape of wheel. That his inefficient giant still ran a mansion for twenty years is a tribute to how much energy the wind holds even when you catch only a sliver of it.

The cube law makes that point vivid, and it is worth carrying as our worked example, because it explains both why wind is so tempting and why bigger, faster machines eventually won. The power available in the wind rises with the cube of the wind speed. Cube means multiplied by itself three times: double the wind speed and you get not twice but two-times-two-times-two — eight times — the power. A breeze that merely doubles from a gentle five meters per second to a brisk ten does not double the harvest; it multiplies it eightfold. This is why a single strong gust carries so much punch, and why the placement of a turbine — on a lakeshore, a ridge, a tall tower up in the faster winds — matters far more than intuition suggests. Brush understood the prize instinctively; what he lacked was a rotor that could take a fair share of it. A modern three-bladed turbine, sweeping the same circle of air, would pull perhaps three or four times more electricity from an identical wind than Brush's slat-covered wheel ever did.

Why, then, did the backyard-wind-plant future in Scientific American not arrive? Because a rival technology was spreading faster: the central power station and the wires that carried its electricity into homes. It was cheaper, once the grid reached your street, to buy power from a big coal-fired dynamo miles away than to raise and maintain a seventeen-meter tower in your garden. Brush himself eventually connected his mansion to Cleveland's growing electrical supply and let the great wheel fall idle; it was dismantled in the 1900s. The turbine did not fail. It was simply outcompeted by the convenience of the grid — the same grid whose absence had made it necessary in the first place. This is a pattern we will see again: wind and solar have repeatedly worked, technically, long before they made economic sense against the incumbent of their day.

Let us clear away the misconceptions plainly. First, wind-generated electricity is not a modern invention; a working, automatic, battery-buffered wind-electric plant lit a Cleveland mansion in 1888, and Scientific American celebrated it in 1890. Second, more blades do not mean more power — Brush's hundred and forty-four blades were a handsome mistake, and the machines that conquered the world would have two or three. Third, early wind power was not some flimsy gadget that barely worked; Brush's ran reliably for two decades. And fourth, a myth we can retire before it forms: that wind and solar are competitors invented in different eras. They were being born, on the same continent, within five years of each other — Fritts's rooftop cells around 1884, Brush's rooftop-high turbine by 1888 — two answers to the same question of how to catch the free power falling out of the sky.

It is worth dwelling for a moment on why Brush bothered at all, because his motive was not the one we might assume. He was not trying to save the planet; the idea of climate change lay generations in the future. He was a wealthy man indulging a technical fascination and, more practically, seeking independence — a private, reliable source of electric light for his home in an age when the public supply was young, patchy, and not yet running down his street when he began. That motive, energy independence and self-sufficiency, is precisely what drives many people to install solar panels and home batteries today, and it is a useful corrective to the notion that renewable energy has only ever been about environmental virtue. For much of its history it has been about freedom from the grid, from the coal merchant, from the distant utility — a very old and very practical human desire. Brush wanted his own light, made in his own garden, on his own terms. Millions of homeowners now want exactly the same thing.

Where do you meet Charles Brush today? In every backyard wind turbine and every off-grid cabin that pairs a small generator with a bank of batteries, you meet his exact architecture. In the water-pumping windmills that still turn on ranches across the American West — those slow, many-bladed steel fans — you meet the tradition he scaled up. And in the great three-bladed giants marching across ridgelines and out to sea, you meet the thing he pointed toward but could not quite build: the electric wind machine, automatic and self-protecting, turning weather into watts. When people speak of wind power as something dreamed up in our own anxious century, the honest reply is that a millionaire inventor was lighting his home with it while Queen Victoria still reigned.

Brush proved that the wind could be tamed into steady electricity and left to run itself. What he could not do was make his machine efficient, because he did not have the science of the blade. That science was taking shape, at that very moment, in the workshop of a Danish schoolteacher who would ask not merely whether the wind could make electricity, but exactly how to make it do so with the fewest, fastest, best-shaped blades — and who, astonishingly, would try to store the wind itself as a bottled gas.

Chapter 14 — Poul la Cour

In a low brick building in the Danish village of Askov, in the flat wind-scoured country of southern Jutland, a schoolteacher was blowing smoke at model windmills. The year was in the mid-1890s. The teacher, Poul la Cour, had built himself a wind tunnel — one of the first ever used to study the machines that catch the wind — and inside it he mounted little rotors, some with many blades, some with few, some flat, some curved, and set air rushing past them while he watched, measured, and recorded. He released streamers of smoke into the flow so he could see, with his own eyes, how the air bent and slid around each blade. Where Charles Brush had built one magnificent machine by instinct and inherited tradition, la Cour did something no one had done before: he turned wind power into a science, testing his ideas against measurement rather than guesswork. Out of that drafty little tunnel in a rural folk school came the discovery that would shape every large wind turbine built from then to now — and it was the exact opposite of what common sense, and Charles Brush's hundred and forty-four blades, would suggest.

Poul la Cour was an unlikely revolutionary. Born in 1846, he was a trained meteorologist and inventor of real distinction — he had done clever work on telegraphy before finding his true calling — but by temperament and circumstance he was above all a teacher. He taught at the Askov Folk High School, part of a distinctively Danish movement of adult education aimed at lifting the rural population through learning. La Cour looked out at the Danish countryside, poor in coal, poor in fast rivers, but rich, endlessly rich, in wind, and saw a way to bring the miracle of electricity to farms and villages that no central power station would reach for decades. Denmark had no coal of its own; every lump was imported. But the wind off the North Sea was free and blew hard. If a farmer could turn that wind into light and power, he could join the modern age without waiting for a distant coal baron to string him a wire. This was not an abstraction for la Cour. It was, quite consciously, a mission of national and social uplift.

To pursue it seriously, la Cour needed to answer a question that Brush had never asked: what is the best shape for a wind wheel? Not the biggest, not the most traditional — the best. And so he built his wind tunnel and began the patient work of comparison. What he found, and demonstrated with data, overturned the ancient wisdom of the windmill. A rotor with a small number of blades, turning fast, extracted more power from the wind than a rotor crowded with many blades turning slowly. Fewer blades, moving quickly and shaped as proper airfoils, were more efficient. It is worth savoring how counterintuitive this was to his contemporaries. Every windmill they had ever seen — the great grain mills of the Netherlands, the water-pumpers of the prairies, Brush's Cleveland giant — packed on blades or sails as if catching the wind were like catching rain in buckets, where more buckets plainly meant more water. La Cour's smoke told a different story, and he trusted the smoke.

Why is he right? We began to see the answer in the last chapter, but la Cour's insight lets us sharpen it into a single beautiful idea that engineers now call the tip-speed ratio — a phrase that sounds forbidding and means something simple. It is just the comparison between how fast the tip of a blade is moving through its circle and how fast the wind is blowing toward it. A many-bladed wheel turns slowly, so its blade tips move only a little faster than the wind — a low tip-speed ratio. A lean, fast rotor whips its blade tips through the air many times faster than the oncoming wind — a high tip-speed ratio. And it turns out that catching the wind's energy well requires the tips to move fast.

Here is an analogy that makes it click. Imagine you are cutting a wide field of grass. You could line up a hundred people shoulder to shoulder, each holding a stationary blade, and walk the whole wall slowly forward — that is the many-bladed windmill, a solid front creeping ahead. Or you could give two people fast-swinging scythes and let them sweep in wide, quick arcs — that is the modern rotor. The two swinging scythes, moving fast, can cut the same field with far less metal and far less drag, because each fast stroke covers a broad swath before the next one comes. In a wind rotor, each slender blade, racing around its circle, sweeps through a fresh column of air and does its work before the next blade arrives — so two or three fast blades can "process" essentially all the wind passing through the disc, while letting the air keep flowing so more can follow. Pack in too many slow blades, and, as we saw, the air piles up and deflects around the whole wheel rather than passing through it. La Cour proved with instruments what Brush had never suspected.

A second analogy comes from the water. A ship's propeller has just three or four blades, not a hundred, even though it lives in dense water that could push on many more. Naval engineers learned long ago that a few well-shaped blades moving fast grip the fluid better than a crowd of them moving slow. A wind rotor is a propeller run in reverse — instead of a spinning blade pushing air backward, moving air pushes a blade around — and it obeys the same logic. Nature had been hinting at the answer for as long as fish had fins; la Cour was the first to read the hint carefully for windmills and write down the rule.

Armed with his findings, la Cour did not stop at models. In 1891 he built full-scale experimental turbines at Askov and used them to generate real electricity. And here we reach the most astonishing thing about the man — a fact that, when I first encountered it, I had to read three times to believe. La Cour used the electricity from his wind turbines to split water. He ran the current through water in a process called electrolysis, which pulls water molecules apart into their two constituent gases, hydrogen and oxygen. He collected the hydrogen and stored it. In other words, in the 1890s, in a Danish village, Poul la Cour was storing the energy of the wind not in batteries but as a bottled, combustible gas — hydrogen — that could be burned later for light or drawn on when the wind was still. He piped the hydrogen to lamps that lit the school. This is, in every meaningful sense, "green hydrogen": hydrogen made by using renewable electricity to split water, with no fossil fuel involved. It is one of the most fashionable ideas in the energy world of the 2020s, discussed in gleaming corporate reports and national strategies as a cutting-edge solution to the problem of storing renewable power for the times when the sun does not shine and the wind does not blow. La Cour did it, at small scale, before the twentieth century began.

It was not without incident. Hydrogen is famously flammable, and impurities of oxygen in the mix make it explosive. La Cour's hydrogen storage reportedly blew out windows at the school on at least one occasion — an early, literal demonstration of the safety challenges that still shadow hydrogen today. But the principle held: he had shown that the wind, that most fickle and unstorable of forces, could be caught, converted, and banked as a fuel to be spent at will. The problem of intermittency — the fact that renewable power comes when the weather offers it, not when we want it — is the central technical problem of this entire book, and la Cour was wrestling with it, and partly solving it, a hundred and thirty years ago.

There is a human detail in la Cour's storage work that I

find quietly moving. Because his hydrogen lamps needed a steady supply, and because the wind would not always cooperate, he was forced to think, decades ahead of anyone else, about the whole chain of a renewable energy system: capture, conversion, storage, and use, each link balanced against the others. He was not merely building a turbine; he was building a system, and he understood that the machine on the tower was only one part of it. The batteries or the gas jars mattered as much as the blades. This systems thinking — the recognition that a variable source demands a way to store and smooth its output — is precisely the insight that grid engineers wrestle with today as wind and solar climb toward half of some countries' electricity. La Cour arrived at it not through theory but through the stubborn practical need to keep a schoolroom lit on a calm night.

He did something else that mattered as much as any machine. La Cour understood that a technology without trained people is a dead end, so he founded a course to teach "wind electricians" — countrymen who could build, install, wire, and maintain wind-electric plants across rural Denmark. He published a journal on the subject. He organized his enthusiasts into a society. He built, in other words, not just turbines but an entire human ecosystem of knowledge and skill around them — engineers, installers, and users who understood the machines from the inside. By the time la Cour died in 1908, dozens of his turbines were generating power in the Danish countryside, and a generation of Danes had learned to think of the wind as a practical servant rather than a curiosity.

This is the seed of one of the most remarkable continuities in the history of technology, and it brings us to a surprising fact that upends a common assumption. Denmark's leadership in wind power is not a recent achievement of clever twenty-first-century policy. It is more than a century old, an unbroken lineage running straight from la Cour's Askov workshop to the present day. His student Johannes Juul carried the ideas forward and built, in 1957, the Gedser turbine

— a three-bladed machine that is widely regarded as the direct ancestor of every modern wind turbine, the moment the "Danish concept" of the three-bladed, upwind rotor crystallized into its lasting form. From that lineage grew the Danish companies that would come to dominate the world: Vestas, today the largest wind-turbine manufacturer on the planet, and the wind business now inside Siemens, which together with a handful of rivals builds the great machines rising across the world's ridgelines and seas. When you read that a Danish or German firm has installed the world's most powerful offshore turbine, you are reading the latest chapter of a story that a schoolteacher began by blowing smoke at models in Jutland in the 1890s. The country that had no coal made itself, instead, the world's teacher of wind.

Let me anchor la Cour's efficiency insight with a worked example we can carry, because it is the number that his fewer-faster blades were reaching toward. Recall from the last chapter that the power in the wind rises with the cube of its speed — double the wind and you get eight times the power. But of the power that actually flows through the circle a rotor sweeps, how much can any wheel, however perfect, hope to capture? Not all of it, because the air cannot be brought to a dead stop — if it were, no more air could pass through behind it, and the flow would simply pile up and go around. There must always be some wind left in the air leaving the blades. The German physicist Albert Betz would prove in 1919, a decade after la Cour's death, that the absolute theoretical ceiling is about 59 percent — sixteen twenty-sevenths, to be exact — of the wind's power. This is the Betz limit, and no rotor ever built can beat it. La Cour did not know the number, but his experiments were pushing his lean rotors toward it, and away from the dismal fraction that Brush's crowded wheel could reach. Modern three-bladed turbines, the direct heirs of la Cour and Juul, capture roughly 75 to 85 percent of that Betz ceiling in good conditions — which is to say, they are close to as good as the laws of physics permit. The whole century of refinement since Askov has been, in a sense, the slow closing of the gap between la Cour's insight and Betz's limit.

Now the misconceptions, named and corrected. The first and largest: that Denmark's world lead in wind energy is a modern political creation, a product of green subsidies in our own time. It is not. It is a continuous tradition more than a hundred and thirty years old, founded by la Cour in the 1890s, carried through Juul in the 1950s, and flowering into Vestas and Siemens in our day. Cut into that lineage anywhere and you find the same DNA. The second misconception is the one we met with Brush and can now bury for good: that more blades catch more wind. La Cour proved with instruments that fewer, faster, well-shaped blades win, which is why the giants on the horizon have three blades and not a hundred and forty-four. The third is that storing renewable energy is a brand-new frontier we are only now beginning to cross. La Cour was banking wind as hydrogen in the 1890s; the frontier is old, even if the engineering is finally maturing. And a fourth, gentler myth: that early renewable pioneers were lone tinkerers whose work died with them. La Cour was the opposite — he built schools, trained a workforce, published journals, and founded a tradition precisely so that his knowledge would outlive him. It did.

Where do you meet Poul la Cour today? Every time you see a three-bladed wind turbine — and there are hundreds of thousands of them now, onshore and off, from Texas to the Baltic — you meet the shape his experiments argued for and his students perfected. Every time a headline announces a Danish firm's latest offshore giant, you meet his lineage. And every time a government minister or a technology company unveils a bold plan to store surplus wind and solar power as "green hydrogen," you meet the idea he was already testing, and occasionally exploding, in a Jutland folk school before the automobile. The wind electrician he invented has a modern descendant in the tens of thousands of technicians who climb inside nacelles today; the tradition he founded is why a small, flat, coal-poor country became the spiritual home of an industry now measured in the hundreds of billions of dollars.

La Cour took the raw fact Brush had demonstrated — that wind can make electricity — and turned it into a discipline: the right blades, the honest measurement, the trained workforce, even the storage of energy for a windless day. He gave wind power its science and its schooling. But while the Danes were mastering the wind, the other half of our story — the Sun — was about to be pursued not with a delicate rooftop cell but on a giant, sweltering, ambitious scale, by an American engineer who dragged mirrors and boilers to the edge of the Egyptian desert and tried to make the Sahara itself run an engine.

Chapter 15 — Frank Shuman's Egyptian Sun Plant

At Maadi, a green suburb on the outskirts of Cairo where the desert meets the irrigated banks of the Nile, a row of enormous curved mirrors lay glittering under the Egyptian sun in the summer of 1913. Each mirror was a long trough, shaped like a rain gutter but polished to a shine, and each ran the length of several railway carriages — five of them, side by side, covering an area larger than a tennis court. They did not lie flat. Through the day they tilted slowly to follow the Sun across the sky, and as they turned they gathered the desert light and threw it, concentrated, onto a black pipe running down the focus of each trough. Inside those pipes, water boiled. The steam it made drove an engine, and the engine drove pumps, and the pumps lifted thousands of gallons of Nile water a minute up onto the surrounding fields. There was no coal in sight, no furnace, no smoke — only sunlight, mirrors, and water. A crowd of dignitaries in linen suits and tarbooshes stood in the heat and watched a machine run on nothing but the sky. The man who built it, an American engineer named Frank Shuman, had constructed the world's first solar thermal power station, and he believed he was looking at the future of human civilization.

Shuman was not a scientist but an inventor-entrepreneur of the classic American type — a man who made his money from a practical patent (he had grown wealthy from a process for making wired safety glass) and poured it into a grander obsession. His obsession was the Sun. He had been building solar engines for years back in Philadelphia, experimenting on his own property, driving small pumps and engines with sunlight, refining his designs failure by failure. He grasped a fact that still startles people when they first hear it, and which is worth stating plainly as the foundation of everything Shuman attempted: the energy that pours onto the Earth from the Sun in a single hour is more than the entire human race uses in a year. The Sun is not a modest supplementary source; it is an ocean of energy, and every other fuel we burn — coal, oil, wood — is just sunlight that plants captured and buried long ago. Shuman reasoned that if the flood was that vast, the only real problem was engineering: how to catch a useful slice of it and turn it into work. He set out to solve that problem at industrial scale.

The physics he was exploiting is different from the photovoltaic effect we met with Fritts, and the difference matters, because it founds a whole separate branch of solar power. Fritts's selenium cell turned light directly into electricity through a subtle quantum interaction between photons and a solid material. Shuman did nothing so delicate. He used sunlight the way you might use a magnifying glass to burn a leaf — as concentrated heat. This is solar thermal power, and its logic is brutally simple: gather sunlight from a large area, focus it onto a small one, get that small spot very hot, boil water into steam, and use the steam to drive an engine, exactly as a coal plant does — only with the Sun doing the burning instead of a firebox full of coal. The clever part, the part Shuman labored over for years, was the gathering and focusing.

Here is where the parabolic trough comes in, and it deserves a first-principles explanation because it is the direct ancestor of the concentrated solar plants operating in the world's deserts today. A parabola is a particular curve, the same shape as the reflector behind a car's headlight or the dish of a radio telescope, and it has a magical geometric property: any ray of light arriving straight-on, parallel to the axis of the parabola, bounces off the curve and passes through a single point — the focus. Turn that curve into a long trough rather than a dish, and instead of a single focal point you get a focal line, a bright hot stripe running the length of the trough. Lay your black pipe of water along that line, and you have concentrated the sunlight falling on the whole broad mouth of the trough onto the narrow surface of the pipe. Shuman's mirrors were roughly two meters across at the mouth; squeezing that width of sunshine onto a slim pipe multiplied the intensity many times over, enough to boil water briskly even accounting for the era's imperfect mirrors and losses.

Two analogies bring it home. The first is the child's cruel trick with a magnifying glass and an ant: the lens gathers sunlight from its whole round face and squeezes it to a pinpoint hot enough to scorch. A parabolic trough is that magnifying glass stretched into a long line, scorching a pipe instead of an ant. The second is a satellite dish, which most of us have seen bolted to a rooftop. A satellite dish is a parabola that gathers faint radio waves from across its whole surface and concentrates them onto the little receiver on the arm at its focus — that is why the dish is big and the receiver is small. Shuman's trough gathered sunlight the same way a satellite dish gathers radio, and put the "receiver" — his water pipe — right at the focus. Once you have seen the parabola do its work in a headlight, a magnifying glass, a satellite dish, and a solar trough, you realize it is the same trick every time: collect over a big area, deliver to a small one.

The Maadi plant's numbers, dated and honest, tell the story of both its ambition and its limits. The five troughs, tracking the Sun through the day, drove an engine that produced somewhere in the range of fifty-five horsepower — call it roughly forty to sixty kilowatts of mechanical power, with sources varying on the exact figure — enough to pump on the order of thousands of gallons of Nile water a minute up onto the fields. For irrigation in a hot, dry, fuel-poor country, that was genuinely useful work. Egypt had almost no coal of its own; every ton was shipped in at cost from Britain. A pump that ran on the one thing Egypt had in ruinous abundance — sunshine — answered a real economic need, not merely a scientific curiosity. This was the point Shuman hammered: solar power did not have to wait for some distant future to be worthwhile. In the sun-drenched, coal-starved lands of the world, it could pay its way immediately, doing the humblest and most vital job of all, lifting water to grow food.

Let us carry a worked example, because it grounds Shuman's whole vision in a number anyone can check. On a clear day with the Sun high, the sunlight striking one square meter of ground in a desert like Egypt's delivers something close to a kilowatt of power — a thousand watts, about the output of a small electric heater, falling free on every square meter, all day. Shuman's five troughs covered on the order of a thousand square meters of collecting area. A thousand square meters, each catching roughly a kilowatt, intercepts on the order of a megawatt — a million watts — of raw solar power at midday. That his engine turned only some tens of kilowatts of that into useful pumping tells you the conversion was inefficient, on the order of a few percent, much of the sunlight's energy lost as heat that escaped the pipes and the boiler before it could be turned into motion. But look again at the raw figure: a megawatt of power, arriving silently and free, on a patch of desert the size of a few tennis courts. Multiply that patch across the Sahara, Shuman argued, and the arithmetic becomes staggering. He was not wrong about the arithmetic. He was wrong, tragically, about the time he had.

For Shuman had grasped the largest implication of his own machine, and he wrote it down with a clarity that reads today like prophecy. He calculated — roughly, in the manner of his age — how much desert would be needed to run the world, and concluded that a modest square of the Sahara, covered in his sun plants, could generate power equal to all the coal then mined on Earth. In an essay and in interviews around 1913 he predicted that after the world's oil and coal were exhausted, "the human race can receive unlimited power from the rays of the sun." He foresaw vast solar stations in the world's sunbelts feeding the energy needs of humanity long after the fossil fuels ran out. Read against the twenty-first century's actual push to build gigawatt-scale solar farms in exactly those deserts — the Sahara, the Arabian Peninsula, the American Southwest — Shuman's forecast is almost eerily accurate. He missed the mechanism, imagining mirrors and steam where we would eventually use photovoltaic panels and molten-salt towers, but he got the destination right: a world ultimately powered by the deserts' sun. That a Philadelphia glassmaker standing in a Cairo suburb in 1913 saw this so clearly is the most surprising fact in this chapter, and the most poignant.

Shuman did not arrive at Maadi in a single leap, and the road there ran through failure, which is worth remembering whenever a technology looks inevitable in hindsight. Back in Pennsylvania he had first tried flat black collectors — simple boxes that soaked up sunlight without concentrating it — and found them too weak to drive a serious engine. He had experimented with low-boiling fluids and with different mirror arrangements, refining each design against the disappointing results of the last. A British physicist, C. V. Boys, contributed a crucial improvement to the Maadi plant: the idea of running the boiler pipe along the focal line of a trough, with backing that made the concentration far more effective than Shuman's earlier flat attempts. The Maadi machine, in other words, was not a flash of genius but the product of years of trial, borrowed insight, and stubborn iteration — the same unglamorous process by which every real technology is ground into working shape.

The demonstration at Maadi was a genuine event. Shuman had attracted British and German backing for the venture, and the opening drew colonial officials, financiers, engineers, and press to stand in the desert heat and watch the mirrors track the Sun and the pumps gush Nile water. Lord Kitchener, the British consul-general who effectively ran Egypt, is said to have taken an interest, seeing in solar pumping a way to irrigate cotton lands cheaply; there was talk of a far larger plant in Sudan, and of the technology spreading across the sun-baked reaches of the British Empire. For a season, Shuman stood at the center of a plausible solar future, courted by empires, his mirrors flashing in the desert as proof that the thing could actually be done. Investors were lining up. Plans were drawn for expansion. The sun plant worked; the economics were closing; the vision had momentum.

Then, in the summer of 1914, the world set itself on fire. The First World War broke out. Shuman's German and British engineers were called home to fight one another. The Maadi plant, a delicate international collaboration, could not survive the severing of its backers across enemy lines. Materials, money, and men were swallowed by the war. The plant was abandoned; some accounts say its metal was scrapped for the war effort. Frank Shuman himself died in 1918, before the war ended, his great project stalled and his fortune poured into mirrors that no longer turned. And then came the second, quieter killer: after the war, the world found vast new oceans of cheap oil — in Persia, in Texas, in Arabia — and coal remained abundant. Against fossil fuel selling for a few dollars a barrel, an expensive field of tracking mirrors could not compete. The economic logic that had briefly favored Shuman's sun plant in coal-starved Egypt was drowned by the flood of cheap petroleum. Solar thermal power went to sleep for sixty years, and would not wake in earnest until the oil shocks of the 1970s reminded the world that cheap fossil fuel was not a law of nature.

Here we should correct the misconceptions, because Shuman's story is a magnet for them. The first is the belief that concentrated solar power is a twenty-first-century invention — that great mirror fields boiling fluids to drive turbines are a product of modern climate policy. They are not. A working solar thermal power station, using parabolic troughs to drive an engine, was pumping water beside the Nile in 1913, and the idea of solar steam engines goes back still further, to the Frenchman Augustin Mouchot in the 1870s. The second misconception is that early solar "never worked" — that these were failed experiments that produced nothing. Shuman's plant genuinely pumped Nile water at industrial scale; it did real work reliably for its short life. It was killed by a world war and cheap oil, not by any failure of the machine. The third misconception is that solar power's problem has always been that "the technology isn't ready." Shuman's fate says otherwise: the technology worked, and was defeated by economics and history — by the availability of something cheaper. That has been the recurring pattern of this whole book, and it cuts against the easy assumption that renewables simply needed more invention. Often what they needed was for the fossil alternative to stop being so cheap.

There is a bittersweet coda worth carrying. Shuman's distinctive advantage — that he was making heat, not electricity directly — turned out to contain the one trick that keeps concentrated solar power alive in the age of cheap photovoltaic panels. Heat is easy to store. You can keep a tank of very hot fluid warm for hours, then use it to raise steam after the Sun has set, and so keep generating into the evening when demand peaks and ordinary solar panels have gone dark. Modern concentrated solar plants exploit exactly this, banking the day's heat in tanks of molten salt to dispatch power on demand. Shuman never built storage into Maadi — he was pumping water in daylight, which needs no nighttime supply — but the thermal branch of solar he founded carried within it the answer to the very problem, storing sunshine for later, that still vexes the photovoltaic panels that overtook him. The road not taken had a hidden virtue.

Where do you meet Frank Shuman today? In the parabolic-trough power plants that still operate in the world's deserts — the long lines of curved mirrors at plants like California's Mojave installations and Spain's solar-thermal stations, which are Shuman's Maadi troughs scaled up and modernized, driving turbines instead of pumps and often storing their heat in molten salt to keep generating after sunset. You meet him in the "power towers," ringed by fields of mirrors that aim the Sun at a single glowing receiver, at places like Morocco's great Noor complex on the edge of the Sahara — very nearly the desert Shuman pointed to. And you meet him, most of all, in the simple modern fact that the world's sunniest, most fuel-poor regions are now racing to build solar power on exactly the logic he laid out in 1913: that where the Sun is fierce and the fuel is dear, catching sunlight is not a dream but a bargain. The mechanism changed; the map he drew did not.

Frank Shuman proved that the Sun could be gathered, concentrated, and made to do heavy industrial work — that a machine could run a nation's pumps on daylight alone. What he could not do was compete with cheap oil, and what neither he nor Fritts nor anyone in their generation could yet do was explain, at the deepest level, how light and matter actually traded energy — why a photon could free an electron in one material and merely warm another. That explanation had, in fact, already been written, in 1905, by an unknown clerk in a Swiss patent office puzzling over a stubborn little effect of light on metal — and it would turn out to hold the quantum key that finally made sense of everything Becquerel, Fritts, and Shuman had been groping toward.

Chapter 16 — Einstein and the Photoelectric Effect

In the spring of 1905 a twenty-six-year-old clerk sat at a wooden desk on the third floor of the Swiss Federal Patent Office in Bern, examining applications for electrical machines he had no particular interest in approving. His name was Albert Einstein. He had a physics doctorate that had not yet been granted, a wife and infant son at home, and a habit of doing his real thinking in the margins of the working day, then carrying it to a café or a friend's apartment in the evening. Over the following months he would send four papers to the German journal Annalen der Physik. One argued that space and time were not the fixed stage everyone assumed. One showed that the jittering of pollen grains in water proved atoms were real. One contained an equation the whole world would eventually recognize. And one — the first he sent, and the one he himself called "very revolutionary" in a letter to a friend — proposed that light does not flow like a smooth stream at all, but arrives in discrete lumps.

That fourth paper is the one this chapter is about, because it is the paper that finally explained the mystery we left hanging with Becquerel and Fritts. When Edmond Becquerel, back in 1839, saw a faint current appear the moment sunlight struck his electrodes, he could describe the effect but not the cause. When Charles Fritts, in 1883, coated selenium with gold so thin you could see through it and watched it make electricity from light, he was convinced he had a rival to Thomas Edison's coal-fired plants — and he still could not say why his little wafer worked. For decades the photovoltaic effect was a laboratory curiosity that no one could account for, a magic trick whose method was hidden. Einstein's 1905 paper handed physics the missing sentence. And here is the fact that surprises almost everyone who hears it: this paper, not relativity, not the famous equation relating mass and energy, is the specific work for which Einstein was awarded the Nobel Prize in Physics. The prize came in 1921, and the citation named "his discovery of the law of the photoelectric effect." Relativity was considered too controversial, too unproven, to honor. The committee reached instead for the quiet paper about lumps of light.

To understand what Einstein saw, we have to start with the puzzle that was driving physicists to distraction at the turn of the century. It went by the name of the photoelectric effect, and it is a cousin of the photovoltaic effect that runs a solar panel. Shine light on a clean metal surface in a vacuum, and under the right conditions the metal spits out electrons. You can catch those electrons, measure them, and study the rules of the game. By the 1890s, thanks largely to the careful experiments of the German physicist Philipp Lenard, those rules were well mapped. And they made no sense.

Here is why they made no sense. Everyone in 1900 was certain that light was a wave. This was not a hunch; it was one of the great triumphs of nineteenth-century science. James Clerk Maxwell had shown that light is a ripple in the electric and magnetic fields, and experiment after experiment had confirmed it. Now, if light is a wave, its energy should depend on its brightness — its amplitude, the height of the ripple. A brighter light is a taller wave carrying more energy, exactly as a taller ocean wave hits the beach harder. So the expectation was simple and obvious. Turn up the brightness, and the electrons flying out of the metal should come out faster, carrying more energy, because you are pouring more energy into them. Turn the light dim, and the electrons should come out sluggish, feeble, barely able to escape.

That is not what happened. What Lenard found was maddening. The energy of each escaping electron did not depend on the brightness of the light at all. You could crank a red lamp up to blazing intensity, flooding the metal with light, and every electron that emerged came out with exactly the same modest energy as before — you just got more of them. But change the color of the light — shift from red toward blue, toward the ultraviolet — and suddenly each electron came out with more energy, even if the light was faint. Worse, below a certain color threshold, nothing happened at all. Red light on many metals produced not a single electron no matter how bright it was, no matter how long you waited. Then you switched to blue, dim as a distant candle, and electrons appeared instantly. Brightness governed how many electrons you got. Color governed how much energy each one carried. The wave picture had no room for this. A wave should not care about color in that way, and it certainly should not have a hard threshold below which a blindingly bright beam does nothing.

Einstein's leap was to stop trying to save the wave. He proposed that when light interacts with matter, it behaves as though its energy is bundled into individual packets — he called them light quanta, and we now call them photons. Each photon carries a fixed, indivisible amount of energy, and — this is the crucial part — the size of that packet is set entirely by the light's color, its frequency. Blue light is made of high-energy photons. Red light is made of low-energy photons. The rule is startlingly simple: a photon's energy equals a constant number, called Planck's constant, multiplied by the light's frequency. High frequency, high energy per packet. Low frequency, low energy per packet.

Brightness, meanwhile, has nothing to do with the size of the packets. Brightness is simply how many photons arrive each second. A bright red lamp is a firehose of small packets. A dim blue lamp is a trickle of large ones.

Now the maddening experiment falls into place with a click you can almost hear. Think of an electron sitting in the metal as a coin-operated turnstile, the kind you push through in an old subway station. It takes a certain minimum payment to release the bar and let the electron through — out of the metal and into freedom. Physicists call that minimum payment the work function, and it is a fixed price set by the particular metal. Now imagine that light pays not in a continuous stream of small change but in coins of a single fixed denomination, and that the denomination depends on the color. Red photons are pennies. Blue photons are quarters. The turnstile costs, let us say, a dime.

Here is the whole mystery, dissolved. It does not matter how many pennies you throw at the turnstile — a million pennies, a truckload — because a single electron can only accept one coin at a time, and no single penny is worth a dime. So bright red light, an avalanche of pennies, frees not one electron. But a single quarter clears the price with change to spare, so even one dim quarter — one blue photon — pops the turnstile instantly, and the leftover value becomes the electron's speed as it flies away. Turn up the brightness of the blue light and you are simply feeding more quarters into more turnstiles: more electrons, each with the same energy as before. Shift toward the ultraviolet and you are handing over silver dollars; each electron escapes with even more energy to spare. The threshold, the color dependence, the indifference to brightness — all of it follows from one idea, that light comes in coins of fixed denomination and an electron must pay its fare with a single coin.

I find it worth pausing on how radical this was, because it is easy now, from the far side of a century, to nod along as if it were obvious. It was not obvious. It was heretical. The wave theory of light was the crown jewel of physics, defended by mountains of evidence. Max Planck, the great German physicist, had introduced the idea of energy quanta five years earlier, in 1900, to solve a different problem about the glow of hot objects — but even Planck regarded his quanta as a mathematical trick, a bookkeeping device, not a real feature of the world. He did not believe light was actually lumpy. Einstein was the one willing to say the lumps were real, that a beam of light genuinely is a hail of particles. For years he stood almost alone in that conviction. When Planck and three colleagues nominated Einstein for membership in the Prussian Academy of Sciences in 1913, they praised his genius and then, in the same breath, apologized for the light-quantum idea, writing that he had sometimes "missed the target in his speculations, as, for example, in his hypothesis of light quanta." His own champions thought this particular idea was his embarrassing misstep.

The man who proved them wrong did not want to. Robert Millikan was an American experimental physicist at the University of Chicago, meticulous, stubborn, and openly hostile to Einstein's photon. He set out around 1912 to demolish it. He built exquisite apparatus, machining fresh metal surfaces inside a vacuum with a tiny knife so that no film of oxide could taint the readings — he later called it a "machine shop in vacuo." He measured, for color after color, exactly how much energy the escaping electrons carried. He spent the better part of a decade trying to catch Einstein's simple straight-line prediction in a lie. And in 1916 he published his results, which showed that Einstein's equation was correct in every particular, the straight line perfect, Planck's constant measured cleaner than ever before. Millikan confirmed the theory to the last decimal, and then, in print, grumbled that the theoretical basis was so unreasonable that he could hardly believe his own confirmation of it. He called Einstein's hypothesis "reckless." He had built the strongest possible case for an idea he despised, and honesty forced him to publish it. It is one of the great examples in science of an experiment triumphing over the experimenter's own wishes.

There is a smaller, human story tucked inside the big one, and it belongs to Lenard, the man whose careful measurements set the whole puzzle up. Lenard won his own Nobel Prize in 1905 for his work on cathode rays, and he had every reason to feel he had done the essential work on the photoelectric effect. But it was Einstein who supplied the explanation, and Lenard never forgave him. In later years Lenard became a bitter figure, an early and vicious supporter of the Nazi movement, and a leader of the grotesque campaign against what he called "Jewish physics" — a category in which he placed relativity and Einstein himself. The tangled genius of the photoelectric effect is that its experimental foundations and its theoretical roof were built by two men who would end up on opposite sides of the darkest fault line of the coming century.

Let me put a number to all this, because the numbers are where the abstraction becomes tangible. The energies involved in freeing an electron are tiny, so physicists measure them in a unit called the electron-volt, abbreviated eV — roughly the energy an electron picks up crossing an ordinary flashlight battery. A photon of red light, near the edge of what our eyes can see, carries about 1.8 electron-volts. A photon of violet light carries about 3.1. That is the whole visible spectrum in a nutshell: from red to violet, the energy per photon not quite doubles. Now consider silicon, the material at the heart of every modern solar cell. It takes about 1.1 electron-volts to knock an electron loose in silicon and let it carry current — that is silicon's threshold, its turnstile price, a quantity we will meet again under the name band gap. Every photon of visible light, red through violet, clears that price; each one can free an electron. This is the deep reason silicon makes a good solar cell and why Fritts's selenium worked at all. The Sun pours photons onto the material, and any photon carrying more than the threshold energy pays its fare and sets an electron moving. Becquerel's faint current, Fritts's glowing wafer — both were forests of turnstiles being cleared, one photon at a time, by coins of sufficient denomination. Neither man had the words for it, but Einstein's 1905 paper is the caption that had been missing from their photographs for decades.

You meet this physics far more often than you might think, and not only on rooftops. The automatic doors at a supermarket, the ones that sense you approaching, often work on a photoelectric principle — a beam of light and a sensor that registers when the electron flow is interrupted. The light meter in a camera, the sensor that dims your phone screen in a dark room, the old smoke detectors that watched for particles crossing a light beam, the solar-powered calculator that has run without a battery on your desk since the 1980s: all of them are children of that Bern patent clerk's revolutionary paper. And every time you see a photovoltaic panel — on a parking meter, a highway sign, a satellite, a suburban roof — you are looking at Einstein's coins of light being spent, one at a time, to free electrons and push a current through a wire.

Because the idea is so counterintuitive, it collects misconceptions the way a magnet collects filings, and three are worth naming and correcting before we move on. The first is the one we have already met but which bears repeating because it is so widespread: that Einstein won his Nobel Prize for relativity. He did not. He won it, in 1921, specifically for explaining the photoelectric effect — the very physics that underlies solar power. Relativity, the work most people associate with his name, was never the subject of a Nobel citation. The second misconception is the one that fooled all of physics before 1905: that brighter light must mean more energetic electrons or more energetic outcomes. It does not. Color, meaning frequency, sets the energy each photon delivers; brightness sets only the number of photons. A dim blue light can free electrons that a blazing red light cannot touch, because energy rides on color, not on intensity — a truth that will matter enormously when we ask why solar cells waste part of the sunlight that hits them. The third misconception is subtler: that Einstein discovered light is a particle and thereby overturned the wave theory for good.

He did not overturn it. Light is stranger than either picture allows. It behaves as a wave when it spreads and interferes, and as a stream of particles when it delivers its energy to a single electron. This double life — wave and particle at once — became the seed of quantum mechanics, the physics that governs the very small, and it remains one of the deepest and most unsettling truths we know. Einstein opened that door, then spent much of his later life uneasy about the strange rooms it led into.

So let us gather the thread. For nearly seventy years the photovoltaic effect worked without anyone knowing why: Becquerel saw the current, Fritts built the cell, and neither could say what light was doing to matter. Einstein, in a single 1905 paper written on the margins of a patent clerk's day, supplied the answer — light arrives in fixed packets whose energy is set by color, an electron must pay a threshold price with one packet to break free, and everything about the maddening experiments falls into place once you see it that way. Millikan, trying to disprove it, proved it instead. The Nobel committee, wary of relativity, honored this quieter idea in 1921. We now have the key: we know why light can free an electron and what determines whether a given color is up to the job. But knowing the physics of a single freed electron is not the same as building a machine that harvests billions of them into useful power — and that machine would be discovered, like so much in this story, entirely by accident, in a cracked rod of silicon on a workbench at Bell Labs.

Chapter 17 — The Silicon Puzzle

The rod of silicon that changed the world was, by every ordinary standard, a failure. It sat on a workbench at Bell Telephone Laboratories in Holmdel, New Jersey, sometime around 1940 — a slim grey cylinder of purified silicon, grown in a furnace and cooled into a solid crystal. It was supposed to be uniform, a clean sample for studying how this odd material carried electricity. Instead it had a flaw running through the middle of it, a crack or seam where the crystal had cooled unevenly, and on either side of that seam the silicon behaved differently. A more disciplined lab might have thrown it out. The man who kept it was Russell Ohl, a quiet, dogged experimental physicist with a fascination for silicon that most of his colleagues at Bell considered a waste of a good mind. Semiconductors were then regarded as dirty, unreliable materials, unworthy of serious attention next to the clean elegance of vacuum tubes. Ohl thought otherwise, and he spent his days coaxing rods of silicon out of a furnace and probing them with electrical contacts to see what they would do.

One day — the details survive in Ohl's own later recollections and in the accounts historians assembled from Bell's records — he was measuring the electrical resistance of this cracked rod when he noticed something absurd. The reading jumped when the light in the room changed. He was near a window, or a colleague passed in front of a lamp, and the needle on his instrument swung. Ohl chased the effect. He shone a flashlight on the rod and watched a voltage appear across it, a real and surprisingly strong voltage, produced by nothing but light falling on the silicon. He is said to have called in his colleagues, including the theorist Walter Brattain, who would later share a Nobel Prize for the transistor, and shown them the trick. The voltage was far larger than anyone expected from a chunk of silicon. Something at that crack in the crystal was turning light into electricity with a vigor that Fritts's selenium and Becquerel's electrodes had never managed. Ohl had stumbled onto the silicon photovoltaic junction — the structure that sits, in more refined form, at the heart of essentially every solar cell manufactured on Earth today. He filed for a patent, which was granted in 1941. And the crucial part, the part that makes this one of the great accidents in the history of technology, is that the whole thing hinged on an impurity gradient in a rod that was supposed to be pure and simple, and was in fact neither.

To understand what Ohl had found, we need to understand what silicon is and why it sits in such an interesting no-man's-land between the materials we usually deal with. The world of everyday electrical materials divides into two camps. There are conductors — copper, aluminum, gold — metals in which electrons roam freely, so that the slightest push of voltage sends a current flooding through. And there are insulators — glass, rubber, dry wood — in which the electrons are locked tightly in place, so that no ordinary voltage can budge them; that is why we sheath wires in plastic and stand power lines on ceramic. Silicon belongs to neither camp. It is a semiconductor, which is exactly what the name says: a material whose ability to conduct sits partway between the two, and — this is the magic — a material whose conductivity we can control with exquisite precision by adding almost unimaginably tiny amounts of other elements. Pure silicon at room temperature is a rather poor conductor, closer to an insulator. Its electrons are mostly tied up in bonds with their neighbors, holding the crystal together in a neat repeating lattice, each silicon atom joining hands with four others.

The controlling trick is called doping, and it is one of the most counterintuitive ideas in all of technology, because it means the material you want is not the pure one but the deliberately contaminated one. Take a crystal of silicon, in which every atom has four outer electrons all busy bonding with four neighbors, and sprinkle into it a trace of phosphorus, whose atoms carry five outer electrons instead of four. Each phosphorus atom slots into the lattice, uses four of its electrons to bond like a good silicon citizen, and is left with one electron over — a spare, loosely held, free to wander. Do this to just one atom in a million and you have flooded the crystal with mobile negative charges. This is called n-type silicon, n for negative, and it conducts far better than the pure material because it is full of loose electrons looking for somewhere to go.

Now take another piece of silicon and dope it instead with boron, whose atoms carry only three outer electrons. Boron slots into the lattice too, but it can only bond with three neighbors; the fourth bond is missing an electron. That missing electron leaves a vacancy, a gap in the bonding structure that physicists, with a nice economy of language, call a hole. And here is the beautiful subtlety: a hole behaves as though it were a positive particle in its own right. An electron from a neighboring bond can slip over to fill the vacancy, but in doing so it leaves a new hole where it came from. Fill that one, and the hole has moved again. The vacancy migrates through the crystal as if it were a real, mobile, positive charge — the way an empty seat appears to travel down a row in a theater as people shuffle over one by one to fill it. Silicon doped this way is called p-type, p for positive, and its current is carried not by spare electrons but by these traveling holes.

By themselves, a slab of n-type silicon and a slab of p-type silicon are not especially interesting; each simply conducts a bit better than pure silicon. The wonder happens when you put them in intimate contact, so that n-type meets p-type across a single boundary within one crystal. That boundary is the p–n junction, and it is the beating heart of the whole enterprise. What Ohl's cracked rod had produced by accident, as the crystal cooled unevenly and impurities segregated to one side of the seam, was exactly such a boundary — n-type silicon on one side, p-type on the other, formed spontaneously by nature in a "bad" ingot no one had designed.

To feel why the junction matters, picture two crowds pressed against opposite sides of a doorway. On the n-side is a crowd of loose electrons, jostling and mobile. On the p-side is a crowd of holes, equally mobile, equally eager. The instant the two sides meet, the loose electrons near the boundary go tumbling across into the p-side, where they drop into waiting holes, and the holes near the boundary drift the other way. But this migration cannot go on freely, because every electron that crosses leaves behind a phosphorus atom that is now short one negative charge — a fixed positive ion, nailed in place in the lattice. And every hole that crosses leaves behind a boron atom that is now carrying an extra electron — a fixed negative ion. Very quickly a thin zone builds up right at the junction: fixed positive charges on the n-side, fixed negative charges on the p-side. Those fixed charges create an electric field across the boundary, a built-in field pointing from the n-side toward the p-side, and that field pushes back against any further crossing. Equilibrium settles in. The junction becomes a one-way street.

That one-way character is the whole point, and it is worth a second analogy because it is the feature that makes a solar cell, a transistor, and a diode all possible. Think of the junction as a turnstile at a stadium — but a turnstile that only turns one way. The built-in field lets charge flow easily in one direction and blocks it firmly in the other. Push current the "easy" way and it passes; push it the "hard" way and the junction resists. This is why a p–n junction, on its own, is a diode: a valve for electricity, letting current through in one direction only. But shine light on it, and the same one-way field becomes an engine. When a photon of sufficient energy strikes the silicon near the junction — a photon carrying more than silicon's threshold of about 1.1 electron-volts, the very physics Einstein handed us in the last chapter — it knocks an electron free and leaves a hole behind, creating a matched pair of charges out of a single packet of light. Ordinarily that electron and hole would simply drift about and recombine, the electron falling back into the hole, the light's energy lost as a whisper of heat. But if the pair is born within reach of the junction's built-in field, the field seizes them. It sweeps the electron one way and the hole the other, pulling them apart before they can recombine, and drives them out to opposite faces of the crystal. Connect a wire between those faces and the separated charges flow through it as an electric current — a current you can use to light a bulb or charge a battery. The junction does not merely let current pass; it actively sorts the light-born charges and pumps them around a circuit. That is the photovoltaic effect made muscular, and it is why Ohl's voltage was so much stronger than anything Becquerel or Fritts had seen. Their cells lacked this clean, powerful, built-in field. The silicon p–n junction supplied it.

There is a surprising fact worth dwelling on here, because it overturns an intuition almost everyone shares. The instinct is that a better solar cell should be made of purer silicon — that impurities are contaminants, defects, things to be scrubbed away. In fact it is precisely the opposite. Without impurities, a silicon crystal is nearly useless for making electricity from light; it has no built-in field, no way to sort the charges, no engine. The impurities are not the problem; they are the solution. The entire art of the semiconductor industry is the art of adding the right impurities, in the right places, in the right vanishingly small amounts — often around one foreign atom for every million or even billion silicon atoms, a concentration so slight that if the silicon lattice were a stadium crowd, the dopant atoms would be a scattered handful of people in a nation of them. The silicon itself must indeed be fabulously pure to begin with, so that no stray unwanted impurities muddy the picture, but the finished device is pure silicon deliberately, precisely, poisoned. Ohl's genius, and his luck, was to have found this out from a rod that poisoned itself.

And that is the second surprise, the one that gives this chapter its shape: none of it was planned. Ohl was not trying to build a solar cell. He was studying the electrical properties of silicon for the purpose of improving radio detectors — the little crystal rectifiers that could pick radio signals out of the air, which the coming war would make urgently important for radar. The photovoltaic effect he discovered was a side road, and the junction that produced it was an artifact of a crystal that had cooled badly. Had the ingot been perfect, uniform, exactly as intended, it would have shown no such dramatic effect and Ohl would have moved on. The one-way field, the sorting engine, the whole architecture of the modern cell, announced itself through a flaw. It is a recurring pattern in this book, from Becquerel's chance observation to the space program's rescue of a technology no one on Earth wanted: the crucial step arrives sideways, through an accident that a careful mind refuses to sweep away.

Ohl's junction did not immediately conquer the world, and it is worth being honest about why. The voltages were real but the efficiency was miserable — his cells converted well under one percent of the light energy that struck them into electricity, no better than Fritts's selenium of sixty years before, and far too feeble to compete with a wire from the power company. The device was a scientific triumph and a practical curiosity. What Ohl had established was the mechanism, the correct blueprint. Turning that blueprint into a cell efficient enough to matter would take another fourteen years, a different set of hands, and a deliberate, systematic effort rather than a happy accident — a story for the next chapter.

You meet Ohl's discovery, refined and multiplied, every single day, and not only in solar panels. The p–n junction he characterized is the fundamental building block of the entire electronic age. Every diode that steers current in your phone charger, every transistor in the processor running the device you may be reading this on, every light-emitting diode in a modern bulb or screen — all are junctions between differently doped semiconductors, sorting and steering charge across a built-in field. The silicon chip industry, the largest and most consequential manufacturing enterprise in human history, is in a real sense the industrialization of what Ohl saw jump on his meter in 1940. When you flip on an LED lamp, you are running a p–n junction in reverse — pouring electrons and holes into the junction so they recombine and release their energy as light instead of harvesting light to separate them. The solar cell and the LED are the same physics run in opposite directions, and both descend from that cracked grey rod.

Two misconceptions deserve correcting before we move on, and both cut against common sense. The first is the belief that what you want in a solar cell is the purest possible silicon, as though purity were the goal. It is not. You want fabulously pure silicon that has then been doped with carefully chosen impurities — phosphorus on one side, boron on the other — because it is the controlled impurities that create the junction and its all-important field. Pure, undoped silicon makes no useful cell at all. The second misconception is that the silicon solar cell was invented, deliberately, as a way to make electricity from the Sun. It was not invented so much as encountered. Russell Ohl was studying radio detectors and grew a flawed crystal; the photovoltaic junction revealed itself through that flaw. The deliberate invention of a practical cell came later, and by a different route, but the foundational discovery was serendipity — a scientist alert enough to notice that his broken sample was doing something extraordinary, and stubborn enough to chase it.

So let us take stock. In a rod of silicon that cooled unevenly on a Bell Labs bench around 1940, Russell Ohl found a natural boundary between differently behaving regions, shone light across it, and watched a strong voltage appear — the silicon p–n junction, patented in 1941, the structure at the core of the modern solar cell. We learned what makes it work: silicon as a semiconductor poised between conductor and insulator; doping with phosphorus to make electron-rich n-type and with boron to make hole-rich p-type; and the built-in one-way field that forms where the two meet, a field that sorts light-born electrons and holes and drives them as current through a wire. We learned that the impurities are the point, not the problem, and that the whole thing was found by accident. What remained was to make it efficient enough, and cheap enough, to be more than a marvel — and that work would begin in earnest at Bell Labs on a spring day in 1954, when three very different men pooled their frustrations into the first solar cell that could actually power something.

Chapter 18 — Bell Labs, 1954

On the morning of April 25, 1954, in a laboratory in Murray Hill, New Jersey, a small painted Ferris wheel began to turn under nothing but the light of a desk lamp. There were no batteries and no wall socket. The only thing feeding the little motor was a strip of silicon wafers the size of a few postage stamps, tilted toward the light. When a reporter's photographer needed a picture, the Bell Labs men obliged with a second demonstration that was, if anything, more striking: a small radio transmitter, its power supplied entirely by sunlight falling on silicon, broadcasting voice and music across the room. The New York Times thought this important enough for the front page the next morning, reporting that Bell scientists had built a device that converted the Sun's rays directly into electricity, and quoting the company's suggestion that the work "may mark the beginning of a new era, leading eventually to the realization of one of mankind's most cherished dreams — the harnessing of the almost limitless energy of the sun." It was, for once, a press release that history would more or less vindicate — though not on anything like the timetable anyone in the room imagined, and not before decades of disappointment.

The three men whose names went on the announcement made an unlikely team, and the accidental way their skills fit together is the real story of that spring. The first was Daryl Chapin, an engineer who had come to the problem not from a love of the Sun but from a mundane and stubborn headache. Bell was the telephone company, and telephones in remote places — jungle outposts, desert relay stations, tropical islands where humidity ate ordinary dry-cell batteries alive — needed a reliable trickle of power to keep working. Chapin had been assigned to find something better than batteries for these far-flung sites, and he had been experimenting with selenium cells, direct descendants of the wafers Charles Fritts had made in 1883. They were hopeless. Chapin measured selenium converting well under one percent of sunlight into electricity, so feebly that a whole rooftop of the stuff could barely light a bulb. He was, by his own account, thoroughly frustrated, on the verge of concluding that solar power was a dead end, when a colleague suggested he look at what two other men down the hall were doing with silicon.

Those two were Calvin Fuller and Gerald Pearson, and they had not set out to build a solar cell either. Fuller was a chemist, a specialist in the young and delicate art of doping silicon — introducing precise traces of impurities to give the crystal the electrical properties you wanted, the very technique we met in the last chapter. Pearson was a physicist, Fuller's frequent collaborator, and the two of them had been probing the behavior of doped silicon as part of Bell's enormous, war-forged effort to master semiconductors. Fuller had learned to make cleaner, more controllable p–n junctions than Russell Ohl's accidental one of 1940 — junctions formed deliberately by diffusing boron and then phosphorus into silicon at high temperature, rather than hoping a crystal would crack in a useful way. When Pearson made one of these junctions and, almost in passing, measured how much current it produced under light, the number stopped him. It was far better than Chapin's tortured selenium. Pearson mentioned it to Chapin. Chapin, who had been about to give up, abandoned selenium on the spot and threw himself into silicon.

What followed over the next months was not a flash of genius but a grind of patient optimization, and it is worth understanding what they were fighting, because it explains the number they eventually reached. A solar cell converts sunlight to electricity, and the fraction it converts is its efficiency. Ohl's cells and the first silicon junctions Fuller and Pearson made hovered under one percent, essentially useless. The three men attacked the losses one by one. They found that the junction had to sit very close to the surface, so that light-born electrons and holes were created within reach of the built-in field before they could recombine and vanish. They discovered that where they put the impurities mattered enormously; an early version doped with lithium worked poorly, and switching the arrangement — a thin phosphorus-doped layer over a boron-doped base — transformed it. They worked out how to attach electrical contacts that collected the current without shadowing too much of the precious light-catching surface. Each improvement was a fraction of a percent, and the fractions accumulated. By the spring of 1954 they had a silicon cell converting about six percent of sunlight into electricity.

Six percent sounds modest, and in a sense it is — a modern rooftop panel does better than three times as well. But the jump from under one percent to six percent was the difference between a laboratory curiosity and a device that could actually do work, and that is the first of this chapter's surprises. It was roughly a sixfold leap in a single concentrated effort, and it vaulted silicon past everything that had come before. Fritts's selenium in 1883, Ohl's silicon in 1940, Chapin's selenium in the early 1950s — all sat around or below one percent, stuck there for seventy years. In a matter of months, three men working in the same building lifted the number to a point where the arithmetic of usefulness finally turned positive. To feel what six percent means, do the sum. At solar noon on a clear day, roughly a thousand watts of sunlight fall on every square meter of ground facing the Sun — a figure worth carrying through this whole book, because it is the budget from which every solar technology must spend. A six-percent cell turns that thousand watts into about sixty watts of electricity per square meter. It is not much, but sixty watts is real; it can charge a battery, run a small radio, turn a toy Ferris wheel. One percent could not. That is the threshold Bell Labs crossed.

The public demonstrations were staged with a showman's instinct, and the anecdote of the Bell demo deserves telling in full because it fixed the technology's image for a generation. Bell called the invention the "solar battery," and after the Murray Hill unveiling they took it on the road. The most theatrical outing came in the fall of 1955 in Americus, Georgia, where the company wired an array of the new cells to a telephone repeater and used the Sun to carry a phone call — the first time in history, they claimed, that a conversation had been powered purely by sunlight. Chapin, ever the practical engineer, also toured with a homely little demonstration kit: the silicon strip, the toy Ferris wheel, sometimes a small fan. The image that ran in newspapers around the world — a whirring toy driven by a lamp through a scrap of silicon — did more to lodge the idea of solar electricity in the public mind than any equation could. It was proof you could see moving. And it planted an expectation that would take half a century to honor: that the Sun might soon run the machinery of ordinary life.

There is a smaller anecdote from those years that captures how uncertain the whole enterprise still felt to the people living it. The Americus telephone experiment ran on a rack of cells mounted on a pole, and it worked the sunlight carried the call — but it ran for only about six months before Bell quietly wound it down, because the cells cost so much and delivered so little that no telephone engineer could justify them against a simple battery and a wire. The demonstration had proved the principle and then bumped, hard, into the price. Chapin himself kept his showman's kit for years and would set the little Ferris wheel spinning for visitors and schoolchildren, delighting in the toy even as he knew, better than anyone, how far it was from lighting a real house. The three inventors received a joint patent for the "solar energy converting apparatus," and Bell, true to the culture of a research laboratory, treated the achievement as a scientific milestone to be published and admired rather than a product to be sold. It is a peculiarly honest failure — a technology that did exactly what its makers claimed and still could not find a single paying job on Earth.

Let us pause and reconstruct, cleanly, how the Bell cell actually turned light into that turning wheel, because it draws together the threads of the last two chapters. Start with the sunlight: a hail of photons, each a packet of energy whose size is set by its color, exactly as Einstein explained in 1905. The cell is a wafer of silicon with a p–n junction just beneath its surface — a thin phosphorus-doped n-type layer sitting on a thicker boron-doped p-type base, with the built-in one-way electric field spanning the boundary between them, exactly the structure Ohl stumbled on and Fuller learned to make on purpose. A photon carrying more than silicon's threshold energy of about 1.1 electron-volts strikes the crystal and knocks an electron free, leaving a hole behind — one packet of light, one pair of charges. If that pair is born near the junction, the built-in field grabs it, sweeps the electron toward the n-side and the hole toward the p-side, and prevents them from recombining. Metal contacts on the two faces collect the separated charges, and a wire strung between them carries the electrons around an external circuit — through a motor, a radio, a Ferris wheel's little engine — before they return to fill the waiting holes. Sunlight in, direct current out, no moving parts, no fuel, no noise. Multiply one cell's half-volt and modest current by wiring many cells together, and you have Chapin's strip powering his toy. That is the entire machine, and in its essentials it has not changed since 1954.

Now the second surprise, and it is the one that punctures the New York Times optimism. The Bell solar battery was astonishingly, absurdly expensive. The materials and painstaking hand-fabrication meant that the electricity it produced cost a fortune. Historians of the technology estimate the early Bell cells at something on the order of hundreds of dollars per watt of capacity — figures commonly cited land around a few hundred dollars for a single watt, at a time when a watt of electricity from the coal-fired grid cost a tiny fraction of a cent to generate. Put concretely: powering an ordinary household with 1954 solar cells would have cost more than the house, more than the neighborhood, a sum running into the millions to replace what the utility supplied for pocket change. To carry the worked example forward, remember that a square meter of six-percent cells made about sixty watts. At a few hundred dollars a watt, that single square meter of silicon represented tens of thousands of dollars — for sixty watts of midday power, and none at all at night. The physics worked. The economics were a catastrophe. This is the crucial and sobering fact about April 1954: the solar cell was born practical in the narrow sense that it could do real work, and utterly impractical in the only sense that mattered for spreading it across the world.

There is a human poignancy in this that the celebratory front pages missed. Chapin, Fuller, and Pearson had solved the scientific problem they set out to solve, and solved it beautifully, and it made almost no immediate difference. Chapin, doing the arithmetic himself, is said to have calculated that generating a family's electricity with his cells would run to an impossible sum, and to have felt the gap between the promise and the price keenly. Bell Labs, a division of a regulated telephone monopoly, had no obvious commercial use for a power source hundreds of times too costly, and quietly let the solar battery drift toward the shelf. The device that the Times had called the dawn of a new era spent its first years as a triumph without a market — a common fate, we will see, for inventions that arrive decades before the world is ready to afford them.

The misconception this chapter exists to correct is therefore the tidy one that the history books sometimes imply:

that the 1954 breakthrough made solar power practical, full stop, and that widespread solar electricity followed naturally from it. It did not. The breakthrough made solar power possible and demonstrated it convincingly, but it left the technology hopelessly uncompetitive on cost — so uncompetitive that for a decade it had essentially no terrestrial use at all. The road from that six-percent cell to the panels now cheaper than coal ran through a ninety-nine-plus-percent collapse in price that would take more than half a century and would depend on manufacturing scale nobody in Murray Hill foresaw. Two related myths are worth naming alongside it. One is that Bell Labs invented the solar cell from nothing; in truth they inherited the photovoltaic effect from Becquerel, the selenium cell from Fritts, and the silicon junction from Ohl, and their genuine achievement was efficiency, not first discovery. The other is that the "solar battery" stored energy, as its name suggests; it did not. It was a converter, making electricity only while the Sun shone, with no capacity to hold power for the night — a limitation that still shapes every debate about solar today, and that Bell's marketers papered over with a comforting but misleading word.

You meet the direct descendants of the Bell cell in more places than you might guess, and in a few of them the ancestry is almost unbroken. The pocket calculator that has run on a sliver of silicon since the 1970s, the solar-powered call boxes on remote highways, the panels on a parking meter or a bus-stop sign — all are the small, low-power niches the technology first crept into once prices began, very slowly, to fall. And every silicon rooftop array on Earth is a Chapin-Fuller-Pearson cell scaled up and cheapened: the same phosphorus-over-boron junction, the same built-in field, the same sunlight-to-direct-current conversion, now churned out by the billion in automated factories rather than assembled by hand in New Jersey. When you see a modern panel rated at, say, four hundred watts, you are looking at what would have cost a fortune in 1954 and now costs less than a good bicycle — the whole arc of this book compressed into one comparison.

So let us gather the thread. On April 25, 1954, Daryl Chapin, Calvin Fuller, and Gerald Pearson announced at Bell Labs the first practical silicon solar cell, about six percent efficient — a sixfold leap over everything before it, dramatized by a toy Ferris wheel spinning under a lamp and a radio transmitter running on nothing but light, and heralded on the front page of the New York Times as the possible dawn of a new era. We saw exactly how their cell worked, drawing together Einstein's photons and Ohl's junction into a single machine, and we saw the brutal catch that the headlines glossed over: at hundreds of dollars per watt it was hundreds of times too expensive to compete, a scientific triumph with no market on Earth. What the invention needed was a customer for whom cost did not matter and for whom nothing else would do — and in the skies above the planet, just four years later, exactly such a customer was about to appear.

Chapter 19 — A Cell for the Space Age

At 7:15 in the morning on March 17, 1958, a slender rocket lifted off from Cape Canaveral carrying a payload the size of a grapefruit. The satellite weighed a little over three pounds, small enough to hold in two hands, and its aluminum skin was studded with six tiny patches of silicon, each no bigger than a large postage stamp. It was called Vanguard 1, and it was only the fourth object humans had ever placed in orbit — following the Soviet Union's Sputnik 1 and 2 and America's own Explorer 1. The Soviet premier Nikita Khrushchev is said to have mocked it as "the grapefruit satellite," a jibe at its modest size next to the heavier Soviet craft. But that grapefruit carried something none of its predecessors had: a set of solar cells, direct descendants of the Bell Labs invention of four years earlier, wired to power one of its two little radio transmitters. And those cells would outlast every other power source in the early space age, turning a mocked afterthought into the first proof that photovoltaics had found the one job on or off the Earth that they alone could do.

Vanguard 1 carried two transmitters. One ran on a conventional chemical battery, the best that 1958 technology could offer, and it did its job for about three weeks before the battery went flat and the signal died, exactly as everyone expected. The other transmitter, drawing perhaps a tenth of a watt from those six silicon patches, kept talking. It transmitted for years. Long after the battery-powered voice had fallen silent, the solar-powered one went on chirping its telemetry back to Earth, tracked by stations around the world, until roughly 1964 when its signal finally faded — brought down not by any failure of the cells but by the slow degradation of the electronics in the harsh radiation of space. Six stamps of silicon had run a radio in orbit for something like six years. No other power source available in 1958 could have come close. This was the killer application that Bell's toy Ferris wheel had promised and could not deliver on Earth: a place where cost was almost irrelevant, where refueling was impossible, and where nothing but sunlight was reliably on hand.

The reason space suits solar power so perfectly is worth spelling out from first principles, because it is the mirror image of everything that makes solar hard on the ground. Down here, sunlight is intermittent and diluted. Clouds pass; night falls for half of every day; the atmosphere scatters and absorbs a good part of the light before it reaches a panel, so that even at noon on a clear day only about a thousand watts land on a square meter, and the daily average is far less. A satellite in orbit escapes almost all of that. Above the atmosphere there is no weather, no haze, no absorption — the full solar constant of roughly 1,361 watts per square meter falls on the panels, more than reaches any point on the ground, and it falls with perfect consistency. A satellite in a high or well-chosen orbit can sit in near-continuous sunlight for long stretches, and even in low orbit its "night," the pass through Earth's shadow, lasts well under an hour rather than the twelve of a terrestrial evening. There are no clouds to wait out and no seasons to endure. And crucially, there is no alternative. You cannot run an extension cord to a satellite. You cannot send a truck of diesel or swap a battery three hundred miles up. Chemical batteries drain in weeks; early nuclear sources were heavy, controversial, and reserved for special missions. Sunlight, converted directly and silently by a device with no moving parts to wear out, was the only power source that could keep working indefinitely in orbit. Everything that makes solar a compromise on Earth — its dependence on the Sun, its lack of stored fuel — either disappears or ceases to matter in space.

The decision to fly solar cells at all was not a foregone conclusion, and the story of how they got aboard Vanguard is the first of this chapter's anecdotes, because it turned on a stubborn argument. Much of the engineering establishment behind the satellite favored proven chemical batteries; they were understood, predictable, and did not depend on an exotic new device that had barely left the laboratory. Solar cells were the unproven newcomer, and there was real doubt about whether they would survive launch and the radiation of space. The push to include them is credited above all to Hans Ziegler, a physicist working for the U.S. Army Signal Corps, who had grasped that batteries alone doomed any satellite to a short life and argued forcefully that solar cells were the only route to longevity in orbit. The compromise that emerged was the two-transmitter arrangement — battery for one, solar for the other — which was, whether anyone intended it or not, a perfectly designed public experiment. When the battery died in weeks and the solar cells ran for years, the argument was settled in the most visible way imaginable. The doubters had, in effect, built the control group that proved them wrong. After Vanguard 1, the question in satellite design was no longer whether to use solar power but how much.

That shift created something new: a real, if tiny, market for photovoltaics at a moment when they had no market on Earth at all. This is the pivotal fact of the whole chapter, and it is easy to underestimate. In 1958 a silicon solar cell was still hundreds of times too expensive to compete with grid electricity, as we saw at Bell Labs; on the ground it was hopeless. But a satellite program does not compare the price of a solar cell to the price of coal power. It compares the price of a solar cell to the price of the entire mission —

the rocket, the launch, the years of engineering — against which a few hundred dollars' worth of silicon is a rounding error. For a customer like that, cost was almost no object and reliability was everything. So the space program became the one buyer willing to pay whatever photovoltaics cost, and in doing so it kept the technology alive through the long decades when nothing on Earth would have. The company that seized the opportunity most aggressively was Hoffman Electronics of California, which had licensed the Bell patents and built much of the early space industry's cell supply. Through the late 1950s and 1960s Hoffman and its competitors pushed silicon cell efficiency upward — from the six percent of the original Bell cell toward ten percent and beyond — and hardened the cells against the radiation and thermal shocks of orbit. Space did not just use photovoltaics; it improved them, funding the incremental gains that would eventually matter on the ground.

The demand grew quickly because the satellites multiplied. The space race between the United States and the Soviet Union poured money into orbit through the 1960s, and satellite after satellite — weather observers, communications relays, spy platforms, scientific probes — needed power for years in a place where only sunlight would serve. Solar panels became standard equipment, the unfolding wings we now recognize on virtually every spacecraft. Telstar, the communications satellite that relayed the first live transatlantic television in 1962, ran on solar cells. So did the deep-space probes, the early space stations, and eventually the great arrays of the International Space Station. A technology that had been a laboratory triumph without a customer became, within a decade, an indispensable and unremarkable part of spaceflight. The market was small in absolute terms — the total worldwide production of solar cells in those years would be dwarfed by a single afternoon's output of a modern Chinese factory — but it was steady, it paid well, and it demanded exactly the reliability improvements that would later serve the terrestrial industry.

Here the surprising facts come thick, and the most delightful of them concerns Vanguard 1 itself. That grapefruit-sized satellite is still up there. Its transmitters fell silent around 1964, but the physical craft remains in orbit to this day, circling the Earth roughly every two hours, and it holds the distinction of being the oldest human-made object still in space — older than any other satellite, older than every crewed spacecraft, a small silver relic of 1958 that will keep orbiting for a century or more before atmospheric drag finally brings it down. The very first satellites, Sputnik and Explorer, long ago burned up in the atmosphere; Vanguard, launched into a higher orbit, outlived them all. A second surprise is how completely solar power conquered its niche: within a few years of Vanguard, running a satellite on anything other than sunlight became the rare exception rather than the rule, and it has remained so ever since — the panels on a spacecraft are now as expected as wings on an airplane. And a third, quieter surprise is the sheer leverage of that tiny early market. The space program never consumed more than a sliver of the world's manufacturing effort, yet by keeping photovoltaics in continuous production and pushing their efficiency and durability for a decade and more, it preserved and nurtured a technology that the terrestrial economy had entirely rejected. Without a buyer, an invention dies on the shelf. Space was the buyer.

Let us carry the worked example a step further to feel the scale. Vanguard's six little cells produced on the order of a tenth of a watt — enough for a whisper of a radio signal, and nothing more. A single modern rooftop panel produces four hundred watts, four thousand times as much, from a surface you could carry under one arm. The International Space Station, at the far end of the same lineage, spreads out roughly an acre of solar arrays generating on the order of a hundred kilowatts — a million times Vanguard's output. The physics linking them is identical: photons above silicon's threshold energy freeing electrons, a p–n junction's built-in field sorting the charges, direct current flowing out through wires. What changed across those decades was not the principle but the efficiency, the reliability, and above all the cost — and the first steps of that improvement were paid for by customers who needed power in a place where sunlight was the only fuel that came for free.

This brings us to the misconception the chapter exists to correct, and it is a subtle one that even sympathetic histories sometimes fall into: the idea that the space program was a detour for solar power, a glamorous but irrelevant sideshow that delayed its real terrestrial destiny. The opposite is true. The space program was not a detour; it was the lifeline. In the years after 1954, solar cells had no economic reason to exist on Earth — they were hundreds of times too expensive, and every rational market rejected them. Had photovoltaics depended on terrestrial demand in the late 1950s and 1960s, the technology might well have withered, its factories closed, its expertise scattered, its slow price decline never begun. Space kept the industry breathing. It provided a customer indifferent to cost, insistent on reliability, and eager for every incremental gain in efficiency — and it did so for the two decades before the oil shocks of the 1970s finally gave solar a reason to come back down to Earth. The satellites did not distract solar power from its future; they carried it there. A second misconception worth naming is the assumption that solar cells were somehow developed for the space program, as a deliberate response to the needs of spaceflight. They were not. They were invented at Bell Labs to solve a telephone company's power problem and then, finding no market, were adopted by the space program as the only available solution to a wholly different challenge — a reminder that technologies routinely find their first real home far from the purpose their inventors imagined. And a third small myth: that the cells failed or degraded quickly in the brutal environment of orbit. On the contrary, they proved remarkably durable, outlasting the batteries and the electronics around them, and it was that unexpected toughness — a radio running for six years on six stamps of silicon — that made the case for solar power more convincingly than any laboratory demonstration ever had.

It is worth lingering on one more anecdote, because it shows how quickly the lesson of Vanguard hardened into doctrine. When American engineers designed the Telstar communications satellite a few years later, they wrapped its spherical body in thousands of small solar cells rather than trusting any battery to last — and when the Soviet Union and the United States began building the heavier, longer-lived satellites of the 1960s, the unfolding solar wing became as standard a feature as the antenna. The cells even drove a quiet arms race of their own: because radiation in orbit slowly degraded the silicon, manufacturers competed to build cells that held their output longer under bombardment by charged particles, adding protective cover glass and refining the crystal until a panel could survive a decade in the harshest environment humans had ever asked a machine to work in. Every one of those improvements — the cover glass, the radiation-hardened silicon, the reliable series wiring of many cells — was paid for by a space program that needed the power and could afford the cost, and every one of them quietly raised the quality of the cells that would, decades later, come down to ordinary roofs. The satellites were, without anyone framing it that way, a lavishly funded research-and-development program for terrestrial solar power, disguised as a space race.

You meet this heritage every time you look up at the right moment and catch the moving point of light that is a satellite crossing the evening sky, and every time you use the things those satellites make possible — the weather forecast drawn from orbiting observers, the television relayed across an ocean, the navigation signal from the constellation of satellites that guides a phone's map, nearly all of them running on solar panels that trace their ancestry to Vanguard. You meet it, too, in the fact that the very reliability improvements the space age demanded — cells that survive radiation, thermal cycling, and years of unattended operation — are part of why the panel on a modern roof is warranted to keep working for a quarter of a century. The satellites hardened the technology that now sits above your head, on the ground.

So let us gather the thread. On March 17, 1958, Vanguard 1 carried six small silicon solar cells into orbit, and while its chemical battery died within weeks, the solar-powered transmitter ran for years — the first proof that photovoltaics had found their killer application in a place where cost scarcely mattered and nothing else would work. We saw why space is the ideal home for solar power: full, uninterrupted sunlight above the atmosphere, no weather, no alternative, and no way to refuel. We saw how the argument over Vanguard's power source became an accidental experiment that settled the question for good; how the space race then made solar panels standard on satellites and kept a tiny but vital industry alive through the 1960s, with companies like Hoffman Electronics pushing efficiency and durability upward; and how that "detour" was in truth the lifeline that carried the technology across the decades when Earth wanted no part of it. Vanguard 1 orbits still, the oldest human object in space, a silent monument to the moment solar power found its first true purpose. But a satellite's panel and a rooftop's are the same machine at heart, and to understand what that machine can and cannot do — how much of the Sun's light it must throw away, and why — we need to walk through it, photon by photon, from first principles.

Chapter 20 — How a Solar Cell Actually Works

Pick up a solar panel — really pick one up, a modern rooftop module — and the first thing you notice is that it is a strangely quiet object. It has no fuel line, no exhaust, no fan, no sound. It is a sheet of glass over a grid of dark blue-black squares laced with thin silver lines, framed in aluminum, with two wires trailing from the back. Set it in the Sun and it does something that would have looked like sorcery to almost every human who ever lived: with no moving part, no combustion, no motion of any kind you can see or hear, it produces electricity. Tilt it toward the light and current flows through the wires. Shade it with your hand and the flow stops. There is no better place to end this part of the book than inside that silent square, following a single ray of sunlight all the way to the wire, because everything we have met so far — Becquerel's current, Einstein's photons, Ohl's junction, the Bell cell, the Vanguard satellite — converges here into one working machine. Let us take it apart, gently, from first principles.

Start with what arrives. Sunlight, as Einstein taught us and as we have carried through these chapters, is not a smooth stream but a hail of photons, each an indivisible packet of energy whose size is set entirely by its color. Red photons are small packets, around 1.8 electron-volts each; violet photons are large ones, around 3.1; and the Sun sends the whole range, along with a great deal of invisible infrared below the red and a little ultraviolet above the violet. At solar noon on a clear day, roughly a thousand watts of this mixed light fall on every square meter facing the Sun — the budget we have carried since Bell Labs, the thousand watts from which every solar cell must try to spend. The question that governs everything is deceptively simple: when this rain of differently sized packets hits a wafer of silicon, how much of it can be turned into electricity, and why not more?

The answer begins with silicon's threshold, which physicists call the band gap, and which is the single most important number in the whole device. In a silicon crystal, most electrons are locked into bonds holding the lattice together — they occupy what is called the valence band, the ground floor, where they cannot move to carry current. To become mobile, an electron must be kicked up to a higher level, the conduction band, the upper floor where it is free to roam. The gap between those two floors — the minimum energy needed to make the jump — is the band gap, and for silicon it is about 1.1 electron-volts. This is silicon's turnstile price, the fare from the last chapter's coin analogy, and it sets the rule for which photons can do useful work. A photon that carries less than 1.1 electron-volts cannot lift an electron across the gap at all. A photon carrying 1.1 or more can, freeing an electron and leaving a hole behind, exactly the paired charges we need.

Now comes the crucial and slightly heartbreaking consequence, the reason a solar cell can never approach a hundred percent efficiency no matter how perfectly it is built. Most of the sunlight's energy is, in a precise sense, the wrong size. Consider the two ways a photon can fail. First, the photons that are too weak: all that invisible infrared light streaming from the Sun, and even the deep-red end of the visible, carries less than 1.1 electron-volts per packet.

These photons simply cannot pay silicon's fare. They pass straight through the crystal, or warm it faintly, and their energy is lost to electricity entirely. A large fraction of the Sun's output — roughly a fifth of it lies in the near-infrared alone — is invisible to a silicon cell in this way, coins too small to clear the turnstile. Second, and more subtly, the photons that are too strong. A violet photon carrying 3.1 electron-volts can certainly free an electron across the 1.1-volt gap — but only 1.1 volts' worth of its energy goes into the useful jump. The extra roughly 2 electron-volts does not make the electron more useful; it makes the crystal a little hotter, then vanishes as waste heat. Physicists call this thermalization, and it is like paying a one-dollar fare with a hundred-dollar bill and being given no change: the excess is simply lost. Every photon above the band gap wastes its surplus this way, and since much of the Sun's light is bluer and more energetic than silicon's threshold, the waste is enormous.

These two losses — the too-weak photons that pass through, and the too-strong photons that squander their excess as heat — are not engineering flaws to be polished away. They are fundamental, baked into the physics of using a single material with a single band gap to catch a broad spectrum of light. Together they impose a hard ceiling on how well any single-junction solar cell can ever perform. That ceiling was calculated in 1961 by two physicists at Shockley Semiconductor, William Shockley — the same Shockley who had helped invent the transistor — and Hans-Joachim Queisser, and it bears their names: the Shockley–Queisser limit. For a single-junction cell under ordinary one-sun illumination, it sits at about 33 percent. No matter how flawless the silicon, how clever the contacts, how pure the crystal, a cell built on one material cannot convert more than roughly a third of sunlight into electricity, because roughly two-thirds of the light's energy is the wrong size for its band gap. This is a genuine law of nature, not a temporary limitation, and it deserves to be set beside the great ceiling we met on the wind side of this book: Betz's limit, the proof that no turbine can capture more than about 59.3 percent of the energy in the wind because the air must keep moving to leave the rotor. Each is a fundamental ceiling derived from first principles; each tells us that a whole category of machine has a maximum it can approach but never exceed. The wind engineer bumps against Betz; the solar engineer bumps against Shockley and Queisser. Knowing where the ceiling is turns out to be as important as knowing how to climb toward it.

With the losses understood, we can walk the working machine in full, and it is a satisfyingly complete story. Our square of silicon is built as a p–n junction, the structure Russell Ohl found by accident and Bell Labs learned to make on purpose: a thin layer of phosphorus-doped n-type silicon, rich in loose electrons, sitting atop a thicker layer of boron-doped p-type silicon, rich in holes, with a built-in one-way electric field spanning the boundary between them. A photon of adequate energy strikes the crystal and knocks an electron free, creating an electron and a hole — a matched pair of charges born from one packet of light. Left alone, that pair would drift about and, within a flicker of time, recombine, the electron dropping back into the hole and the energy lost as a whisper of heat. The junction's field prevents this. It sweeps the freed electron toward the n-side and the hole toward the p-side, hauling them apart before they can rejoin. Now there is a surplus of electrons at one face of the cell and a surplus of holes at the other — which is to say, a voltage, a difference in electrical pressure between the two faces. Connect a wire between them and the electrons flow through it, from the n-side around the external circuit and back to the p-side to fill the waiting holes, doing work along the way. That flow is direct current, DC, electricity moving steadily in one direction — the same form that comes out of a battery, and the reason every solar array needs an inverter, which we will come to, to turn it into the alternating current the grid uses.

Two everyday analogies make the sorting engine at the heart of this vivid. The first we have used before: the one-way turnstile. The junction's built-in field is a turnstile that lets charge through in only one direction. Light-born electrons and holes, created in a jumble, are herded through it and cannot come back — which is exactly why they accumulate on opposite faces and build up a voltage rather than simply recombining where they were born. The second analogy is water over a dam. Picture the Sun's photons as a pump lifting water — the electrons — from a low reservoir up to a high one; the energy of the photon is the lift. The band gap is the height of the dam wall: a photon must supply at least that much lift to get the water over the top, and any lift beyond the wall's height is wasted, the water simply splashing down the far side with energy you never capture. Once over the wall, the water — the electrons — flows downhill through your wire, through your turbine, doing useful work, before returning to the low reservoir to be lifted again. The cell is a pump driven by sunlight, tirelessly lifting electrons over a dam and letting them run back down through your circuit.

This picture also explains, at last, the little visual mysteries of the panel you picked up. Why is it blue-black? Because a bare silicon wafer is actually shiny grey and reflects away roughly a third of the light that hits it — light reflected is light wasted, photons that never enter the crystal to free an electron. So manufacturers coat the cell with an anti-reflection layer, a thin film tuned so that the light bounces the way that lets most of it in rather than out, and that coating happens to reflect a little blue while admitting the rest, giving the cell its characteristic dark blue or, in the best modern cells, near-black appearance. A black cell is simply a cell that has been engineered to reflect as little sunlight as possible — its color is the visible signature of a photon trap. Why the thin silver lines gridding the surface? Those are the metal fingers, the contacts that collect the freed electrons and carry them to the wire. They must touch the silicon to gather current, but every line also casts a shadow, blocking the light beneath it, so their design is a careful compromise: thin enough and spread wide enough to shadow as little of the surface as possible, yet dense enough to collect the charge before it is lost. And why is the cell so thin — a wafer well under a fifth of a millimeter, thinner than a business card? Partly to save silicon, which is costly to purify, but also because silicon absorbs light well enough that a thin slice catches most of what it can catch; making it thicker would waste material without gathering much more light. Every visible feature of the cell — its color, its silver grid, its wafer thinness — is a solution to the problem of getting the maximum number of photons to free the maximum number of electrons and then collecting them before they slip away.

Let us put numbers to the whole machine, carrying our worked example to its conclusion, because the arithmetic is where the physics becomes real. A thousand watts of sunlight fall on a square meter at noon. A good modern commercial silicon module converts about 20 to 22 percent of that into electricity — comfortably below the roughly 33-percent Shockley–Queisser ceiling, because real cells suffer additional losses beyond the fundamental ones: some reflection despite the coating, some shadowing by the fingers, some resistance in the wires, some recombination of charges before they are collected. Call it 20 percent to be conservative. That square meter therefore yields about 200 watts of electricity in full Sun — more than three times what the six-percent Bell cell of 1954 managed from the same area, a measure of seventy years of refinement. A typical rooftop panel is a little under two square meters and is rated around 400 watts, which fits. A house needing, say, six kilowatts of peak capacity would want roughly fifteen such panels, covering perhaps twenty-five to thirty square meters of roof — an entirely ordinary suburban array, and a direct scaling-up of Chapin's toy-powering strip. The Sun's thousand-watt budget, spent through a fifth-efficient cell, is what powers the modern rooftop.

Three misconceptions cluster around this machine, and clearing them is the best way to fix what we have learned. The first is the hope, perennial and understandable, that solar cells could someday be made a hundred percent efficient, or nearly so, if only we were clever enough. They cannot — not a single-junction cell. The Shockley–Queisser limit of about 33 percent is a ceiling set by nature, the direct consequence of matching one band gap to a broad spectrum in which most photons are the wrong size. This is not defeatism; it is the same honest realism that Betz's limit brings to wind. Cleverness can climb toward the ceiling, and — as we will see when this book reaches the frontier of tandem and perovskite cells that stack two band gaps to catch more of the spectrum and have already passed 34 percent in the laboratory — cleverness can even build a taller ceiling by changing the game. But within the rules of a single silicon junction, a third of the sunlight is the practical best, and any claim of a simple ninety-percent silicon panel is a claim to have repealed a law of physics. The second misconception is that solar cells need direct, bright sunlight to work at all, and are useless under cloud. They are not. A cell responds to any light that reaches it, including the diffuse, scattered light of an overcast sky — which is why panels still generate on a grey day, just less, and why solar works in cloudy climates like Germany's, which became one of the world's great solar nations despite weather no one would call sunny. Diffuse light is weaker, so output drops, but it does not vanish; the cell is not a switch that clicks off when a cloud passes, but a meter that reads lower. The third misconception is the belief, sometimes offered as a knock against the technology, that the visible color of a panel is arbitrary or merely cosmetic. It is not: the dark blue-black is the deliberate result of anti-reflection engineering, the color of a surface tuned to swallow light rather than bounce it, and a shinier, prettier cell would be a worse one.

So let us gather the thread, and with it this part of the book. A solar cell is a silent machine that turns sunlight directly into electricity with no moving parts, and we have now followed the whole process from first principles: a hail of photons, each a packet whose energy is set by its color, falls on a p–n junction of doped silicon; any photon carrying more than silicon's band-gap threshold of about 1.1 electron-volts frees an electron and leaves a hole; the junction's built-in field sweeps the two apart before they can recombine; and the separated charges flow out through the metal fingers and the wires as direct current. We saw why so much sunlight must be thrown away — the too-weak photons that pass through and the too-strong ones that waste their surplus as heat — and how those unavoidable losses set the Shockley–Queisser ceiling near 33 percent for a single junction, a fundamental limit as real as Betz's for the wind. We saw why the cell is blue-black, why it wears a grid of silver fingers, and why it is made so thin, each feature a solution to the problem of catching and collecting the most light. And we ran the arithmetic: a thousand watts on a square meter, a fifth of it captured, two hundred watts out, fifteen panels for a house. But a single cell makes only about half a volt, far too little to be useful on its own — and so the next step in our story is how these silent squares are wired together, by the dozen and the hundred, into the modules and arrays that actually power a home and a grid.

Chapter 21 — From Cell to Module to Array

Hold a single solar cell up to the light and you are holding something almost comically underwhelming. It is a thin blue-black square, roughly the width of your palm, cool to the touch, with a faint grid of silver lines printed across its face like the veins of a leaf. Clip a small motor to it, carry it to a sunny window, and the motor will spin — lazily, but it will spin. Then measure the voltage across the cell's two contacts and you will read something that sounds like a rounding error: about half a volt. Not five volts, not twelve, not the hundred and twenty or two hundred and thirty that hums in your walls. Half a volt. A single alkaline battery from the kitchen drawer outmuscles it three to one. And here is the detail that trips up almost everyone the first time they hear it: that half a volt does not change if you make the cell bigger. Cut the wafer in half and each half still reads about half a volt. Grow it to the size of a dinner plate and it is still about half a volt. Size buys you current, not voltage. The voltage is a property of the material itself — of silicon and the way a photon knocks an electron across the band gap we followed in earlier chapters — and it is stubbornly fixed at that meager number no matter how grand the ambitions of the person holding the cell.

So the whole engineering problem of turning sunlight into usable electricity begins with a question of arithmetic. How do you get from half a volt to something a house can use? The answer is the same trick that every builder of electrical things has relied on since the nineteenth century, and it has two moves. To add voltage, you wire cells in series — the positive contact of one to the negative of the next, like people joining hands in a line, so that the small pushes add up end to end. To add current, you wire cells (or whole strings of cells) in parallel — all the positive contacts joined on one side, all the negatives on the other, like several lanes of a highway merging their traffic. Series stacks the pressure; parallel widens the flow. Every solar panel you have ever seen is an exercise in these two moves, repeated with quiet discipline.

There is a homelier way to feel the difference between series and parallel, and it is worth carrying with you, because it recurs everywhere in this book. Think of water. Voltage is like water pressure — the push behind the flow — and current is like the volume of water actually moving. Wiring cells in series is like stacking pumps one behind another so their pressures add up, building a taller head of water; wiring them in parallel is like laying several pipes side by side so more water flows at the same pressure. A rooftop needs both a decent pressure and a decent flow, and so a panel needs both moves, in the right proportion. Get the proportion wrong and you have either a trickle at high pressure or a flood that goes nowhere, neither of which lights a house.

Consider the most common arrangement. Take about sixty cells, each contributing its half-volt-ish push, and wire them all in series. Half a volt times sixty gives you thirty-something volts — now we are in the range of real electrical work. String them at seventy-two cells instead and you climb past forty volts. This is why, if you climb a ladder and count the little squares on a residential rooftop panel, you will almost always arrive at sixty or seventy-two. It is not an aesthetic choice or a coincidence of manufacturing. It is arithmetic made visible: the number of cells is chosen to build the panel's voltage up to a level that inverters and wiring can handle efficiently. A panel, in other words, is not a single device. It is a committee of small identical devices, each too weak to matter alone, wired into a majority that can.

But a committee of delicate silicon wafers has a problem the arithmetic ignores: silicon is brittle, and the world is hostile. A bare cell is thinner than three sheets of paper stacked together, sharp-edged, and cracks if you look at it wrong. It has to survive on a roof for a quarter of a century — through hail, ultraviolet light, sub-zero nights, hundred-degree afternoons, driving rain, and the endless expansion and contraction of temperature swings that would fatigue almost any material. So the cells are entombed. This is the part of a panel most people never think about, and it is where much of the real cleverness lives. The cells are sandwiched between layers in a process called encapsulation. On top goes a sheet of tempered, low-iron glass — low-iron because ordinary window glass has a faint green tint that would steal a sliver of the incoming light. Below the glass, and again beneath the cells, go layers of a clear polymer, usually a plastic called EVA (ethylene-vinyl acetate), which is laid in as a sheet and then melted in a vacuum oven until it flows around the cells and sets into a transparent, waterproof cushion. At the very back is the backsheet, a tough weatherproof film (or a second pane of glass in the newer double-glass designs), and the whole laminated slab is clamped into an aluminum frame that gives it rigidity and something to bolt down. The result is a module — the proper name for what most people call a panel — and it is engineered to shrug off decades of weather while losing only a whisper of performance.

It is easy to underrate this laminated sandwich as mere packaging, but the encapsulation is arguably as hard-won an achievement as the cell it protects, and the industry learned its importance the painful way. In the early decades of terrestrial solar, panels routinely failed not because the silicon wore out but because moisture crept in at the edges, corroded the delicate metal contacts, and clouded or delaminated the plastic, turning a working panel into a fogged, useless slab within a few years. The reason today's modules can be warrantied for a quarter of a century is that manufacturers spent decades perfecting the glass, the polymer, the edge seals, and the lamination process — the vacuum oven that squeezes every last bubble of air out from around the cells so no pocket of trapped moisture can begin its slow rot. There is a direct inheritance here from the space program: the first great users of solar cells were satellites like Vanguard 1, which had to survive the savage temperature swings and radiation of orbit, and the discipline of building cells that could not be repaired once launched taught engineers to build for brutal longevity. Some of that hard-won ruggedness came back down to Earth and now sits on suburban roofs.

Bolted to the back of that frame, usually in one corner, is a small plastic box that deserves far more attention than it gets. This is the junction box, and it is where the panel's wiring comes to a head and where a set of humble components called bypass diodes are hidden. To understand why they matter, you have to appreciate the great vulnerability of a series string, and the best way in is an image everyone of a certain age carries: the old-fashioned Christmas lights. In the cheapest kind, the bulbs were wired in series, so the current had to pass through every bulb in turn to complete the circuit. Which meant that when one bulb burned out, the entire string went dark — and then came the miserable evening of testing bulbs one by one to find the culprit. A series string of solar cells has exactly the same weakness. The current that flows out of the panel must pass through every cell in the series, one after another, so the string as a whole can only carry as much current as its weakest member allows. And nothing weakens a cell faster than shade.

Here is the surprising and genuinely counterintuitive fact: shading one corner of a panel does not cost you one corner's worth of power. It can cost you the whole panel. If a single cell falls into shadow — the shadow of a chimney, a vent pipe, a fallen leaf, a bird dropping — that one cell stops generating and instead becomes a bottleneck, a resistor, choking the current that all the other sunlit cells are trying to push through it. Worse, all that thwarted energy piles up as heat in the shaded cell, which can develop a "hot spot" hot enough to damage the encapsulation. A panel is only as strong as its shadiest cell. This is where the bypass diode earns its keep. A diode is a one-way valve for electricity, and the bypass diodes in the junction box are wired so that if a group of cells is shaded and its voltage collapses, the current can detour around that group entirely — flowing through the diode instead of fighting through the dead cells. A typical sixty-cell panel has three such diodes, each guarding a third of the cells. When a leaf lands on one corner, that third of the panel is bypassed and drops out, but the other two-thirds keep working. You lose a third, not everything, and the hot spot is defused. It is a small, cheap, elegant piece of protection, and it is the direct answer to one of the most common misconceptions about solar panels: that a single broken or shaded cell kills the whole module. Without bypass diodes, that would very nearly be true. With them, the panel simply routes around the injury and carries on.

Now let us build a system, because a single module — even a healthy one wringing thirty-odd volts and four hundred watts out of the noon sun — is still not a rooftop's worth of power. Modules are themselves wired in series into what installers call a string, stacking their voltages up to the several hundred volts an inverter likes to see, and strings are wired in parallel to add their currents together. The whole assembly of modules on a roof or in a field is an array. The same two moves — series for voltage, parallel for current — repeat at every scale, from the cells inside one panel to the panels inside one string to the strings inside one array, like a fractal of the same simple idea. Where you put the array shapes the engineering: a rooftop array is cheap because the roof and its structure already exist, but it must live with whatever orientation and shading the house happens to have. A ground-mount array in an open field can be aimed at the ideal angle, spaced to avoid self-shading, and even set on motors that tilt to follow the sun across the sky, but it needs land, foundations, and its own supporting steel.

Let me make it concrete with the question every homeowner actually asks: how many panels do I need? Suppose a household uses something in the neighborhood of thirty kilowatt-hours of electricity a day — a reasonable figure for a mid-sized home with the usual appliances. A single modern panel rated at, say, four hundred watts, sitting in a decently sunny climate that delivers the equivalent of four to five hours of full "one-sun" output a day, will produce roughly one and a half to two kilowatt-hours daily once you account for the real world's imperfections. Divide the household's thirty by the panel's rough two and you land at somewhere around fifteen to twenty panels — a system of perhaps six to eight kilowatts, the sort of installation that now blankets suburban roofs by the millions. The exact count depends on the roof, the latitude, the weather, and how much the household actually uses, but the shape of the answer is always the same: a few dozen modules, a few kilowatts, arithmetic all the way down from that single stubborn half-volt cell.

Before we leave the panel itself, there is a counterintuitive quirk of its behavior worth pinning down, because it corrects a third widespread misconception: the belief that solar panels love heat, that the hotter and sunnier the day the better they perform. In fact the opposite is true. A solar cell is a semiconductor, and semiconductors lose voltage as they warm up. On a blazing summer afternoon a rooftop panel can easily reach sixty or seventy degrees Celsius, far above the twenty-five degrees at which it is rated, and at that temperature its output sags by ten percent or more compared to the same sunshine on a cold day. This is why some of the most productive solar conditions on Earth are not steamy tropical noons but cold, bright, high-altitude days — crisp winter sunshine on a snow-dusted mountainside, where the light is strong and the panels stay cool. It is also why the deserts that host the world's great solar farms pay a quiet penalty for their heat, and why engineers leave an air gap behind rooftop panels so they can shed warmth. Sunlight the panel wants; heat it merely tolerates.

A related misconception is that panels do nothing unless the sun is blazing directly on them. They do less on a cloudy day, certainly — perhaps a fifth to a half of their clear-sky output — but they do not switch off. A solar cell responds to the diffuse light scattered across an overcast sky as well as to direct beams, which is exactly why solar power works, and works well, in famously gray places. Germany, no one's idea of a sun-baked country, became for years the world's leading installer of solar panels, generating serious quantities of electricity under skies that Californians would call miserable. The panel is not a sunbather waiting for a perfect day; it is a patient collector of whatever light the sky sends down.

The panels will not stay quite as strong as the day they were installed, and honesty about that is part of understanding them. Silicon modules degrade slowly, typically losing around half a percent of their output per year — a slow fade driven by ultraviolet light, thermal cycling, and the gradual yellowing of the encapsulant. This is why manufacturers offer that reassuringly specific warranty: they guarantee the panel will still produce something like eighty to eighty-five percent of its original power after twenty-five years. Read that promise carefully, because it contains a genuinely surprising fact. A solar panel is one of the very few pieces of technology you can buy that is warrantied to still be doing its job, at better than four-fifths strength, a quarter of a century after you unbox it. There are panels installed in the 1980s that are still generating today, decades past their nominal design life. This is the quiet counterpoint to the second great misconception — that solar panels are fragile, high-maintenance things needing constant cleaning and fussing. In reality a rooftop array has no moving parts, and rain handles most of the cleaning; in all but the dustiest climates, the gain from washing panels is small enough that the labor rarely pays for itself. The maintenance schedule for most home solar arrays is, honestly, closer to "check that nothing has fallen on it" than to anything resembling a chore.

There is one part of the system, though, that will almost certainly not last the full twenty-five years — the box that all this direct current is flowing toward, sitting on the wall of the garage, converting the array's DC into the alternating current your home and the grid actually speak. The panels are the marathon runners of the system; that box is the part that tends to tire first, and it is also, as we are about to see, no longer the cheap part of the bargain. We have followed one small cell's half a volt all the way up through series and parallel, through glass and diodes, into a rooftop array humming with several hundred volts of direct current. Now we have to translate that current into a language the world can use — and that is the job of the inverter.

Chapter 22 — The Inverter and the Balance of the System

On the wall of a suburban garage in California, next to the water heater and the recycling bins, hangs a gray metal box about the size of a carry-on suitcase. It hums faintly when the sun is up. Most homeowners could not tell you its brand, and many do not know it is there at all. Yet if you unbolted every panel from the roof and laid them in the driveway, and then unbolted this one box and set it beside them, you would be looking at a curious modern truth: the box on the wall, and the wires and racks and permits and labor that go with it, quite possibly cost the household more than all the panels combined. The glamour of solar power is up on the roof, catching the light. The intelligence — and, increasingly, the expense — is down here in the garage.

The box is an inverter, and to see why it is indispensable we have to confront a mismatch built into the whole enterprise. A solar cell produces direct current, or DC — a steady, one-directional flow of electrons, the same kind of current that comes out of a battery, pushing always in one direction from minus to plus. But the electrical grid, and every socket in your home, runs on alternating current, or AC —

a current that reverses direction rhythmically, sloshing back and forth fifty or sixty times a second depending on which country you are in. This is not a trivial incompatibility you can wire your way around. It is the residue of a famous nineteenth-century fight, the so-called War of the Currents between Thomas Edison, who backed direct current, and George Westinghouse and Nikola Tesla, who backed alternating current. Alternating current won, and it won for a good reason: its voltage can be stepped up and down with a transformer, which let power companies push electricity across hundreds of miles of transmission line at high voltage and then tame it down to safe levels at the neighborhood. The entire planet's grid speaks AC. Your solar array speaks DC. Something has to stand between them and translate. That something is the inverter.

The simplest way to picture what it does is to think of it as a very fast, very precise switch — or a whole bank of them — flipping the direction of the panels' direct current back and forth to manufacture the wave-like rhythm of alternating current. Early inverters did this crudely, producing a blocky, stair-stepped approximation of a wave. Modern ones switch thousands of times a second and use clever filtering to sculpt an output so smooth and clean it is indistinguishable from grid power, a flawless sine wave rising and falling exactly in step with the utility's own. But calling the inverter a mere switch undersells it badly. A better word is translator, or perhaps gearbox, and the gearbox analogy is worth dwelling on. A car engine produces the most power at a particular range of engine speeds, and the transmission's job is to keep the engine in its sweet spot no matter how fast the car is going. A solar panel, too, has a sweet spot — a particular combination of voltage and current at which it delivers the most power — and that sweet spot drifts constantly as clouds pass, as the sun climbs and sinks, as the panels heat up and cool down through the day. If the inverter simply drew power at a fixed voltage, it would spend most of the day off the panel's peak, leaving free energy on the table.

So the inverter is always hunting for that peak, using a technique with a name that sounds more forbidding than the idea behind it: maximum power point tracking, or MPPT. Picture the relationship between a panel's voltage and the power it delivers as a hill — push the voltage too low and you get little power, push it too high and the current collapses and you again get little power, but somewhere in between is a summit where the power is greatest. The inverter is like a hiker in fog who cannot see the summit but can feel the slope underfoot, constantly taking a small step, checking whether the power went up or down, and adjusting — nudging the operating voltage this way and that, many times a second, to stay perched on the peak of the hill as the whole landscape shifts beneath it with every passing cloud. This relentless little dance is why a good MPPT inverter can wring several percent more energy out of the same panels than a naive design, and it is the single most important reason to think of the inverter as the brain of the system rather than just its translator.

If the gearbox analogy captures the tracking, a second everyday image captures the conversion itself. Think of the inverter as a translator standing between two people who cannot understand each other — the panels, who only speak in the steady monotone of direct current, and the grid, which speaks in the rhythmic rise and fall of alternating current. A bad translator renders the meaning crudely, in broken phrases; the old blocky inverters were like that, and appliances fed on their choppy output ran hot and hummed. A great translator renders it so fluently that the listener cannot tell the words were ever in another language, and that is what a modern inverter does, sculpting an output so smooth that the grid accepts it as its own. The translation is not free — a few percent of the energy is always lost as heat in the conversion, which is why the inverter needs cooling fins and sometimes a fan — but a good one gives back better than ninety-five percent of what the panels hand it.

The inverter has two more jobs, both about being a good

citizen of the grid. The first is synchronization. To feed power into the grid, the inverter's manufactured AC wave must line up with the utility's wave with almost perfect timing — same frequency, same phase, cresting and troughing in lockstep. Pushing an out-of-step wave onto the grid would be like shoving someone on a swing at exactly the wrong moment: instead of adding energy you fight the motion, and equipment suffers. So the inverter listens to the grid's rhythm and matches it continuously. The second job is safety, and it carries a slightly ominous industry name: anti-islanding. If the grid goes down — a storm brings a line down, a crew is dispatched to repair it — you emphatically do not want your rooftop array cheerfully back-feeding electricity into what the line workers have every right to assume is a dead wire. So the inverter is required to detect the grid's absence within a fraction of a second and shut itself off. This is a genuine feature, but it is also the root of one of the most disappointing surprises solar owners discover, usually during their first blackout: the sun is shining, the roof is covered in working panels, and yet the house is dark. Because a standard grid-tied inverter shuts down the instant the grid fails, an ordinary solar system does not keep the lights on in a power cut. To ride through a blackout you need a battery and an inverter specifically designed to disconnect from the grid and form its own little island of power — hardware most homes do not have. The panel on the roof and the sunshine falling on it are, in that moment, useless, held hostage by a safety rule that exists to protect the people fixing the line.

Now to the money, because this is where the modern story genuinely upends people's assumptions. For most of solar's history the panels were the expensive part; everything else was an afterthought bolted around them. Then, over roughly fifteen years, the price of the panels themselves fell off a cliff — a collapse we will trace to Chinese manufacturing scale and the relentless logic of the learning curve — until modules became almost a commodity, priced by the watt at a small fraction of their old cost. But the other stuff did not fall nearly as fast. Everything that is not the panel is lumped together under a slightly bureaucratic phrase: the balance of system, or BOS. That means the inverter, the mounting rails and roof anchors, the wiring and conduit, the combiner boxes and disconnects and the meter. And in the case of home solar it also means all the intangible costs — the permits, the inspections, the sales and marketing, the design engineering, the crew's labor and the truck they drove in. Add those up and you reach a fact that still startles people: for a typical American residential solar installation, the panels are a minority of the total cost. The great majority is balance of system, and a large slice of that is the so-called soft costs — the permits, paperwork, and labor — which vary wildly from country to country. This is why an identical rooftop system can cost dramatically more to install in the United States than in Australia or Germany, even though the panels came off the same production line. The hardware has been commoditized; the friction has not.

Let me make that concrete with the kind of breakdown an industry analyst would recognize, keeping the numbers as fractions rather than as prices. For a typical American home installation, the panels themselves account for only something like a sixth of the total outlay. The inverter and the electrical hardware take another slice; the racking and mounting another; the wiring, conduit, and small parts another. And then, larger than any single piece of hardware, comes the pile of soft costs — the permit fees and inspection appointments, the customer acquisition and sales commissions, the system design and engineering, the overhead, and the labor of the crew on the roof. In the United States these soft costs alone can rival or exceed the entire cost of the hardware. Ship the identical pallet of panels, inverter, and racking to a rooftop in Australia, where permitting is streamlined and installation crews are lean and plentiful, and the finished system can cost less than half what it costs in America — same sunshine, same silicon, a wildly different price, and almost the whole difference is friction rather than physics. That gap is one of the most important and least appreciated facts in solar economics: the frontier of cheap solar in a rich country now runs through the permitting office, not the factory.

So the old misconception — that the panels are the expensive part of going solar — is now simply wrong, and understanding that reframes the whole economics of the technology. The frontier of cost reduction is no longer the silicon. It is the mounting hardware, the permitting office, and the hour of a licensed electrician's time.

Within the balance of system, the inverter itself comes in two main philosophies, and the choice between them is one of the few real decisions a homeowner's installer will make. The traditional approach is the string inverter: one central box, back in the garage, that all the panels feed into, wired in the series strings we met in the last chapter. It is cheaper and simpler, but it inherits the string's weakness — because the whole string shares one MPPT brain, shading or dirt on one panel drags on all of them, and the array as a whole tends to perform at the level of its weakest string. The alternative is to distribute the intelligence: a microinverter is a small inverter bolted to the back of each individual panel, converting that one panel's DC to AC right there on the roof. Microinverters cost more in total, but each panel gets its own MPPT brain and its own place in the sun, so a shaded or underperforming panel no longer drags its neighbors down, and the system tolerates complicated roofs with dormers and chimneys far more gracefully. A close cousin, the power optimizer, splits the difference — a small box on each panel that does the tracking but hands off the actual DC-to-AC conversion to a single string inverter. There is no universally right answer; there is only the trade of upfront cost against resilience to shade and the peace of mind of per-panel monitoring.

Whichever kind you choose, there is an uncomfortable fact to plan around: the inverter is the part of a solar system most likely to fail and need replacing. The panels, as we saw, are warrantied for twenty-five years and often outlast that; the inverter, packed with fast-switching electronics and heat-stressed components, is typically warrantied for something more like ten to fifteen. Over the life of an array, many owners will replace the inverter at least once. It is the beating heart of the system and, like a heart, the part most likely to wear out. Microinverters shift this calculus somewhat — with dozens of small units, a single failure costs you one panel's worth of output rather than the whole system, though replacing one means a trip up onto the roof.

There is one more number the inverter is quietly responsible for, and you meet it on your electricity bill. In many places, a rooftop system tied to the grid is governed by an arrangement called net metering: when your panels produce more than your home is using, the surplus flows backward through your meter into the grid, and the utility credits you for it, effectively spinning your meter in reverse. At night, or on a dull day, you draw from the grid as usual and the credits are drawn down. The meter tracks the net — hence the name — and the grid serves as a kind of vast, effortless battery that you never have to buy or maintain. Net metering has been one of the great engines of rooftop solar's spread and also one of its great political battlegrounds, because how generously the utility credits that exported power decides whether a home installation pays for itself in seven years or seventeen. The rules are being rewritten in state after state, generally in the direction of crediting exports less generously, and the terms of that fight will shape the economics of home solar for a generation.

Net metering also quietly exposes another misconception, the mirror image of the blackout surprise: many people assume that a home with solar panels is somehow independent of the grid, off on its own, unplugged from the utility. In truth the ordinary rooftop system is more tethered to the grid than a conventional home, not less. It leans on the grid as its battery, its backup, and its reference signal all at once — drawing power at night and in bad weather, dumping surplus during the day, and taking from the grid the very rhythm the inverter must synchronize to. Cut a standard system loose from the grid and it does not become a proud little independent power station; it goes dark the moment a cloud passes or the sun sets. True independence — keeping the lights on through a multi-day blackout, unplugging from the utility altogether — requires batteries, and batteries are their own chapter in the story of how these systems are changing. For now the point is simply that the humble grid connection, so easy to take for granted, is doing an enormous amount of invisible work behind every rooftop array.

We have now assembled the whole chain: a half-volt cell multiplied into a module, modules strung into an array, the array's direct current translated by a hunting, synchronizing, self-protecting brain into grid-ready alternating current, and the surplus metered out to the neighborhood. That is the modern rooftop, and it rests on a century of accumulated cleverness. But solar is only half of this book's story, and the wind half has been waiting, on a cold and windy hilltop in Vermont, for a man with an outsized ambition and two eight-ton stainless-steel blades. To find the birth of the modern wind turbine we have to go back to 1941, to a farmer's summit called Grandpa's Knob.

Chapter 23 — Palmer Putnam's Giant

On the night of March 26, 1945, a caretaker named Harold Perry was alone in the control house at the foot of a Vermont hill called Grandpa's Knob when he heard a sound he would remember for the rest of his life — a deep, wrenching bang, followed by a lurch that shook the entire steel tower above him. One of the two great blades of the Smith–Putnam wind turbine, each of them sixty-six feet long and weighing eight tons, had torn loose from the hub while the rotor was spinning and hurled itself off the mountain. It cartwheeled through the dark and landed some distance down the slope, digging a furrow in the frozen ground. No one was killed. But the largest wind turbine the world had ever seen, a machine that had been feeding electricity into the Vermont grid on and off for three and a half years, had just thrown a blade, and it would never turn again. The strangest part of the story is not that it failed. It is that nothing of its size would be built again for nearly forty years.

To understand how a 1.25-megawatt wind turbine came to stand on a Vermont hilltop in 1941 — a megawatt being a million watts, enough to supply hundreds of homes — you have to understand Palmer Cosslett Putnam, and Putnam was not a wind man at all. He was an MIT-trained engineer, restless and persuasive, a geologist by first training and an entrepreneur by temperament, and his interest in wind power began not with a grand vision of clean energy but with a personal irritation. Having built a house on Cape Cod in the early 1930s, he found himself appalled by two things: the electricity bill, which was high, and the wind, which was constant. It struck him as absurd that so much free energy was blowing past his windows while he paid dearly for power delivered down a wire. He began to wonder, with an engineer's arithmetic, whether the wind could be harnessed at a scale that would actually matter to a utility — not a farm windmill charging a few batteries, which was old news, but a machine large enough to feed a serious quantity of electricity into the public grid at a price that could compete with coal.

This was a genuinely radical reframing, and it is worth pausing on because it is the intellectual birthplace of every modern wind farm. The windmills of history — the Persian grain-grinders, the Dutch polder-drainers, Charles Brush's battery-charging dynamo of 1888, the wind-chargers dotting American farms in the 1920s — were all, in a sense, private machines, sized to a local task and a local store. Putnam's idea was different in kind. He wanted a utility-scale, grid-connected wind turbine, a machine that would not charge a battery or grind a farmer's grain but would synchronize its output to the alternating current of the public electricity network and pour power into it, indistinguishable from the output of a coal or hydro plant. That single conceptual leap — from private windmill to grid generator — is the reason Putnam belongs in this book. He was the first to build the thing we now take for granted, and he built it decades before the world was ready to want it.

It is worth dwelling for a moment on just how large a leap the numbers represent, because the scale is what makes the story astonishing rather than merely interesting. Charles Brush's celebrated Cleveland wind dynamo of 1888 — the machine that had impressed the readers of Scientific American as the first genuinely automatic wind-electric plant —

produced about twelve kilowatts, twelve thousand watts, enough to charge a bank of batteries and light one large house. Putnam was proposing to build a machine of 1.25 megawatts, which is to say one thousand two hundred and fifty kilowatts, more than a hundred times Brush's output, and to do it not to charge batteries in a cellar but to feed alternating current straight into a regional grid alongside coal and hydro plants. It was as if someone in the age of the propeller biplane had proposed to skip straight to the jumbo jet. The gap between the largest existing wind machine and the one Putnam wanted to build was not incremental; it was a chasm, and crossing it in a single stride is exactly what makes the Smith–Putnam turbine one of the boldest engineering gambles in the history of energy.

Putnam had ambition and connections but no money and no manufacturing base, so he did what able promoters do: he found partners. He brought his idea to the S. Morgan Smith Company of York, Pennsylvania, a firm that built water turbines for hydroelectric dams and was looking to diversify. They agreed to finance and fabricate the machine, and the collaboration gave the turbine its double-barreled name, Smith–Putnam. Putnam also assembled an extraordinary brain trust of consultants — among them the great aerodynamicist Theodore von Kármán and engineers from MIT and General Electric — to work out the punishing physics of a rotor that would be, by an enormous margin, the largest ever attempted. Everything about the machine was at the outer edge of what 1940 could build. The rotor stretched a hundred and seventy-five feet across, mounted atop a steel tower some hundred and ten feet tall, on a summit chosen after a survey of Vermont's windiest peaks. The two blades were built of stainless steel and could change their pitch — rotate about their long axis — to regulate the machine's speed and shed load in a gale, a feature we still find on turbines today. And the whole rotor was hinged in a clever way to flap and relieve the ferocious, uneven stresses that a wind gust throws at a blade sweeping through a hundred-and-seventy-five-foot circle.

On October 19, 1941, on Grandpa's Knob in the town of Castleton, the machine turned in earnest and did something no wind turbine had done before: it synchronized to the grid of the Central Vermont Public Service Corporation and began delivering electricity to paying customers. Read that sentence again and let its date sink in, because it contains a genuinely startling fact. In 1941 — before the transistor, before the first digital computer, in the same weeks that the United States was being drawn into the Second World War — an American hilltop carried a wind turbine rated at 1.25 megawatts, feeding utility power into the public grid. That capacity would not be matched by another wind machine anywhere in the world until around 1980. For four decades the high-water mark of wind engineering was a wartime experiment that had already been dismantled. When people tell you, as they often do, that giant wind turbines are a recent invention of the environmental age, this is the fact to set gently in front of them. The megawatt turbine is older than the pocket calculator.

To grasp why synchronizing to the grid was such an achievement, picture what it actually demands. The grid is a vast chorus of generators all singing the same note in perfect time — in the United States, sixty cycles a second, the wave cresting and falling sixty times every second in lockstep across thousands of miles of wire. For a new generator to join that chorus it cannot simply blurt out its own tune; it must match the grid's frequency and phase exactly, cresting when the grid crests, or the mismatch will jolt the machine and the network alike. A steady coal plant, spinning at a rock-solid speed, finds this easy. A wind turbine, whose rotor speeds up and slows with every gust, does not. Putnam's engineers had to hold their giant rotor's output in step with the Vermont grid despite the wind's endless variation — a genuinely hard control problem in an age before digital electronics, solved with mechanical governors, gearing, and the pitchable blades that spilled excess wind to keep the speed in bounds. That they managed it at all, and repeatedly, is a fact that tends to get lost beneath the drama of the blade that flew off.

The machine's working life was neither triumphant nor a farce; it was the messy middle. It ran intermittently through the war years, and running it taught its engineers, in the hardest possible way, the lessons that only a full-scale prototype can teach. The bearings gave trouble. The control systems had to be nursed. And the blades, those eight-ton stainless-steel wings, flexed and vibrated under loads that the era's fatigue science could only partly predict. There is a revealing anecdote from the early testing: the machine's main bearing failed in 1943 and the turbine sat idle for two full years, not because the engineers did not know how to fix it but because it was wartime, and the priority list for scarce steel, skilled machinists, and heavy castings did not have room for an experimental windmill in Vermont. A replacement bearing that might have been fabricated in weeks in peacetime took two years to obtain. The turbine, in other words, was as much a casualty of the war economy as of any flaw in its design.

And that context is essential to reading the blade failure correctly. During those two idle years of waiting for the bearing, the crew had noticed a crack developing near the root of one blade. In ordinary times they would have taken the blade down and repaired or replaced it. But it was 1944, then 1945, and the materials and shop time to do the job properly simply were not available; the country's metalworking capacity was consumed by tanks and ships and aircraft. So the engineers made a calculated gamble. They ran the machine anyway, watching the crack, hoping to gather more operating data before the war's end freed up resources for a proper overhaul. The gamble failed. On that March night in 1945 the cracked blade let go and flew off the mountain. It was not a mysterious act of God or a sign that the wind itself was untamable. It was a known crack, in a known blade, that could not be fixed on time because of a war. The distinction matters enormously, and it is the correction to the second great misconception about the

Smith–Putnam turbine: that it proved wind power "doesn't work." It proved no such thing. The machine had already demonstrated, over three and a half years, that a large turbine could reliably synchronize to a grid and deliver power. What killed it was wartime metal shortage and, ultimately, cold economics.

Because after the blade failure came the reckoning, and here the story turns on money rather than metal. Smith–Putnam's engineers studied the wreck, understood the fatigue problem, and concluded that a fixed, stronger blade could be built. The company even worked out what a production version might cost. But the sums did not close. In the flush of the postwar years, fossil fuel was astonishingly cheap and getting cheaper; coal-fired and hydroelectric power could be delivered for well under what a fleet of these enormous, finicky, hand-built turbines would require. The S. Morgan Smith Company ran the numbers, concluded that wind electricity could not compete with the going price of conventional power, and in 1945 quietly ended the project. The tower on Grandpa's Knob was dismantled and its steel sold. A machine that had been decades ahead of its time was scrapped not because it could not generate electricity but because it could not generate electricity cheaply enough to matter in a world drowning in inexpensive fossil fuel.

The engineers left behind one more artifact that deserves to be remembered: a book. Palmer Putnam himself wrote up the whole adventure in 1948 in a volume titled Power from the Wind, a candid, technical, unsparing account of what the project had attempted, achieved, and failed to do. It became something close to a sacred text for the small band of wind engineers who came after, precisely because Putnam did not varnish the failures — he laid out the fatigue problems, the vibration, the bearing troubles, and the economics, so that anyone who picked the thread back up would not have to relearn every lesson from scratch. When the wind revival came in the 1970s, the engineers who did know their history reached for Putnam's book. In a field where so much institutional memory had evaporated, that single honest volume was one of the few bridges across the forty-year gap, and it is a reminder that writing down your failures clearly can be as valuable a legacy as any working machine.

There is something almost unbearably modern in that verdict, and it is the through-line that connects Putnam's giant to everything in this book. Wind power did not fail on physics. Betz's limit, which we met earlier, told us that a rotor can in principle capture up to about fifty-nine percent of the energy in the wind passing through it, and the Smith–Putnam machine, for all its troubles, captured a respectable fraction of that. The kinetic power available in the wind — proportional, as we saw, to the cube of the wind speed and the square of the rotor's diameter — was there for the taking, and Putnam's engineers took a real slice of it. What defeated the turbine was the price of the alternative. For the next thirty years, whenever anyone proposed building large wind machines, the same answer came back: why bother, when oil and coal and gas are this cheap? The economics of wind would not turn until the price of the alternative did, and the price of the alternative would not turn until a October morning in 1973 when the oil-exporting nations of the Middle East changed the arithmetic of the entire industrial world overnight.

The forgetting was not total, and where you meet Putnam's legacy today is precisely in the fact that his central bet turned out to be right, just early. Every utility-scale wind farm now feeding the grid — every array of megawatt-and-larger turbines synchronized to the alternating current of a national network — is the vindication of the idea he staked his fortune and reputation on in 1941: that wind could be harnessed not as a private curiosity but as a serious contributor to the public power supply. When a modern offshore machine of a dozen megawatts pours its output into the grid off the coast of England or China, it is doing, at ten times the scale, exactly what the Smith–Putnam turbine did on a Vermont hilltop before most of the engineers who build them were born. Putnam did not live to see it become ordinary. He died in 1984, just as the wind revival he had prefigured was finding its feet in California and Denmark.

You can still find Grandpa's Knob on a Vermont map, a modest wooded summit near Rutland, and for years almost nothing marked what had stood there. The turbine left behind no monument and no successor — only a foundation, some old photographs of two vast blades turning against a New England sky, and a lesson that the industry would have to learn again from scratch a generation later. When engineers in Denmark and Germany and California finally returned to large wind turbines in the late 1970s, many of them did not even know that an American had already built a megawatt machine, synchronized it to the grid, and run it through a world war. Putnam's giant had been so thoroughly forgotten that its central achievement had to be reinvented.

But Putnam's failure planted one idea that would not die, and it is the idea that carries us into the next chapter. The Smith–Putnam blades had been huge, heavy slabs of stainless steel, and their fatal weakness had been exactly that heaviness — eight tons of metal flexing and cracking under loads no one could fully predict. What if the blade did not have to be a massive steel wing at all? What if it could be light, slender, and shaped by the same aerodynamic science that was lifting airplanes into the sky? In postwar Germany, an engineer named Ulrich Hütter was about to ask precisely that question, and his answer would look far more like the wind turbines outside your window today than anything on Grandpa's Knob.

Chapter 24 — Ulrich Hütter and the Slender Blade

In the hills near Stuttgart in the late 1950s, on a windswept rise called the Stötten, a wind turbine turned that looked wrong to almost everyone who saw it. It had only two blades, and those blades were absurdly thin — long, tapering, glossy things that swept through the air with none of the muscular bulk people expected of a machine meant to wrestle power from the wind. Farm windmills of the age were dense fans of many broad slats, built to look like they were pushing hard against a strong breeze. This machine, by contrast, looked delicate, almost fragile, more like the wing of a glider stood on end than like any windmill in living memory. And that resemblance was not an accident. The man who designed it, Ulrich Hütter, had spent the war years around aircraft, and he had brought to wind power a single heretical conviction that would quietly reorganize the entire field: a turbine blade is not a sail. It is a wing.

That sentence sounds simple, but it overturns the intuition almost everyone starts with, and correcting that intuition is the whole point of this chapter. Watch a sailboat run before the wind and it is obvious what is happening — the wind pushes on the sail, the sail drags the boat along, and the harder the wind blows the harder it shoves. It is natural to assume a windmill works the same way, that the moving air simply pushes against the blades and shoves the rotor around like a hand pushing a revolving door. Machines that work this way do exist; engineers call them drag machines, and the old multi-bladed farm water-pumper is a rough example, presenting a broad face for the wind to push against. But drag machines are fundamentally limited. A surface being pushed by the wind can never move faster than the wind that pushes it — you cannot outrun the thing shoving you from behind — and a blade that only captures the wind's push leaves most of the wind's energy untouched. To get more, you have to stop pushing and start flying.

There is a homely demonstration of the difference that anyone can do with a hand out a car window, and it captures the whole idea. Drive at a steady speed and hold your flat palm out the window, fingers forward, face-on to the wind. You feel the air shove your hand backward — that is drag, the sail's kind of force, pure push. Now tilt your hand slightly, thumb up, so it becomes a little wing cutting through the flow at an angle. Suddenly your hand wants to climb, to rise on its own, lifted by a force that points not backward with the wind but upward, across it. You have just felt lift, and you have felt that it is stronger and springier than the mere push you felt a moment before. That upward yank on your tilted palm is, in miniature, exactly the force that drives a modern turbine — not the wind shoving the blade along, but the blade, shaped and angled like your tilted hand, being pulled sideways around its circle.

Here is the beautiful counterintuitive core of it. A modern turbine blade does not get shoved by the wind; it gets pulled by lift, exactly as an airplane wing is pulled upward. Recall how a wing works: it is shaped so that air flowing over its curved upper surface travels in a way that lowers the pressure above the wing relative to below, and the resulting pressure difference sucks the wing toward the low-pressure side. Tilt that wing and mount it so the lift points not up but sideways — around the circle the rotor sweeps — and the wing is pulled along its path, dragging the rotor around with it. This is why a good turbine blade is a carefully sculpted airfoil, curved and tapered and twisted, and not a flat paddle. And it explains a fact that startles people the first time they hear it: the tip of a large modern turbine blade is not lumbering along at wind speed. It is moving far faster than the wind. The blade, pulled by lift, races ahead of the air that drives it, sometimes at seven or eight times the wind's own speed — so that in a brisk wind of forty or fifty kilometers an hour, the blade tips can be slicing through the air at well over two hundred and eighty kilometers an hour, faster than a Formula One car on a straight. A sail can never outrun the wind. A wing routinely does.

Engineers capture this in a single number that Hütter understood better than almost anyone of his generation: the tip-speed ratio, the ratio of how fast the blade tip moves to how fast the wind blows. A slow, draggy, many-bladed windmill has a low tip-speed ratio — its blade tips barely keep pace with the wind. A slender, fast, lift-driven rotor has a high one. And the deep insight, which reaches back to the Danish schoolteacher Poul la Cour and his wind-tunnel experiments of the 1890s, is that the faster and more airfoil-like your blades, the fewer of them you need. This is not obvious, and it is worth making concrete. If a rotor turns slowly, each blade has to be broad and there have to be many of them, because the air passing through the disk between the slow-moving blades slips by untouched; you need a lot of blade to intercept it. But if the rotor spins fast, each blade returns to sweep through a patch of air just moments after the previous blade left it, before that air has had time to recover and flow on unimpeded — so a fast rotor can harvest the whole disk of wind with only a handful of thin blades. Speed substitutes for surface area. La Cour had grasped this in principle. Hütter, armed with modern aerodynamics, turned it into a design philosophy: few blades, slender, fast, and shaped like wings.

The fact that the tips outrun the wind so dramatically has a consequence you can hear, and it is where most people actually meet this physics today. Stand near a large modern turbine on a windy day and the sound it makes is not the creaking, groaning labor you might expect of a giant machine; it is a soft, rhythmic whoosh, the swish of a blade tip cutting the air at highway speed. That sound is the signature of a lift machine running fast and clean, and it is the reason engineers obsess over the exact shape of the outer third of the blade, where the speed and therefore the aerodynamic action are greatest. It is also why blade tips are the part that erodes — at nearly three hundred kilometers an hour, a raindrop or a speck of grit hits the leading edge like a tiny bullet, and offshore operators now bond special protective tape or coatings to the tips to keep them from being sandblasted smooth over the years. The wing that outruns the wind pays for its speed at its fingertips.

None of this violates the fundamental ceiling we established with Albert Betz. Betz proved in 1919 that no rotor, however clever, can extract more than about fifty-nine percent of the energy in the wind flowing through it, because the air has to keep moving to get out of the way of the air behind it — bring the wind to a complete stop and you would dam the flow and capture nothing. Hütter was not trying to beat Betz's limit; nobody can. He was trying to approach it efficiently and, crucially, cheaply — to capture a large fraction of that theoretical maximum with the least possible material, weight, and cost. A draggy multi-bladed rotor wastes not only energy but metal and money on all that blade area. A slender high-speed rotor gets close to the Betz ceiling with a fraction of the material. Efficiency and economy pointed the same way, toward fewer and thinner blades, and that convergence is why the modern turbine looks the way it does.

Let me make the tip-speed idea concrete, because a single worked number fixes it in the mind. Imagine a large turbine with a rotor a hundred meters across, its blades stretching fifty meters from hub to tip, turning at a leisurely-looking twelve revolutions a minute — one full turn every five seconds. That sounds slow, almost stately, and to the eye it is. But follow the tip: in one revolution it travels around a circle a hundred meters in diameter, which is more than three hundred meters of arc, and it does so five times a minute, so the tip is covering better than fifteen hundred meters a minute — around ninety kilometers an hour just from that unhurried-looking rotation, and far more in a strong wind driving the rotor faster. The hub barely moves; the tip screams. That enormous difference in speed along a single blade is why the blade must be twisted along its length, presenting a steeper angle near the slow hub and a shallower one near the racing tip, so that every section meets the oncoming air at the correct angle to make lift. A turbine blade is not a simple wing but a wing that changes its shape and angle continuously from root to tip — a subtlety Hütter understood and built into his designs when many contemporaries were still bolting on flat paddles.

But slender, fast-spinning blades create a brutal engineering demand, and this is where Hütter's real genius lay. A long thin blade whipping around at two hundred kilometers an hour and bending in every gust must be simultaneously light, strong, and stiff — three properties that fight one another in most materials. Make it of steel, as Palmer Putnam had, and it is strong but crushingly heavy, and its own weight and the fatigue of flexing will eventually crack it, as Putnam learned on Grandpa's Knob. Hütter's answer came from a material that was, in the 1950s, still novel: glass-fiber-reinforced plastic, what we now simply call fiberglass — thin glass filaments embedded in a resin that could be molded into the exact compound curves an airfoil demands, yielding a blade that was feather-light for its size yet strong and springy enough to bend with the gusts instead of snapping. Hütter pioneered the techniques for building these composite blades and for attaching them to the hub, and his experimental turbine on the Stötten, the machine known as the W-34, with its two slender fiberglass blades and a rotor about thirty-four meters across, became a working proof that the wing-not-sail philosophy could be built and run. It was not a commercial success in its own time — postwar Germany, like postwar America, was awash in cheap fossil energy, and there was no market for wind electricity — but it was a laboratory for ideas, and the ideas outlived the machine.

There is a quieter human dimension to Hütter's story that an honest account should not skip. Like a number of German engineers of his generation, his career had roots in the aviation work of the Nazi era, and his path into wind ran through the same aerodynamic science that had served military aircraft. The postwar decades let him turn that expertise toward the wind, and it is a recurring pattern in this history that the deep knowledge of how air flows over a wing — knowledge developed at great expense for aviation — became the intellectual inheritance that made modern wind power possible. The turbine blade is, in a real sense, a piece of aeronautical engineering that came home to roost on a tower.

It is worth naming the misconception this whole chapter is built to dislodge, because it is astonishingly durable. Ask most people how a windmill works and they will tell you, with complete confidence, that the wind blows against the blades and pushes them around — the sail picture, the revolving-door picture, the intuition that a windmill is a thing being shoved. It is wrong, and Hütter's life work is the correction. The blades of the machine on the ridge are not being pushed; they are being pulled, flown around their circle by lift, in exactly the way a wing is flown through the sky. The wind does not so much drive the blade as the blade flies through the wind. Once you see it that way, a dozen puzzling facts click into place at once — why the blades are so few and so thin, why they can spin faster than the wind that powers them, why their shape matters so much more than their bulk, and why an airplane engineer, not a millwright, turned out to be the right person to design them.

Hütter's influence spread by example and by the students and engineers who absorbed his thinking, and it flowed in two directions that this book will follow. One channel ran to the United States, where in the 1970s, after the oil shock finally made wind worth a second look, government-funded programs at NASA and the Department of Energy reached back to Hütter's principles — lightweight composite blades, few in number, high tip-speed ratios — and built a series of large experimental turbines. Several of those American machines, like Putnam's giant before them, chose two slender blades in the Hütter tradition. The other channel, more consequential in the end, ran to Denmark, where engineers were absorbing the same aerodynamic lessons but would fold them into a design of stubbornly different temperament — sturdier, simpler, and, fatefully, three-bladed rather than two.

For it turns out that Hütter's philosophy, right in almost every particular, was not right in every particular, and the one place the modern turbine parted company with him is instructive. Hütter favored two blades, because two blades use less material than three and, in pure aerodynamic terms, a well-designed two-bladed rotor can be nearly as efficient. But two-bladed rotors have a vibration problem that is subtle and stubborn: as the rotor turns, the two blades pass in front of the tower and swing through regions of differing wind speed in a way that sets up an uneven, shuddering load, a dynamic imbalance that stresses the machine and unsettles it. A third blade, though it adds cost and weight, smooths all of this out, giving the rotor a rotational steadiness — a kind of gyroscopic poise — that two blades cannot match. And there is an aesthetic and psychological dimension too, which sounds soft but proved real in the marketplace: three slowly turning blades simply look calmer, more balanced, more trustworthy to the human eye than two blades scissoring around, and public acceptance turned out to matter for a technology that lives in full view on the landscape. The modern turbine took Hütter's slender, fast, lift-driven, composite blade — and then added one more of it.

So we arrive at the threshold of the machine we recognize today, assembled from parts contributed by everyone we have met. From la Cour, the insight that fewer, faster blades win. From Betz, the ceiling that tells us how much we can hope to capture. From Putnam, the proof that a large turbine can feed the grid, and the cautionary tale of the heavy steel blade that cracked. And from Hütter, the decisive reconception of the blade as a wing — light, slender, aerodynamic, composite — that made large turbines practical rather than punishing. What remained was for someone to combine these threads into a single robust, reliable, buildable design and prove it could run for years without flying apart. That someone was a Danish engineer named Johannes Juul, and the machine he raised beside a small town on the Danish coast in 1957 would turn out to be the direct ancestor of nearly every wind turbine spinning on Earth today.

Chapter 25 — The Gedser Turbine

On a spit of land at the southernmost tip of Denmark, where the island of Falster runs out into the cold gray Baltic near the ferry town of Gedser, a wind turbine went into service in 1957 that almost nobody outside Denmark noticed at the time. It had three blades, it faced into the wind, it stood on a squat concrete tower, and it turned with an unglamorous, almost agricultural steadiness. It was rated at two hundred kilowatts — a fraction of Palmer Putnam's wartime giant, a tenth of a megawatt and then some, enough for a village rather than a city. By the standards of ambition, it was a modest machine. And yet, of all the wind turbines built in the twentieth century, this is the one whose shape you would recognize instantly if you drove past a wind farm tomorrow. The Gedser turbine is the machine that everything since has been a scaled-up copy of. When we talk about "the modern wind turbine," we are, whether we know it or not, talking about a design that a Danish engineer named Johannes Juul finalized on a Baltic headland in 1957.

Juul was not a young man chasing a novelty. He was in his late sixties when he built the Gedser machine, and he had, in a sense, been preparing for it his entire adult life. Back in 1904, as a young man, he had attended a course taught by Poul la Cour — the Danish schoolteacher-scientist we met earlier, the father of wind electricity, the man who had run the world's first wind-power laboratory and shown that fewer, faster blades were more efficient. Half a century later, Juul, by then a senior engineer at a Danish utility, returned to his old teacher's subject and set out to build the reliable, grid-connected wind turbine that la Cour had dreamed of and Putnam had failed to make economic. There is a lovely continuity in that: a direct chain of teaching and influence stretching from la Cour's experiments in the 1890s, through the student who sat in his classroom in 1904, to the machine that would define the industry for the next hundred years. Denmark had never really stopped thinking about wind — the country is flat, windy, poor in coal, and long accustomed to windmills — and that unbroken cultural attention is a large part of why the winning design emerged there rather than in America or Germany.

It is worth being clear about what kind of achievement this was, because it was not the achievement of a lone inventor with a flash of insight. Juul invented almost nothing at Gedser in the sense of a brand-new principle; the upwind rotor, the induction generator, the airfoil blade, the idea of stall — all of these existed already, scattered across the work of earlier engineers. What Juul did was harder and rarer than invention: he made the right choices and fit them together into a whole that worked. Engineering history is full of brilliant components that never cohered into a reliable machine, and full of clever machines that ran once for the cameras and then broke. Juul's genius was integration and judgment — knowing which of the available options to trust, how to make them reinforce rather than undermine one another, and how to build the result rugged enough to survive a decade of Baltic weather. It is a quieter kind of genius than the inventor's, and in engineering it is often the decisive one.

What Juul assembled at Gedser was not a single new invention but a particular combination of choices, and nearly every one of those choices became the industry standard. To see why they won, it helps to walk through the design decisions the way an engineer would, because each one embodies a principle worth understanding in its own right.

Start with the most visible choice: which way the rotor faces. A wind turbine can be built with its rotor on the downwind side of the tower, so the wind hits the tower first and then the blades, or on the upwind side, so the wind meets the blades first. Downwind machines have a seductive advantage — they can, in principle, swing around to follow the wind on their own, like a weather vane, with the blades trailing behind. But they have a nasty flaw: every time a blade passes behind the tower, it moves through the tower's wind shadow, a pocket of disturbed, slackened air, and that repeated flick of the blade in and out of the shadow sets up a shuddering pulse that fatigues the machine and, on a still day, makes it thump audibly. Juul chose upwind — blades in front of the tower, meeting clean, undisturbed air on every pass. It costs you something: an upwind machine cannot free-vane, so it needs a mechanism to actively turn and hold the rotor into the wind, a system engineers call yaw. But the payoff in smoothness and longevity is decisive, and upwind is what virtually every large turbine uses today.

Then there is the question of how you stop the machine from destroying itself in a storm. Recall the iron law of wind power, the cube relationship we established earlier: the energy in the wind rises with the cube of its speed, so that doubling the wind speed delivers not twice but eight times the power. That cube law is a gift in a good breeze and a mortal threat in a gale — a wind twice as strong is trying to feed your machine eight times the power it was built to handle, and something has to give. Putnam's turbine had used pitchable blades that rotated about their long axis to spill the excess, an elegant but mechanically complex solution, and the failure of that machinery was part of his machine's troubles. Juul chose a different, brutally simple approach called stall regulation. He fixed his blades rigidly to the hub — no pitching mechanism at all — and shaped them so that when the wind grew too strong, the airflow over the blade would break away and become turbulent, exactly the way an airplane wing "stalls" and loses its lift when tilted too steeply. A stalled blade stops accelerating the rotor; the machine passively refuses to over-speed, with no moving parts, no sensors, no control system needed to save it. The wind's own excess violence triggers the very stall that tames it. It is the kind of solution engineers call beautiful — robustness achieved by geometry rather than by machinery that can fail.

Stall regulation is worth one more moment, because it is such a clean illustration of good engineering, and there is an everyday image that captures it. Think of holding a dinner plate flat out of a fast-moving car window, its face angled slightly up into the wind. At a gentle angle the air streams smoothly over it and the plate rides steadily on a cushion of lift. Tip it too steeply, though, and the smooth flow suddenly tears away into a churning, useless wake — the plate shudders, the lift collapses, and it just batters against the wind. That collapse is a stall, and it feels like a failure. Juul's insight was to turn that failure into a safety feature. By fixing the blades at just the right built-in angle, he arranged for them to ride smoothly in ordinary winds but to stall automatically, all by themselves, exactly when the wind rose high enough to threaten the machine — the flow breaking away, the lift bleeding off, the rotor quietly refusing to be driven any faster. No sensor decides this and no motor executes it; the physics of the airflow does the whole job. A gale that would tear a lesser machine apart instead trips the very stall that saves the Gedser turbine, and it does so with not one moving part.

The third choice was the generator, and here Juul made a decision whose importance is easy to miss. He used what is called an asynchronous, or induction, generator, and he connected it directly to the grid. The significance is this: an induction generator, once the grid is up and running, is effectively locked to the grid's own frequency — it draws its rhythm from the network it is feeding, spinning in sympathy with the fifty-cycle alternating current of the Danish grid. This solved, almost for free, the synchronization problem that had made grid-connected wind so fiddly. There was no need for delicate machinery to match the turbine's output to the grid's wave, phase for phase, the way Putnam's engineers had struggled to do; the induction generator simply fell into step with the grid and stayed there, feeding power in whenever the wind turned the rotor fast enough. It was a self-synchronizing, forgiving, robust way to marry a variable wind machine to a steady electrical network, and it, too, became a template.

And then there were the three blades. Juul settled on three where Hütter, across the border in Germany, had argued for two. The choice looks small and turned out to be foundational, and the reasons gather together everything the previous chapters have taught. Three blades give the rotor a rotational steadiness that two cannot — the load stays balanced as the rotor turns, without the shuddering imbalance that plagues a two-bladed machine as its blades sweep past the tower and through the uneven winds at different heights. Three blades cost more material than two but far less than the dense multi-bladed fans of old, and paired with the slender, fast, lift-driven airfoil shape that la Cour and Hütter had championed, three was the sweet spot where stability, efficiency, and cost all pointed the same way. There is even the human factor we noted before: three slowly wheeling blades read to the eye as calm and balanced, and a technology that lives on the skyline in full public view cannot afford to look agitated. Stability, economy, and even aesthetics converged on the number three, and the industry has scarcely wavered from it since.

Put all these choices together — three blades, upwind, stall-regulated, asynchronous generator directly grid-connected, on a sturdy tower — and you have what came to be called the Danish concept, and it is no exaggeration to say it is the genetic blueprint of the modern turbine. But a blueprint is only as good as the machine's ability to survive, and this is where Gedser truly earned its place in history. Putnam's giant had thrown a blade in three and a half years. Juul's machine ran, reliably and largely untended, feeding power into the Danish grid, for eleven years, from 1957 until 1967, when it was finally shut down not because it had failed but because the cheap oil of the 1960s had once again made wind electricity look economically pointless. It had proven the one thing that mattered most and that every earlier large turbine had failed to prove: that a big wind machine could be genuinely reliable, running year after year through Baltic storms with the sturdy indifference of a well-built tractor. Reliability, not peak efficiency, was the prize, and Gedser won it. Consider what that decade of steady running actually meant in the arithmetic of the thing: eleven years is roughly four thousand days, and on a windy Baltic headland the rotor turned through the great majority of them, its stall-regulated blades weathering gale after gale without a control system to nurse them and without a catastrophic failure to halt them. Putnam's giant had managed a few years and thrown a blade; Juul's humbler machine simply kept its appointment with the wind, morning after morning, until cheap oil rather than any fault of its own finally sent it into retirement.

The most surprising twist in the story came a decade after the machine fell silent. In the 1970s, when the oil shocks finally made the world care about wind again, engineers scrambling to build large turbines went looking for a design that was actually known to work — and found that the answer had been standing quietly at Gedser all along. The American space agency, NASA, then leading a United States government effort to develop big wind turbines, took a direct interest in Juul's old machine. At their request, and with American funding, the Gedser turbine was refurbished and restarted in the late 1970s so that its performance could be carefully measured with modern instruments and its data fed into the new generation of designs. A Danish turbine built by a man in his late sixties in 1957, shut down as uneconomic in 1967, was brought back to life in the 1970s to teach the world's most advanced space agency how to build wind turbines. There is no better illustration of how far ahead of its time the Danish concept was.

There is a deeper lesson buried in Gedser's triumph, and it is one the whole history of technology keeps teaching: the best design is rarely the most efficient one on paper. Hütter's two-bladed rotors were, in narrow aerodynamic terms, arguably more elegant than Juul's; a sleek two-blader wastes less material and can capture a whisker more of the Betz-limited energy. Yet it was the sturdier, heavier, three-bladed, stall-regulated Danish machine that conquered the world, because it was the one that kept running. Juul optimized not for peak performance in a wind-tunnel but for reliability, simplicity, and forgiveness — for the ability to stand on a storm-lashed coast for a decade and just work, tended by ordinary people with ordinary tools. In an industry where a machine has to earn its keep spinning through gales, in salt air, hundreds of times a day, year upon year, robustness beats brilliance. This is the same verdict that had doomed Putnam's dazzling giant, only read the other way: the machine that survives is the machine that matters, and Juul built for survival. It is a principle worth carrying forward into every chapter that follows, because it explains a great deal about why the technologies that win are so often not the cleverest ones.

All of which corrects the last and most persistent misconception in this stretch of the story — the belief that the modern wind turbine is a recent, clean-sheet invention of the environmental age, a product of twenty-first-century engineering. It is nothing of the kind. The three-bladed, upwind, stall-or-pitch-regulated machine turning on a hillside near you is not a new design; it is the Danish concept of 1957, refined and scaled up almost beyond recognition in size but essentially unchanged in its architecture. The blades have grown from a handful of meters to well over a hundred; the generators have grown more sophisticated; controls that Juul never dreamed of now pitch the blades and steer the nacelle by computer. But the DNA is his. When Danish companies like Vestas and Bonus began building commercial turbines in the late 1970s and 1980s, and when Denmark rode that head start to the front of the global wind industry, they were building on Juul's template, and through him on la Cour's, in an unbroken Danish line of wind engineering stretching back to the 1890s. The country that had never stopped thinking about wind ended up defining what wind power would become.

We have now traced the long road to the modern machine, from a Vermont hilltop through a German glider-maker's slender blades to a Baltic headland where all the threads finally twisted into one durable design. What we have not yet done is take that machine apart. The turbine on the hillside is a marvel of subsystems most people never see — the nacelle packed with gearing and generators, the pitch motors, the yaw drives, the anemometer that tells it where the wind is coming from — and each of them is a small engineering story in its own right. Having followed how the modern turbine was born, we are ready at last to climb inside it and examine the anatomy of the machine itself.

Chapter 26 — Anatomy of a Modern Turbine

The maintenance technician clips her harness to the ladder and begins to climb. Above her, inside a steel tube as tall as a forty-story building, a single yellow safety line disappears into the gloom. It is roughly a hundred and thirty meters to the top — the height of the tower — and even with the small service lift that runs up the inside of many modern machines, the last stretch is a hand-over-hand haul through a hatch into a windowless steel room the size of a city bus. This room, bolted to the top of the tower and swiveling slowly to follow the breeze, is the nacelle. It weighs, on a large onshore machine, something in the neighborhood of two to three hundred tons — heavier than a fully loaded jumbo jet, hoisted into the sky and balanced on a pole. When the technician finally pushes open the roof hatch and stands atop the nacelle, the three blades wheel past her, each one longer than the wing of any airliner ever built, and the whole structure hums underfoot with the low, patient torque of the wind. Let us climb up there with her, and take the machine apart piece by piece.

Start with what your eye lands on first: the blades. A modern utility turbine has three of them, and on today's onshore machines each is typically sixty to ninety meters long, with the very largest offshore blades now stretching past a hundred and fifteen meters. To grasp that, picture an American football field, goal line to goal line, at a hundred meters plus the end zones; the biggest blades are longer than the whole field. Yet each one weighs surprisingly little for its size — often just twenty to fifty tons — because it is not solid. A blade is a hollow, precisely curved shell, and it is a marvel of materials engineering. The outer skin is fiberglass, a fabric of spun glass fibers set in resin, chosen because it is stiff, cheap, and can be molded into a smooth aerodynamic curve. Down the inside runs a load-bearing spar, the blade's backbone, and here the engineers reach for stronger stuff: a sandwich of lightweight balsa wood or foam faced with glass, and, in the most demanding designs, carbon fiber — the same black, ferociously stiff material used in racing bikes and jet parts. Balsa, the featherlight wood beloved of model-airplane builders, turns out to be an ideal core precisely because it is light yet resists crushing. A blade must be light enough to lift and cheap enough to mass-produce, but strong enough to survive twenty years of being bent back and forth by every gust, roughly ten million times over its life. It is a compromise written in glass and wood.

The three blades bolt onto a central casting called the hub — a rounded, faintly comical dome, often painted the same off-white as everything else, that sits at the very nose of the machine facing into the wind. The hub is where the blades and their controls all come together, and it rotates as one piece with them. Behind the hub, that whole spinning assembly of hub-plus-blades has a name: the rotor. Now, here is the first of several facts that surprise people. That enormous rotor turns slowly. Watch a big turbine and count: the tips sweep around only about ten to fifteen times a minute, one lazy revolution every four to six seconds. It looks stately, almost drowsy. The illusion of speed comes only from the sheer size — the blade tip, despite the leisurely rotation, is actually slicing through the air at well over two hundred kilometers an hour, because it has so far to travel around the circle. But the shaft itself, the thing the blades are trying to spin, turns at a walking pace.

That slow rotation is the central engineering problem of the whole machine, and solving it is the job of everything packed inside the nacelle behind the hub. Electricity is easiest to generate when you spin a magnet past a coil of wire quickly — the generators in a coal or gas plant whir along at fifteen hundred or three thousand revolutions per minute. Our rotor offers barely a dozen. So the drivetrain, the chain of machinery from rotor to generator, has two ways to bridge that gap, and the choice between them is one of the great forks in turbine design. The first way is a gearbox: a set of meshing gears, much like the gears of a bicycle in reverse, that takes the slow, powerful turn of the main shaft and steps it up to the fast, light spin a compact generator wants — a ratio of roughly a hundred to one. Gearboxes are compact and let you use a small, cheap generator, but they are also the part most likely to fail; they run hot, they wear, and swapping one out a hundred meters up is a helicopter-and-crane nightmare. The second way dispenses with gears entirely. In a direct-drive machine, the slow rotor is bolted straight to a very large generator with a great ring of magnetic poles, so that even at a dozen revolutions a minute it produces usable electricity. Fewer moving parts, less to break — but the generator must be huge and heavy, and to make it work at that size the designers reach for powerful permanent magnets.

And here we brush up against a story this book will return to. The strongest permanent magnets we know how to make are built from rare-earth elements — neodymium and praseodymium, with a pinch of dysprosium or terbium to help them keep their strength when hot. A single large direct-drive generator can contain hundreds of kilograms of these magnets. The elements are not, despite the name, especially rare in the Earth's crust, but the mining and — above all — the refining of them is concentrated overwhelmingly in China, which makes every wind turbine quietly dependent on one country's chemical industry. It is a vulnerability the whole business is aware of and working to trim, and we will follow that supply chain in full when we reach it later in the book. For now, simply note the magnets are there, spinning in the dark inside the nacelle, and that they are why your turbine has a thread running back to a mine in Inner Mongolia.

However the drivetrain is arranged, its output is the same: alternating electric current, produced as the magnets sweep past the generator's copper coils and shove electrons back and forth. But this raw generator current is not yet grid electricity. It comes out at the wrong voltage and, in many machines, at a frequency that wobbles with the wind, so it passes through power electronics — banks of transistors that clean it up and lock it to the grid's steady rhythm, fifty or sixty cycles a second depending on the country — and then through a transformer that steps the voltage up high, into the tens of thousands of volts, so it can travel down the tower and across the countryside without frittering itself away as heat in the cables. On some machines the transformer sits in the base of the tower; on others it rides up in the nacelle. Either way, the journey of the power is now complete in outline: wind pushes blades, blades turn rotor, rotor spins generator, generator makes messy electricity, electronics and transformer tidy it into grid-ready current, and it flows away underground to join the power that lights your kitchen.

Hold the whole assembly up, and you have the tower — the part everyone underestimates, because it looks like a mere pole. It is nothing of the kind. A modern tower is a tapering tube of rolled steel, eighty to a hundred and fifty meters tall, built in sections that are trucked to the site and bolted together on the ground before the whole column is craned upright. At the base it is wide enough to walk into standing up, its steel walls tens of millimeters thick, and it flares outward at the bottom into a foundation of hundreds of cubic meters of reinforced concrete sunk into the earth — a hidden iceberg of ballast that keeps the whole two-hundred-ton nacelle-and-rotor from levering the tower over in a gale. The engineering challenge of the tower is not mainly holding the weight up, which is easy; it is resisting the enormous overturning force of the wind pushing on the rotor a hundred meters overhead, a force that tries, every second of every windy day, to tip the machine like a tree. The tower must also be tuned, like a tuning fork, so that its natural sway does not fall into step with the rhythm of the passing blades, which would shake it to pieces through resonance — the same effect that lets a singer shatter a wine glass, or a platoon marching in step bring down a bridge. Designers compute that sway with great care, and it is why a big turbine, seen from below on a windy day, visibly nods and flexes at the top: it is meant to.

Two more systems keep the whole thing pointed and safe, and both are easy to miss. The first is the yaw system: a ring of gears and motors where the nacelle meets the tower, which slowly rotates the entire two-hundred-ton nacelle so that the rotor always faces into the wind. A small wind vane and an anemometer — a spinning-cup wind-speed gauge, the same instrument you have seen on weather stations — sit on the nacelle roof and tell the controller which way the wind has shifted; the yaw motors grind the machine around to meet it, a few degrees at a time. The second is the pitch system, tucked inside the hub, which can rotate each blade about its own long axis, turning the blade like a screwdriver in its socket. In a gentle wind the blades are angled to grab as much air as possible; as the wind rises toward the machine's limit, they twist edge-on to spill the excess and hold the power steady; and in a storm they turn fully sideways — "feathered," in the old aviation word — so the wind slides past doing almost no work, and the rotor coasts safely to a near stop. Overseeing all of it is the controller, a computer in the base or nacelle that reads the anemometer, the temperatures, the vibrations, and the grid, and decides moment by moment how to yaw, how to pitch, and when to shut down. A modern turbine is, in a real sense, a robot: it starts itself, aims itself, protects itself, and calls for help when a bearing runs hot.

Let me make this concrete with a single machine. Picture a fairly typical modern onshore turbine rated at somewhere between three and five megawatts — a megawatt being a million watts, enough instantaneous power for a few hundred homes. Give it a rotor about a hundred and forty meters across, so each blade is roughly seventy meters, and set the hub at about a hundred meters up, on a tapering steel tower that at its base is wide enough to drive a small car into. Over a year, in a decent wind, a machine like this will not run flat out — the wind is too fickle for that — but it will deliver on the order of fifteen million kilowatt-hours of electricity. An average household in a wealthy country uses very roughly four or five thousand kilowatt-hours a year, so one such turbine, spinning on one patch of farmland, quietly powers something like three thousand homes. The newest offshore giants, with rotors past three hundred meters, can each supply tens of thousands. It is worth pausing on that image: a single object, made mostly of steel and spun glass, turning too slowly to seem impressive, keeping the lights on for a small town.

There is a subtlety hidden in that yearly figure, and it is worth drawing out because it trips up almost everyone. A turbine rated at, say, four megawatts does not produce four megawatts most of the time. It produces four megawatts only when the wind blows in the sweet band between its rated speed and its cut-out speed; the rest of the time — in light airs, in dead calms, in the shutdowns of a storm — it makes less, or nothing. Engineers capture this with a single number, the capacity factor: the actual energy a machine produces over a year divided by the energy it would have made had it run flat out the whole time. For onshore turbines that number is typically somewhere between a quarter and a little over four-tenths; offshore, in the stronger, steadier sea winds, it climbs toward half. Think of it as the difference between a car's top speed and its average speed on a real journey full of traffic lights and slow zones. The rating tells you what the machine can do at its peak; the capacity factor tells you what it actually delivers across a windy, gusty, sometimes-calm year. When we said one machine powers three thousand homes, we already folded that reality in — the fifteen million kilowatt-hours is the real, capacity-factored harvest, not the fantasy of a rotor spinning at full power every hour of every day.

It is worth measuring the modern machine against its ancestors for a moment, because the leap is easy to take for granted. When Palmer Putnam bolted his pioneering turbine to a Vermont hilltop called Grandpa's Knob in 1941, it was a wonder of its age: a 1.25-megawatt machine, the first ever to reach the megawatt scale, feeding the local utility. In 1943 one of its two great steel blades cracked and flew off in the night, and wartime shortages killed the project before it could be rebuilt. Putnam's machine had two blades of solid steel and a fraction of the sophistication of what we have just toured; yet its rated power would be unremarkable on a modern farm, exceeded by a single one of today's mid-sized onshore turbines, and dwarfed many times over by the offshore giants. Everything the modern machine does that Putnam's could not — the pitching blades, the yaw that tracks the wind, the controller nursing the drivetrain through every gust, the composite blades that flex instead of snapping — is the accumulated answer to problems his cracked blade first posed.

Now let me clear away the misconceptions people most often carry up the ladder with them. The first is that a wind turbine has a motor — that something drives the blades around. It is exactly backwards. A motor takes electricity and makes motion; a turbine takes motion and makes electricity. The proper word for the thing in the nacelle is a generator, and the only motors on the machine are the small ones that yaw and pitch, servants to the main event, not its engine. The blades are turned by nothing but moving air. The second misconception is the one we have already met: that the blades whip around fast. They do not. That drowsy ten-to-fifteen revolutions a minute is deliberate, because spin the tips much faster and they begin to roar, to erode in the rain, and to lose efficiency — a limit we will unpack in the next chapter. A third, quieter misconception is that the machine is simple, a child's pinwheel scaled up. Standing in that steel bus in the sky, deafened by the transformer's hum and surrounded by hydraulic lines, pitch bearings, cooling fans, and a controller running a hundred safety checks a second, you understand that it is nothing of the kind. It is one of the most sophisticated pieces of machinery a farmer will ever have standing in a field.

Where do you meet all this today? Almost anywhere with an open horizon. Drive past a wind farm on a motorway and you are looking at the anatomy we have just dissected: the off-white towers, the slowly wheeling three-blade rotors, the nacelles turning almost imperceptibly through the day to track the wind, the small red lights blinking on top for aircraft. Blade transporters — trucks with steerable rear trailers, or in some places purpose-built trains and ships — creep along back roads at night, each carrying a single seventy-meter blade like a colossal white bone, because no bridge or roundabout was designed for such a load. Every one of those blades, before it was trucked out, was laid up by hand and machine in a mold longer than a swimming pool, glass and balsa and resin cured into a shape computed to a fraction of a degree.

And that shape is the real secret of the machine — the thing that turns a wall of moving air into a hundred meters of graceful, twisting curve. We have toured the turbine as a collection of parts: blades, hub, rotor, nacelle, gearbox or generator, yaw, pitch, tower, transformer, controller. But we have not yet asked the deepest question of all. When the wind meets that beautiful, twisted blade, why does the rotor turn at all — and why is the blade shaped like an airplane's wing rather than the flat sail you might expect? To answer that, we have to leave the nacelle, climb out onto the blade itself, and follow a single parcel of air across its surface.

Chapter 27 — The Aerodynamics of a Blade

Cut a wind-turbine blade in half — not lengthwise but straight across, the way you'd slice a cucumber — and lay the cross-section on a table. What you see is not a plank and not a paddle. It is a teardrop: rounded and fat at the front, tapering to a thin trailing edge at the back, with the top surface bulging more than the bottom. An aircraft engineer would recognize it at a glance, because it is the same shape as a wing. This is the single most important fact about how a wind turbine works, and it is the one almost everyone gets wrong. The wind does not push the blades around like air shoving a sail. The blade is a wing, and it is pulled around by the same invisible force that holds an airliner in the sky: lift.

To see why that matters, imagine the crude alternative. Picture a child's pinwheel, or the flat vanes of an old farm windmill: flat boards, tilted so the wind catches them and shoves them along. That is drag-driven, and it works — barely. A flat plate dragged along by the wind can never move faster than the wind itself, wastes most of the air's energy in turbulence, and produces feeble torque for its size. Poul la Cour, the Danish schoolteacher-scientist who built the first real wind-power laboratory in the 1890s, worked out something that ran against every farmer's intuition: fewer, slimmer, faster blades shaped as airfoils beat a great many flat ones. His insight, refined over the following century, is why a modern rotor has just three slender blades rather than the forty or a hundred wooden slats of a nineteenth-century water-pumper. The flat blade is the obvious design, and it is the wrong one.

So what is lift, and how can a moving fluid pull a wing sideways to its own flow? The everyday way to feel it is to stick your hand out of a moving car window — cautiously — and tilt it. Held flat and level, your hand just slices the air. But cock it a few degrees, thumb up, and your whole arm is yanked skyward with startling force. You have not pushed against anything; you have made a wing. The air rushing past is deflected downward by your tilted hand, and — by Newton's old rule that every action has an equal and opposite reaction — the air pushes your hand up in return. Physicists tell the same story a second way, in terms of pressure: the air flowing over the longer, curved top of a wing ends up moving faster and at lower pressure than the air beneath, and the wing is sucked toward the low-pressure side. Both descriptions are true; they are two languages for one event. The upshot is a force, lift, that acts roughly perpendicular to the oncoming wind — sideways to the flow, not along it. And crucially, lift can be many times larger than the drag that comes with it. That is the whole game. A blade angled to the wind feels a great sideways tug of lift and only a small backward nag of drag, and the lift, pointed the right way, hauls the rotor around.

The angle you cock your hand to — the tilt between the wing and the oncoming air — has a name: the angle of attack. Increase it and lift grows, up to a point. Push it too far and the smooth flow over the top of the wing peels away into churning turbulence, lift collapses, and the wing stalls — the same stall that pilots fear, and, as we will see, the same one turbine engineers deliberately use to tame a storm. So the angle of attack is the knob that governs everything, and a good blade holds every part of itself at the sweet angle where lift is generous and stall is still safely distant.

Now comes the subtlety that gives a turbine blade its strange, sculpted beauty, and it follows from something we noted last chapter: the rotor turns slowly, but the tip travels far. Think about the blade as a spinning line anchored at the hub. The root, near the center, creeps around a tight little circle. The tip, seventy meters out, races around an enormous one — and in the same few seconds, so it must be moving vastly faster through the air. On a big machine the root might be sauntering along while the tip is screaming through the air at two hundred and fifty kilometers an hour or more. But every part of the blade also feels the actual wind, blowing in from the front at, say, forty kilometers an hour. And here is the key idea, the one that makes blade design a genuinely three-dimensional puzzle: what each section of the blade actually "feels" is not the wind alone and not its own motion alone, but the two combined — the apparent wind.

Let me make the apparent wind vivid. You already know it, though you may never have named it. Ride a bicycle on a still day and you feel a breeze in your face although the air is dead calm; that breeze is the apparent wind, created entirely by your own motion. Now ride on a day with a real crosswind, and the breeze you feel is slanted — a blend of the true wind and your riding wind, coming from somewhere between the two. A sailor knows it intimately, which is why a sailboat can, astonishingly, sail faster than the wind and even angle upwind: the boat's own speed adds to the true wind to make an apparent wind, and the sail, working as a wing, generates lift from that blend. A turbine blade is doing exactly this. Each section flies through an apparent wind that is the sum of the true wind coming from the front and the "riding wind" coming from the side, created by the blade's own rush around the circle.

Because the tip rushes around so much faster than the root,

the apparent wind at the tip is slanted far more steeply — it comes at the blade from a much shallower angle, almost edge-on, dominated by the blade's own motion. At the root, where the blade barely moves, the apparent wind comes nearly straight from the front. If the blade were a simple flat twist-free board set at one angle, only one short stretch of it could ever meet its apparent wind at the ideal angle of attack; the rest would be stalled or barely working. So the designers twist the blade. Run your eye from root to tip and you can see it: the fat root sits at a steep pitch, cocked well back; as you travel outward the blade slims and flattens, rotating gradually until the tip is nearly aligned with the plane of rotation. The whole blade is wrung like a slow corkscrew so that every station along its length meets its own local apparent wind at the same efficient angle of attack. It is one continuous, carefully computed twist — perhaps twenty degrees or more from root to tip — and it is the reason a blade seen edge-on looks like a frozen ribbon caught mid-turn rather than a straight plank.

Let me follow a single parcel of air across one section of the blade, as this book likes to do. Our parcel arrives from the open field, part of the true wind, and meets the blade at a point midway out, where the blade is already sweeping sideways at a good clip. To our parcel, the blade seems to loom up out of a slanted apparent wind. It splits at the rounded leading edge; part of the parcel flows over the bulging top, speeding up and thinning into low pressure, part slides along the flatter underside. In passing, the parcel is bent — nudged off its course, deflected toward the back of the rotor — and in bending it, the blade feels the reaction: a firm sideways pull of lift, angled so that a large component of it points the way the rotor is already turning. Our parcel, a fraction slower and more turbulent than when it arrived, tumbles off the thin trailing edge and drifts away downwind, having handed over a sliver of its motion. Multiply that tiny handshake by the millions of air parcels crossing all three blades every second, and you have a rotor turning a generator.

There is a third analogy that makes the whole thing click for people who have felt it in their bodies, and it is the ceiling fan run backwards. A ceiling fan is a motor turning shaped blades to push air down at you — electricity in, wind out. A wind turbine is precisely that machine running in reverse: wind in, electricity out. If you could grab a ceiling fan's blades and shove them around by hand against a stiff breeze, you would be doing, clumsily, what the wind does to a turbine — and you would notice at once that the blades are pitched, angled, not flat, because a fan with flat blades barely moves any air. The same shaping that lets a fan throw a strong draft with a gentle spin lets a turbine catch a strong pull from a gentle wind. Engineers even speak of a turbine and a fan as the same device solving inverse problems, and the mathematics that describes one describes the other.

Poul la Cour, whom we met a moment ago, discovered the superiority of the fast, few-bladed airfoil rotor the hard way, by building wind machines behind his Danish folk school in the 1890s and measuring them in a homemade wind tunnel — one of the first anyone had built for the purpose. His finding ran so hard against common sense that it took decades to be fully accepted: that a rotor with a handful of slim, fast-moving blades harvests more of the wind than one crowded with many slow ones. The reason lies in the apparent wind. A slow, crowded rotor blocks the air like a wall and lets little through; a fast, sparse rotor of true airfoils slips through the flow, each blade riding its own steep apparent wind and generating clean lift, disturbing the air just enough to take its energy and no more. It is a lovely instance of less being more — three blades beating thirty — and it is why every large turbine on Earth today wears la Cour's slim signature rather than the crowded fan of a Kansas water-pumper.

How much can the blade take from the wind in total? Not all of it — and here we meet an old friend from earlier in this book. If a rotor extracted every scrap of the wind's energy, the air would have to stop dead behind it, and stopped air cannot get out of the way of the air still coming. The flow would jam. In 1919 the German physicist Albert Betz proved that the most any rotor can ever capture is sixteen twenty-sevenths of the wind's power — about 59.3 percent, the Betz limit. It is a ceiling set by physics, not engineering, and it applies no matter how clever the blade. Real turbines, with their beautifully twisted airfoils, reach perhaps 75 to 85 percent of that ceiling — a power coefficient around 0.4 to 0.5 — which is close enough to the theoretical best that the remaining gains are hard-won. The airfoil blade is not just better than the flat sail; it is nearly as good as nature permits.

There is a hard limit at the other end too — a speed limit for the tips. Why not simply spin the rotor faster and wring out more power? Three reasons, all lurking near a tip speed of eighty to ninety meters per second, roughly three hundred kilometers an hour. First, noise: the faster a tip cuts the air, the louder its swish, and since the noise rises steeply with speed, a modest increase in tip speed can turn a quiet farm turbine into an audible nuisance. Second, erosion. At three hundred kilometers an hour a raindrop is no longer soft; it hits the leading edge like a fistful of gravel, and over years the rain sandblasts the blade's front edge, pitting and peeling the coating until the airfoil is spoiled and the blade must be repaired mid-air by a technician dangling on ropes. Leading-edge erosion from ordinary rain is one of the wind industry's quiet, expensive headaches, and it is a direct consequence of tip speed. Third, and most fundamental, the approach to the speed of sound. Sound travels through air at about three hundred and forty meters a second; as any surface nears a fraction of that, the air begins to compress and misbehave, drag climbs sharply, and the tidy lift-to-drag bargain sours. Turbine designers keep the tips comfortably short of that regime. So the leisurely rotation you see is not laziness or weakness; it is the blade held deliberately below the speed at which it would start to howl, erode, and fight the air.

The pitch system we met in the last chapter is where all this aerodynamics becomes a control lever. Because lift depends on the angle of attack, the operator can manage the machine simply by rotating the whole blade a few degrees on its axis, changing the angle at which it meets the apparent wind. In light winds the blades are pitched to a fat angle of attack, greedily harvesting lift. As the wind strengthens past the machine's rated power, the controller feathers the blades a little — twisting them to a shallower angle so they generate less lift and shed the surplus, holding the output steady and sparing the drivetrain from overload. And in a true storm the blades feather almost fully, turning nearly edge-on so the apparent wind slips past with little grip, deliberately courting the stall that a pilot dreads, and the rotor coasts down to a safe idle. The same physics that pulls the rotor around in a breeze is used, inverted, to protect it in a gale. Every angry gust that would tear a flat sail apart is quietly bled away by a few degrees of twist.

Let me name the misconceptions plainly, since they are so common. The first is that the wind pushes the blades round. It does not; lift pulls them round, sideways to the wind, which is why the tips can outrun the wind that drives them, exactly as a sailboat outruns the breeze. The second is that a flat blade would do just as well — that the airfoil is a refinement, a bit of polish. On the contrary, the flat blade is feeble and turbulent, capturing a small fraction of what a shaped wing captures; the airfoil is not the polish but the whole point. A third, subtler error is to imagine the blade could be a simple, untwisted shape. It cannot, because the apparent wind slants ever more steeply from root to tip, and only a twisted blade can keep its whole length working. The twist is not decoration; it is the geometry of moving air made solid.

Where do you meet this today? In the smooth, unhurried arc of every turbine on the horizon — and, if you look closely at a stationary one, in the visible corkscrew of the blade, fat and steep at the root, slim and flat at the tip. You meet it too in the airliner that carries you on holiday, whose wing is the same device solving the same problem; and in the sailboat leaning upwind against a following breeze. All three — wing, sail, and blade — are lift machines, cousins under the skin.

We now understand the single blade: why it is shaped like a wing, why it is twisted, how lift hauls the rotor around, and why the tips must be reined in. But a blade is only as good as it is big, because the wind's power depends on the area the rotor sweeps. That simple fact has driven turbines to grow, decade after decade, from farm-yard pinwheels to machines taller than cathedrals — and it is the relentless logic of that growth that we turn to next.

Chapter 28 — Bigger, Taller, Farther

In October 2024, in a factory in the Chinese port city of Fuzhou, engineers rolled out a wind turbine so large it strains the imagination. The machine, built by the state-owned firm Dongfang, is rated at 26 megawatts. Its rotor is about 310 meters across — wider than the tallest building in most cities is tall, wider than three football fields laid end to end. Its hub sits some 185 meters above the sea. A single one of these machines, standing in a good offshore wind, is expected to generate on the order of a hundred million kilowatt-hours a year — enough, by rough reckoning, to power something like twenty thousand homes. Set that machine beside the turbine that many historians count as the ancestor of the whole modern industry, Denmark's Gedser turbine of 1957, and the contrast is almost comic. Gedser had a rotor of about 24 meters and produced 200 kilowatts — a fifth of a megawatt. In sixty-seven years, the rotor grew roughly thirteenfold in diameter and the power grew about a hundred and thirtyfold. Why? What force has been pushing turbines relentlessly upward and outward, decade after decade, with no ceiling yet in sight?

The answer is written in a single equation we met earlier in this book, and it is worth pulling back out into the light. The power available in a stream of wind is given by a formula that multiplies together the density of the air, the area the rotor sweeps, and — most fiercely — the cube of the wind speed. Two of those three terms are the engines of gigantism. Look first at the swept area. A rotor sweeps out a circle, and the area of a circle grows with the square of its diameter. That little word "square" is the whole story of turbine scaling. Double the length of the blades, and the rotor diameter doubles, but the area it sweeps — and so the power it can capture — goes up not twofold but fourfold. Triple the diameter and you get nine times the capture. The wind is a resource spread thin across a wide sky, and the only way to gather more of it is to reach a bigger net into it. A longer blade is a bigger net, and the payoff rises faster than the cost of the blade.

Let me make that concrete with the worked example this chapter turns on. Suppose you have a turbine with a rotor 100 meters across, and you replace it with one 200 meters across — double the diameter. Your intuition, trained on everyday objects, might whisper "twice as much." But the swept area of the small rotor is the area of a circle 100 meters wide, and the big one is a circle 200 meters wide, and because area goes with the square, the big rotor sweeps four times the area. Same wind, same day, but four times the power harvested, from one machine instead of, in effect, four. This is the quiet violence of the square law, and it is why doubling rotor diameter is such an irresistible bargain. Every meter you add to a blade buys you disproportionately more wind.

The third term in the equation — wind speed cubed — drives the other dimension of growth: height. Here the key fact is that wind is not the same at all altitudes. Near the ground, the air is dragged and slowed by friction against trees, hills, buildings, and the rough skin of the land itself; climb higher and the wind grows faster and steadier, less snarled by obstacles. Meteorologists call this gradient wind shear, the increase of wind speed with height. And because the power in the wind depends on the cube of its speed, even a modest gain in wind speed pays enormous dividends. A tower that reaches air moving, say, ten percent faster does not gather ten percent more energy but roughly a third more, because the cube law magnifies every gain. That is why towers have marched upward from Gedser's stubby perch to the 80-to-150-meter steel columns of today, and why the offshore giants push their hubs past 180 meters. The taller the tower, the stronger and smoother the wind it drinks — and the cube law turns a small climb into a large reward.

Put the two laws together — power rising with the square of diameter and with the cube of the wind that taller towers reach — and you have a machine that wants, at every turn, to be bigger and taller. The history of wind power over the past half-century is very largely the history of engineers obeying that arithmetic. From Gedser's quarter-megawatt in the 1950s, machines crept up through the hundreds of kilowatts in the 1980s wind rush, passed a megawatt in the 1990s, reached two and three megawatts in the 2000s, five and eight in the 2010s. Today the largest offshore machines are rated at 15, 16, 18 megawatts, and the Dongfang prototype at 26 — with rotors sweeping past 300 meters. The line on the graph has bent upward for seventy years and has, so far, refused to flatten.

It helps to feel the square law in a homelier setting before trusting it with turbines. Think of a pizza. A pizza thirty centimeters across does not give you half the food of one sixty centimeters across; it gives you a quarter, because the eating is in the area and the area goes with the square of the width. Anyone who has ordered pizzas for a party has run into this arithmetic, usually to their delight — the large is a far better deal than it looks. A rotor is a circle too, and it obeys the same rule; a blade that is twice as long feeds on four times the wind, the same bargain the party-planner discovers. Now stack the height effect on top. The wind at 150 meters is not merely a little stronger than the wind at 50; it is meaningfully stronger and, just as important, steadier and less gusty, because it flies above the rough tumble of the surface. And since the energy in wind climbs with the cube of its speed, a tower that reaches a stream ten percent faster harvests not ten but roughly thirty-three percent more energy from every square meter of rotor. The two multipliers compound: a taller tower drinks faster wind, and a bigger rotor drinks more of it, and the product of the two is the steep upward curve of turbine history.

The cube law has a mischievous corollary worth pausing on, because it explains why wind sites are so unequal. If doubling the wind speed gives eight times the power, then a site whose average wind is merely a bit breezier than its neighbor's can be worth dramatically more — not a little more, but a lot. This is why developers hunt so obsessively for the windy ridge, the exposed coast, the open plain, and why a turbine that would be a marginal investment in a sheltered valley becomes a goldmine a few kilometers away on a hilltop. The same physics that rewards height rewards location, and both rewards are magnified out of all proportion by that unforgiving cube.

But there is a second, colder reason turbines keep growing, and it has nothing to do with physics and everything to do with money. A wind farm is not just turbines. It is foundations poured in concrete, access roads bulldozed across the land, cranes hired by the day, electrical cables trenched to each machine, substations, permits, and the endless paperwork of connecting to the grid. Many of those costs are fixed per turbine, not per megawatt. It costs nearly the same to pour a foundation, dig a road, and run a cable to a machine whether that machine makes three megawatts or fifteen. So if you can get fifteen megawatts from one tower instead of five megawatts from each of three, you have poured one foundation instead of three, cut one road instead of three, run one cable instead of three, and hired the crane once instead of thrice. The electricity from the bigger machine is therefore cheaper per unit — the levelized cost of electricity, the LCOE, falls — even though the single machine costs more. Fewer, larger turbines beat many small ones on the metric that ultimately decides everything: the cost per kilowatt-hour delivered. Bigness is not vanity; it is economics.

This is the moment to correct a misconception that dogs the subject, because it seems so reasonable. Surely, people say, a bigger machine must be less efficient — you cannot keep scaling up without losing something. In fact the opposite is usually true. Larger turbines tend to be more efficient, not less. Their longer blades reach into faster, smoother air; their greater size lets them run at higher, steadier tip speeds within the aerodynamic limits we discussed; and the losses in a big generator are proportionally smaller than in a tiny one. The efficiency with which a well-designed rotor captures the wind's energy has crept up, not down, as machines have grown, and all of them remain bounded by the same Betz limit — that ceiling of about 59.3 percent of the wind's energy that no rotor can exceed. Bigger has meant cheaper and, if anything, slightly better, not worse.

The second misconception is that there must be an obvious ceiling — some size beyond which turbines simply cannot go. Engineers have been predicting that ceiling for forty years, and the ceiling has kept moving. In the 1990s, serious people doubted turbines would ever usefully exceed a megawatt or two. Then it was five. Then ten. Each supposed limit fell as materials improved, as carbon fiber stiffened the longest blades, as manufacturing and logistics caught up. There are real constraints, and we will name them, but "the obvious maximum size" has proven to be a mirage that recedes as you walk toward it.

There is an anecdote that captures how fast the ceiling has moved. When the German engineer Ulrich Hütter designed elegant, lightweight composite-bladed turbines in the 1950s and 1960s, he was working near what everyone then took to be the sensible limits of size, wrestling with blades a few tens of meters long and worrying whether they could be built light enough to survive. The whole later industry inherited his insight that the blade should be a slender composite airfoil rather than a heavy metal plank — the very principle that lets today's blades reach a hundred meters. Hütter would likely have regarded a 300-meter rotor as science fiction; and yet the line from his modest composite blades to the Dongfang giant is unbroken, each generation of designers quietly pushing past the limit the previous one had assumed was final. The lesson repeats so often it ought to be a law of the field: whenever an expert declares a definite maximum size for a wind turbine, the safest bet is that a bigger one is already on a drawing board.

That said, the constraints are real, and they explain why the very biggest machines now go to sea. The first is transport. A blade seventy, eighty, a hundred meters long is a nightmare to move over land. It cannot bend around a tight corner or fit under a low bridge; it must be hauled on specialized trailers with steerable rear axles that can swing the tail wide, sometimes creeping through the night with police escorts, and in some places on purpose-built railcars or, increasingly, simply shipped by sea. There is a reason blades are among the largest single objects routinely moved on the world's roads. The second constraint is the crane. To bolt a nacelle weighing hundreds of tons onto a tower 150 meters up, you need one of the tallest, heaviest-lift cranes on Earth, and such cranes are scarce, expensive, and themselves hard to transport and erect. The third is materials: as blades lengthen, they sag and flex more under their own weight and the wind's load, and keeping them stiff without making them too heavy pushes designers to the limits of glass, balsa, and carbon fiber. Every meter added fights physics a little harder than the last.

Offshore, several of those walls fall away. On the water there are no bridges to duck under, no roundabouts to negotiate, no back roads; a blade or a whole pre-assembled nacelle can be floated out on a barge or a purpose-built installation vessel, and giant floating cranes do the lifting. The wind at sea is stronger and steadier than over land, so the reward for going big is greater still, and there are no neighbors to trouble with the swish of a fast tip or the flicker of a shadow. This is why the records are set offshore, why the 26-megawatt Dongfang and its 15-and-18-megawatt rivals are marine machines. The sea is where the square law and the cube law can run furthest before logistics reins them in. It is also, as later chapters will explore, where the costs are highest and the engineering hardest — floating foundations for deep water, cables run for miles along the seabed, maintenance crews ferried out by boat and helicopter — so the offshore frontier is not a free lunch, only a bigger table.

Let me gather the surprising facts into one place, because they bear repeating. A single modern offshore turbine can now power roughly twenty thousand homes — a machine that a person could, in principle, walk up to and touch, keeping the lights on for a small town. The largest prototype as of late 2024 is rated at 26 megawatts, a hundred and thirty times the output of the 1957 Danish machine that pioneered the three-blade design still used today. The blades of these giants are longer than a football field and are moved, in some cases, by dedicated trains and ships because no ordinary vehicle can carry them. And the reason for all of it collapses into two lines of grade-school geometry: capture rises with the square of rotor diameter, and the wind you reach rises with height, its energy magnified by the cube of speed. Double the rotor, quadruple the harvest.

Where do you meet this today? On any long drive past a wind farm, watch how much bigger the newest machines are than the ones beside them from a decade earlier — the older turbines suddenly look like toys. Watch, on a coastal road, for the blade transporters and the sections of tower, each a rolled steel cylinder tall enough to stand a bus inside. And if you ever pass a port that services offshore wind — Esbjerg in Denmark, Grimsby in England, Fuzhou in China — you may see the components staged on the quay: nacelles like beached whales, blades racked in rows longer than the ships that will carry them. The scale is genuinely hard to hold in the mind until you stand beside it.

We have followed wind power up its steepening curve — bigger rotors, taller towers, the march out to sea — and traced that curve back to a few lines of physics and a colder line of economics. But none of this happened in a vacuum of pure engineering. The reason the world suddenly began pouring money and talent into wind and solar in the last third of the twentieth century was not, at first, a graph of swept area or a worry about the climate. It was a shock — a political and economic earthquake that hit the West in the autumn of 1973, when the oil stopped flowing and the price of everything changed overnight. To understand why the modern renewable-energy era began when it did, we have to leave the turbine hall and stand, for a moment, in a line of cars stretching around the block, waiting for gasoline.

Chapter 29 — The Oil Shock

On a cold morning in the winter of 1973, an American motorist pulled up to a gas station in New Jersey and found the pumps already ringed by cars, engines off, drivers slumped behind the wheel with newspapers and thermoses, the line stretching back around the corner and out of sight. A hand-lettered sign hung on the pump: NO GAS. Or, on a luckier day, a limit — ten gallons, then five, sold only to cars whose license plates ended in an odd number, on odd-numbered days. Fistfights broke out. Some stations sold out by mid-morning and shut; some raised their prices by the hour; a few closed on principle, unable to guarantee their regulars a tankful. To a nation that had spent the postwar decades assuming energy was as endless and cheap as air, the sight of those lines was a genuine shock, in the fullest sense of the word — a jolt to a system that had never imagined it could run short. For thirty years the West had built its suburbs, its highways, its whole way of life on the quiet certainty of cheap oil. In a matter of weeks, that certainty evaporated, and it never fully returned.

The trigger was war and politics, not geology. On the sixth of October 1973, on the Jewish holy day of Yom Kippur, Egypt and Syria launched a surprise attack on Israel. The United States and several European nations moved to resupply Israel with arms. In retaliation, the Arab members of the Organization of the Petroleum Exporting Countries — OPEC, the cartel of oil-producing states — announced an embargo: they would cut production and halt oil shipments to the United States and other nations seen as supporting Israel. The oil-exporting countries had discovered their leverage. For years the price of crude had been set largely by the big Western oil companies; now the producing nations seized that power for themselves, and they used it. The effect on price was immediate and brutal. A barrel of oil that had sold for around three dollars before the embargo climbed to nearly twelve dollars within months — a roughly fourfold increase, an inflationary hammer-blow that rippled through every economy that ran on imported oil, which was to say nearly all of them.

To understand why a squabble over crude shipments could paralyze the United States, you have to see how deeply oil had soaked into modern life. Oil is not merely fuel for cars, though the gas lines made that vivid. It heats homes, powers factories, runs trucks and trains and ships, and in the 1970s it generated a meaningful share of electricity too. It is the feedstock for plastics, fertilizers, asphalt, and countless chemicals. An economy on oil is like a body on a single, irreplaceable nutrient: choke the supply and everything, everywhere, sickens at once. Prices rose not just at the pump but in the supermarket, the factory, the utility bill. Economists coined an ugly new word for what followed — stagflation, the previously unthinkable combination of stagnant growth and galloping inflation. The comfortable assumption of the postwar West, that energy would always be abundant and cheap and that the only question was how fast to consume it, cracked and fell away.

The political response was swift and, in retrospect, historic. In November 1973, in a televised address, President Richard Nixon announced what he called Project Independence, a national goal — grand, and as it turned out wildly optimistic — for the United States to meet its own energy needs by the end of the decade and never again be held hostage by a foreign cartel. He invoked the Manhattan Project and the Apollo program, the great national mobilizations of bomb and moon, as the model for a new crash effort on energy. Whatever one thought of the rhetoric, the machinery of government began to turn. Washington created new institutions to marshal the effort: the Energy Research and Development Administration, stood up in the mid-1970s to coordinate the nation's energy research, which within a few years would be folded, in 1977, into a brand-new cabinet department, the Department of Energy — the DOE that still funds and shapes American energy research today. The federal government, which had treated energy as a matter for private industry, suddenly owned it as a matter of national security.

And here is the fact this chapter most wants you to carry away, because it corrects a misconception so widespread it has become a kind of folk history. Ask most people today why the world began to care about solar panels and wind turbines, and they will tell you it was concern about climate change — greenhouse gases, a warming planet, melting ice. That concern is real and it matters, but it is not why the modern renewable-energy era began. It began a decade and a half before climate change entered mainstream politics, and it began not out of environmental idealism but out of hard-nosed national fear. The men who wrote the first big checks for solar and wind research in the 1970s were not, for the most part, worried about carbon dioxide. They were worried about being held to ransom by OPEC. They wanted energy that came from within their own borders, that no cartel could embargo — energy security, in the language of the day. The sun that fell on Arizona and the wind that blew across the Great Plains had one overwhelming political virtue in 1974: no foreign government could turn them off. Renewables were born, in their modern form, as a response to an oil crisis, not a climate crisis.

Money followed the fear, and it followed fast. Federal research budgets for solar energy, which had been a rounding error, swelled by orders of magnitude in the years after the embargo. To house the new effort, the government established, in 1977, the Solar Energy Research Institute — SERI — in Golden, Colorado, a national laboratory dedicated wholly to harvesting sunlight. SERI would grow, be renamed in 1991 the National Renewable Energy Laboratory, and become, as NREL, the single most important American institution in the field — the keeper of the famous efficiency-record charts we have quoted throughout this book, the place where perovskite tandems and silicon cells are certified and celebrated. But it was founded in the shadow of the gas lines, a direct institutional child of the oil shock. It is a genuinely surprising thing to sit with: the laboratory that now tracks the march of solar efficiency toward and past 34 percent exists because, in 1973, Americans could not reliably buy gasoline. The same crisis funded the first serious American wind-turbine research programs, big federal machines built through the late 1970s to see whether the dream of Palmer Putnam's wartime megawatt turbine could be revived at scale.

It is worth dwelling for a moment on how strange and new all this felt, because the shock was as much psychological as economic. For a generation raised in the long postwar boom, cheap energy had been simply a fact of the world, like gravity or weather — something you never thought about because it never failed. Gasoline had cost about the same for years; home heating oil was an afterthought on the household budget; the family car had grown steadily larger and thirstier precisely because no one had any reason to care what it drank. The embargo shattered that innocence in a single autumn. Suddenly Americans were being asked to lower their thermostats, to carpool, to obey a new national speed limit of fifty-five miles an hour imposed largely to save fuel, to sit in those humiliating lines. The very idea that a great industrial nation could be made to queue for gasoline by a decision taken in the Persian Gulf was, to the public, almost unbelievable — and unforgettable. That psychological wound, more than any spreadsheet, is what made energy a permanent fixture of national politics thereafter, and what gave the sun and the wind their first real constituency.

Europe and Japan, even more dependent on imported oil than the United States, felt the blow at least as hard, and their responses foreshadowed much of what was to come. In the Netherlands, Sunday driving was banned outright for a stretch, and photographs from that winter show families strolling and cycling on empty motorways, an eerie preview of a world without cheap petroleum. Denmark, which imported nearly all its energy, was hit so hard that the crisis reoriented the whole country's thinking, planting the seeds of the wind-power leadership that would make tiny Denmark a giant of the turbine industry in the decades to follow. Japan, almost wholly without domestic fossil fuels, launched sweeping efficiency and alternative-energy programs. The oil shock was a global schooling, and different nations drew different lessons from it — but nearly all of them came away convinced that depending utterly on imported oil was a strategic mistake they must, somehow, begin to correct.

The story did not end with one embargo, and the second act deepened every lesson of the first. In 1979, revolution swept Iran, one of the world's great oil exporters. The Shah fell, the oil fields convulsed, exports collapsed, and the world price of crude, already high, roughly doubled again — a second oil shock on the heels of the first. Once more the gas lines formed; once more inflation surged; once more energy security dominated the front pages and the halls of Congress. Two shocks in six years hammered home the point that the first alone might have let the West forget: oil was a political weapon, its price hostage to distant wars and revolutions, and a nation that ran entirely on it had handed a measure of its sovereignty to others. The case for homegrown energy — for the sun and the wind, which no ayatollah and no cartel could embargo — had never looked stronger.

The wind side of the story deserves its own note, because

it too was reborn out of the embargo. Wind-electricity research in America had gone quiet since the failure of Palmer Putnam's pioneering megawatt machine on a Vermont hilltop in the 1940s; for a generation the idea of large turbines had been a curiosity. Now, flush with crisis money, the federal government revived it in earnest, commissioning a series of large experimental machines through the late 1970s and into the 1980s — big, government-funded turbines built to test whether utility-scale wind power could be made to work with modern materials and controls. Many of these experiments were clumsy, and some were outright failures, plagued by the very blade-fatigue and drivetrain problems that the previous chapters showed the industry would spend decades learning to solve. But the point was that the money and the attention had returned, and with them a generation of engineers who would carry the lessons — good and bad — into the commercial wind boom that followed. The turbines on the horizon today have a lineage that runs straight back through those government test machines to the anxiety of the gas lines.

Let me pause on a first-principles way of seeing what had happened, because it recurs throughout the history of energy. A society's energy system is also its political anatomy. When your power comes from a fuel that must be dug up in a few places and shipped across oceans, then whoever controls those places and those shipping lanes holds a knife to your throat, and your foreign policy, your economy, and your daily comfort all bend to that fact. When your power comes instead from a resource spread across your own territory — sunlight on your deserts, wind over your plains — the knife is gone, or at least the hand holding it is your own. The oil shocks taught the West this lesson in the most concrete possible way: through empty pumps and quadrupled prices. Everything that followed — the research institutes, the tax credits, the first commercial wind farms, the panels on the White House roof that we will meet in the next chapter — flowed from that hard schooling in the geopolitics of energy. It is worth remembering, too, that the lesson has a modern echo: the concern today about who controls the minerals and the factories for solar panels and turbine magnets is, at bottom, the same anxiety about energy security, wearing new clothes.

There is a bitter coda to the 1970s boom, and honesty requires telling it, because it sets up a theme that will run through the rest of this book. The surge of research money was enormous, but it was not durable. It rode the political emotion of the crisis, and when the crisis eased — when oil prices, having spiked, began in the mid-1980s to fall again, when the lines vanished and the pumps flowed and cheap energy returned — the emotion drained away, and with it much of the money. Solar research budgets that had been swollen in the late 1970s were slashed in the early-to-mid 1980s, some by half or more; programs were cancelled, the young field starved of the funds it had briefly been promised. The scale of that collapse surprises people who imagine that support for renewables has climbed steadily upward. It has not. It has lurched — a boom born of fear in the 1970s, a bust when the fear faded in the 1980s, and much stop-and-start since. The technology's fortunes have been tied, again and again, less to its own steady progress in the laboratory than to the price of oil and the mood of politics.

Let me name the misconceptions squarely, as this book likes to do. The first, and most important, is the belief that interest in renewable energy began with the fight against climate change. It did not; it began with the fight for energy security in the 1970s, a decade and a half earlier, out of fear of OPEC rather than fear of carbon. The second is that government support for solar and wind, once it began, grew steadily. It did not; it boomed and then collapsed within a single decade, the first of many such swings. And a third, quieter misconception is that the oil shocks were about oil physically running out — "peak oil," the well running dry. They were not. The oil was still in the ground in 1973; the crisis was political, a deliberate embargo and a cartel flexing its power, which is precisely why it was so frightening. A shortage caused by geology might at least be predictable. A shortage caused by a foreign government's decision could come at any time, over any war, with no warning at all.

Where do you meet the legacy of the oil shock today? You meet it every time a politician utters the phrase "energy independence," a slogan minted in Nixon's Project Independence and worn smooth by half a century of use. You meet it in the very existence of the Department of Energy and of NREL, institutions that shape the field to this day and that were forged in the crisis. You meet it in the strategic petroleum reserves that many nations built afterward, vast underground stores of crude kept against the next embargo. And you meet it, most of all, in the basic political logic that still drives so much renewable-energy policy: the desire, older than any climate treaty, to make your own power on your own land, where no one else can switch it off.

The crisis produced not only institutions and budgets but symbols — and the most famous symbol of all was about to be bolted to the most famous roof in the world. In the summer of 1979, in the trough of the second oil shock, a president who had trained as an engineer and who felt the energy crisis as a moral as much as an economic emergency climbed onto the roof of the White House to make a promise to the American people about the sun. It is to Jimmy Carter's panels, and to the whiplash of politics that followed them, that we turn next.

Chapter 30 — Carter's Panels

On the twentieth of June 1979, President Jimmy Carter stood on the West Wing roof of the White House, squinting in the summer sun, and behind him a crew of workmen bolted down the last of thirty-two solar panels. It was the depth of the second oil shock; the Iranian revolution had convulsed the oil markets, gas lines had returned to American streets, and the mood of the country was sour and anxious. Carter — a former naval officer trained in nuclear engineering, a man who genuinely understood a heat exchanger and a flow rate — had come up to the roof to make the installation an event, a piece of political theater in the best sense. Squinting into the cameras, he delivered a short speech that has echoed down the decades. The panels above him, he said, could be looked at in two ways: as a curiosity, a museum piece for future Americans to shake their heads over, or as the leading edge of something enduring. He wanted them to be the latter. Solar energy, he declared, was part of "one of the greatest and most exciting adventures ever undertaken by the American people." He set the country a target: twenty percent of American energy from renewable sources by the year 2000. And then he climbed back down, and the panels stayed up on the roof, quietly doing their humble work, for the next seven years.

Now, there is a detail about those panels that almost everyone gets wrong, and it matters enough to stop and set straight, because it goes to the heart of a confusion that still muddles conversations about solar energy. Carter's panels were not the blue-black photovoltaic cells you picture on a modern rooftop, the silicon wafers that turn light directly into electricity. They were solar thermal panels — solar hot-water panels. This is a completely different technology, and the difference is worth understanding from first principles, because the two are forever being mixed up.

A photovoltaic panel, the kind this book has spent many chapters on, performs a subtle quantum trick: a particle of light, a photon, strikes a semiconductor and knocks an electron loose, and that liberated electron becomes electric current. Light goes in one end; electricity comes out the other; no moving parts, no boiling, no heat required — indeed heat is a nuisance to a PV cell. A solar thermal panel does something far older and far simpler, something you could explain to a child. It uses sunlight to warm water. Picture a garden hose left out on a summer lawn: the water that first comes out is scaldingly hot, because the sun has been beating on the black rubber all afternoon. A solar thermal panel is that principle engineered into a proper collector — a dark absorbing plate under glass, with water or another fluid running through pipes bonded to it, so the sun heats the plate, the plate heats the fluid, and the hot fluid is piped down into the building to provide hot water for washing and, sometimes, warmth for the rooms. There is no electricity in it at all. It is a way of catching the sun's heat directly, the same lineage that runs back through this book to Augustin Mouchot's solar steam engines and Frank Shuman's parabolic troughs in the Egyptian desert. Carter's thirty-two panels heated water for the White House — for the staff cafeteria, by most accounts. They were, in the plainest terms, a very fancy set of the same kind of collectors you can still see on rooftops in Israel, Greece, or China, where solar water heating remains enormously common.

Why does the confusion matter? Because it flatters and it distorts. To remember Carter's panels as photovoltaic is to imagine the White House generating its own clean electricity in 1979, a more futuristic and glamorous thing than what actually happened. What actually happened was more modest and, in a way, more honest: a president chose the mature, affordable, sensible solar technology of his day — hot water, which genuinely paid for itself — and made of it a symbol of a larger ambition. The symbol was the point. The panels themselves were never going to power the nation; they were a flag planted on the most visible roof in America, a statement that the country's leader took the solar adventure seriously.

The confusion between the two solar technologies is so persistent that it is worth a second everyday picture, because sorting it out will serve you for the rest of this book. Imagine two ways to use the sun to make a cup of tea. The first way: put a kettle of water out in a black pot under a magnifying arrangement of mirrors, let the sun boil it directly, and pour. That is solar thermal — sunlight to heat, straightforwardly. The second way: use a solar panel to generate electricity, run that electricity into an electric kettle, and let the element boil the water. That is photovoltaic — sunlight to electricity to heat, a longer and more roundabout chain, but one that gives you electricity along the way, which you can send down a wire to do anything electricity can do. Carter's panels were the black-pot method, applied to the White House hot-water supply. They were simple, robust, and, in the sunny economics of the late 1970s, they actually saved money — solar water heating was then, and in many countries still is, one of the most cost-effective uses of sunlight there is. The glamorous, universal technology — light straight to electricity — was in 1979 still absurdly expensive, confined mostly to satellites and remote outposts, which is exactly why a practical engineer like Carter reached for hot water instead.

Alongside the flag, Carter planted real policy, and this is

the part that lasted longer than the hardware. His administration expanded federal tax credits for solar and other renewable installations, letting homeowners and businesses subtract a slice of the cost of a solar system from their taxes — a direct financial nudge to get collectors and panels onto roofs. Combined with the research money that had begun flowing after the 1973 shock and the founding of the Solar Energy Research Institute we met in the last chapter, the late 1970s were, briefly, a golden hour for American solar ambition: budgets up, institutes founded, tax credits enacted, a target set, and the president himself up a ladder for the cameras. For a moment it looked as though the United States might make itself the leader of the coming solar age.

It did not last. In 1981 Ronald Reagan succeeded Carter, arriving with a very different philosophy — a conviction that the market, not the government, should pick energy winners, that the crisis of the 1970s had passed, and that the subsidies and mandates of the Carter years were an overreach to be rolled back. Oil prices, which had spiked so frighteningly, were beginning their long slide back down through the 1980s; the gas lines were a receding memory; the sense of emergency that had funded the solar boom was draining away, exactly the boom-and-bust dynamic the previous chapter described. Under Reagan the swollen solar research budgets were cut sharply — by large fractions, in some programs more than half — and the renewable-energy tax credits Carter had championed were allowed to lapse in the mid-1980s. The golden hour ended.

And then, in 1986, during a routine repair to the White House roof, the thirty-two panels were quietly taken down and not put back. There was no ceremony, no speech, no counter-theater; the panels simply came off and went into storage, and their removal became, almost instantly, one of the most potent symbols in the whole history of energy policy. Whatever the mundane practicalities of a roof repair, the optics were unmistakable: the solar panels a president had installed with fanfare, another administration removed without a word. If putting them up had said the sun is our future, taking them down said, or seemed to say, never mind. Historians and journalists have returned to that image again and again, because it captures in a single gesture the whiplash that has dogged renewable energy ever since — the lurching from enthusiastic support to abrupt abandonment and back, driven less by the technology's own steady progress than by the price of oil, the color of the party in power, and the mood of the moment. The panels' rise and fall are a parable of the whole field's stop-and-start century.

It is easy, from a distance of decades, to be glib about that reversal, so let me be fair to both sides of it, as this book tries to be. Reagan's people were not cartoon villains twirling their mustaches at the sun. They held a genuine and defensible view: that the government had no business subsidizing particular technologies and picking winners, that markets allocate capital better than bureaucrats, and that the emergency conditions which had justified the Carter-era spending had passed once oil grew cheap again. On that reading, propping up expensive solar hardware with taxpayers' money was a distortion that should end, and the panels on the roof were an embarrassing bit of symbolism best quietly retired. Carter's defenders held an equally genuine and defensible view: that energy is a matter of national security and long-term public good, that young technologies need patient support to cross the valley between the laboratory and the market, and that abandoning solar the moment oil got cheap was a short-sighted surrender that would cost the country its early lead. Both cases have force, and the argument between them — market purity versus strategic patience — is not a relic of the 1980s but the very same argument that rages over renewable-energy subsidies today. The panels came down; the debate never did.

The afterlife of those panels is one of the strangest and most charming footnotes in energy history, and it is worth telling in full, because it is so improbable. After their removal, the panels sat forgotten in a government storehouse in Virginia for years. Then, in the early 1990s, a small environmental college in Maine — Unity College — acquired a batch of them and put them to work, of all places, heating water for the cafeteria on its own campus, so that for years students at a rural Maine college washed their dishes with hot water warmed by the sun on collectors that had once sat on the roof of the White House. In time the panels became too valuable as artifacts to keep on a cafeteria roof, and they scattered. One went to the Smithsonian in Washington. Another traveled across the Pacific to a solar science museum in Dezhou, China — a city that styles itself China's "Solar Valley" — where a piece of American energy history now sits in the country that would go on to dominate global solar manufacturing, an irony almost too neat to be true. Others found homes in museums and collections elsewhere. A single set of hot-water panels, installed on one summer day in 1979, thus ended up dispersed across a Maine college, the Smithsonian, and a Chinese museum — a physical diaspora tracing the arc of solar power itself, from an American ambition, through neglect, to a future increasingly written in China.

The whiplash was not confined to a roof; it played out across the whole American solar landscape, and nowhere more visibly than in California. The Carter-era federal tax credits, stacked on top of generous state credits, had touched off a genuine boom in the early 1980s — not only in rooftop collectors but in the first great fields of wind turbines strung across the mountain passes at Altamont, Tehachapi, and San Gorgonio, the famous "wind rush" that put California, briefly, at the center of the renewable world. Thousands of small turbines went up in a few frantic years, many of them flimsy and hastily built, chasing the tax breaks. When the credits lapsed in the mid-1980s, the boom collapsed almost overnight; developers folded, half-built projects were abandoned, and fields of stalled turbines stood rusting in the desert wind as monuments to a subsidy switched off. It was the same lesson the White House panels taught, written across whole hillsides: a young industry built on incentives that can vanish with an election or an oil price is an industry building on sand. Denmark and Germany, which would soon overtake America in wind, drew precisely that lesson from the California bust and designed their own support to be steadier and more durable — a story this book will return to.

Let me draw out the surprising facts this story holds, because they are easy to miss inside the narrative. The first is that whole "boom-bust" shape of 1980s incentives: support for solar did not simply fade, it was actively dismantled within a few years of its peak, budgets cut and credits killed while the technology itself was still improving in the lab. The second is the sheer improbability of the panels' journey — from the West Wing to a college cafeteria in Maine to a museum in China, an itinerary no one could have scripted. And the third, quieter than the others, is that Carter's twenty-percent-by-2000 target, so easily mocked as naive, was not absurd in spirit even if the number proved far out of reach; it named an ambition that much of the world is only now, decades late, beginning to approach.

The misconceptions to correct are two, and we have already met them both. The first is the persistent belief that Carter's White House panels were photovoltaic, generators of electricity; they were solar thermal, heaters of water, a different and older technology, and getting that right is part of understanding what solar energy actually was in 1979. The second is the comfortable assumption that public support for renewables, once won, moves steadily forward — that the arc bends reliably toward more. Carter's panels are the standing refutation. Support has been whiplash, not a ramp: up under Carter, down under Reagan, and lurching ever since, in the United States and, in different rhythms, elsewhere. The technology's progress in the laboratory has often been far steadier than the politics that funds it.

Where do you meet Carter's panels today? You meet them, in a sense, on the roof of the Smithsonian's collections and in that museum in Dezhou, preserved as relics. You meet them in the phrase, still current, of a president "putting solar on the White House" — a gesture repeated, in fact, by later administrations, one of which installed modern photovoltaic panels on the residence in the 2010s, closing a loop Carter had opened. And you meet the deeper lesson every time renewable-energy policy swings hard with an election or an oil price: the reminder, written on that West Wing roof, that this technology's fortunes have always been political as much as physical.

But symbols and subsidies, however dramatic, could not by themselves make solar electricity cheap enough to matter. For photovoltaics to escape the space program and the government demonstration project and climb down onto ordinary rooftops, someone had to attack the brutal cost of a silicon cell and drag it, order of magnitude by order of magnitude, toward the price of everyday power. That work had already begun, in a laboratory funded by an oil company, in the hands of a chemist named Elliott Berman — and it is to his quiet campaign to make the sun affordable that we turn next.

Chapter 31 — Elliott Berman and the Cheaper Cell

In a rented corner of a Massachusetts industrial park in the early 1970s, a chemist named Elliott Berman was doing something that would have horrified the engineers who built the solar cells aboard America's satellites. He was buying silicon that the electronics industry had thrown away. Cracked wafers, off-spec ingots, the swept-up remainders of chip factories — material stamped "reject" and sold for a fraction of the pristine stuff. To the space program, a solar cell was a jewel: hand-selected, flawless, priced without apology because the alternative was a dead satellite orbiting uselessly forever. To Berman, that mindset was the whole problem. He was not trying to power a spacecraft. He was trying, quite deliberately, to build a solar cell cheap enough that someone on Earth, standing on ordinary ground under an ordinary sky, might actually choose to buy one.

It sounds like a modest ambition. It was in fact a heresy, and it changed everything.

To understand why, you have to remember where solar cells had come from. As we saw earlier in this book, the first practical silicon cell was born at Bell Labs in 1954, the work of

Chapin, Fuller, and Pearson, converting about six percent of sunlight into electricity. It was a marvel and, for terrestrial purposes, a curiosity. The trouble was money. A cell that could power a transistor radio in a magazine photograph cost so much that no rational person would use it to power anything that a wire could reach. The number usually quoted for the early 1970s is around a hundred dollars per watt of generating capacity. Let that figure settle for a moment. A watt is not much — a single old-fashioned Christmas-tree bulb might draw several of them. To assemble enough solar cells to run a modest household refrigerator, you would have been looking at a sum that could instead have bought a car, or two. Solar electricity was, quite literally, the most expensive electricity human beings had ever produced.

And yet it had found one customer who did not care about the price at all: outer space. A satellite has no extension cord. It cannot be refueled. Whatever powers it must arrive folded up at launch and then work, silently, for years. For that job, solar cells were not merely competitive; they were the only sane option. So through the late 1950s and 1960s, the entire solar-cell industry — such as it was — served the space program and the space program alone. Prices stayed astronomical because the customer's budget was, in effect, unlimited, and because no one saw a reason to change. Why cheapen a product whose only buyers gladly paid full freight?

Berman saw the reason. He had come to solar by an unusual route, a chemist who had worked on photographic and imaging materials and who thought, correctly, that the people making solar cells were asking the wrong question. They kept asking: how do we make the best possible cell? He wanted to ask: how do we make the cheapest acceptable one? Those are not the same question, and the gap between them is where an entire industry was hiding.

Here is the first-principles insight that guided him, and it is worth stating plainly because it governs the whole story of solar's descent from luxury to commodity. The cost of a solar cell is not one number; it is a stack of separate costs, and most of them have nothing to do with the physics of turning light into electricity. There is the silicon itself. There is the elaborate purification. There is the way the cell is cut and shaped. There is the encapsulation — the glass, the frame, the sealing that lets a delicate wafer survive weather. And there is the accumulated habit of an industry that had never once been asked to economize. Berman's genius was to attack the whole stack at once, and to accept a worse cell in exchange for a far cheaper one, because for his intended customers a worse cell was still miraculously good.

Think of it this way. Imagine two watchmakers. One makes a chronometer for a naval expedition to the Arctic: it must survive salt, ice, and a decade at sea, and its owner will pay anything for reliability. The other makes a wristwatch for a commuter who just needs to catch the 8:15. If the second watchmaker insists on Arctic-grade everything, his watch will cost a fortune and no commuter will buy it. The trick is not to build a worse watch out of carelessness; it is to build exactly the watch the commuter needs and not one atom more. Berman was building the commuter's watch of solar cells. Reject-grade silicon, cheaper encapsulation, a module designed for a fence post rather than a rocket — each substitution shaved the price, and the sum of the substitutions was transformational.

His second analogy is even closer to what he actually did. Consider bottled water. The water is nearly free; almost the entire price is the bottle, the label, the cap, the shipping. If you wanted to make bottled water cheaper, you would not try to invent cheaper water — you would redesign the packaging. Berman grasped that in a solar module of that era, the "water" (the silicon that does the physics) was only part of the cost, and the "bottle" (the glass, the frame, the assembly, the profit margins inherited from the space business) was ripe for demolition. He replaced expensive space-grade cover materials with cheaper industrial ones. He designed modules that could be manufactured without the fussy hand-labor of a satellite contractor. Cost by cost, he brought that hundred-dollars-per-watt figure crashing down — toward twenty dollars per watt, a fivefold cut, over just a few years. Solar was still not cheap. But it had crossed from "impossible" to merely "expensive," and that is the crossing that matters, because expensive things find niches, and niches fund the next round of cost-cutting.

Now for the part of the story that reliably surprises people. Berman's crusade was bankrolled by Big Oil. His company, Solar Power Corporation, founded in 1968 and brought to real scale in the early 1970s, was funded by Exxon — then one of the largest oil companies on Earth. The image many of us carry, of a scrappy solar movement pitted against a monolithic petroleum industry, does not fit the founding facts. In the years around the first oil shock of 1973, several major oil companies looked at the coming decades and decided, sensibly enough, that they were in the energy business rather than merely the oil business, and that a small bet on photovoltaics was cheap insurance. Exxon's research managers gave Berman room and money to chase his cheaper cell. For a crucial stretch of the 1970s, the company doing the most to make solar electricity affordable on Earth was a subsidiary of an oil giant. History is rarely as tidy as our slogans.

But a cheaper cell is only half a business. Berman still needed customers — people who would pay twenty dollars a watt when the wall socket delivered power for pennies. Who on Earth would do that? The answer reveals the deepest and most durable idea in this chapter, and it is worth slowing down for.

The grid is a wire, and a wire has to be built. Running electricity to a new location means stringing poles, laying cable, and maintaining that line forever after. Over short distances in a city, this is trivial and cheap. But the cost of a wire climbs with every kilometer, and in remote or inconvenient places it climbs steeply — sometimes to tens of thousands of dollars per kilometer of line, sometimes to figures that make a small remote installation simply not worth connecting at all. This is the hidden truth that Berman built a company on: the relevant comparison for solar was never the price of grid electricity. It was the price of the wire you would otherwise have to run, plus the diesel generator you would otherwise have to fuel and service, plus the technician you would otherwise have to send into the wilderness to keep it all going. Against that comparison, a silent panel that needed no fuel and no wire could win handsomely — even at twenty dollars a watt.

So Berman and his rivals went hunting for places with no grid and no patience for one. They found them, and the list is gloriously unglamorous. Offshore oil and gas platforms needed warning lights and navigation aids, and stringing a cable across the seabed to a lonely platform, or ferrying out diesel and a mechanic, was absurdly expensive. A small solar panel bolted to the rig, charging a battery, running a lamp — that was cheap by comparison, and it needed no boat trips. Here is a delicious irony to file away: some of the very first paying terrestrial customers for cheap solar were offshore petroleum installations, oil rigs powering their safety lights with the sun. The fossil-fuel industry did not merely fund early solar; in places, it was early solar's best customer.

Then there were the buoys. All along the world's coasts and shipping lanes float navigation buoys — the bobbing markers that keep ships off the rocks — and each one needs a light, and each light needs power, and there is no wire to a buoy. For a century these had run on batteries that had to be swapped by crews in boats, or on gas systems that had to be refilled, an endless, costly maintenance chore across thousands of markers. A solar panel on top of a buoy, trickling charge into a battery, could run for years with almost no attention. Marine navigation buoys became one of solar's first genuine killer applications — an unheralded but enormous market, precisely because the alternative was so miserably expensive. Lighthouses and offshore beacons followed. So did remote telecommunications: microwave repeater stations perched on mountaintops, relaying phone calls across empty country, where hauling fuel to a diesel generator up a dirt track cost more each year than a solar array cost once. Railroad signaling in remote stretches, cathodic-protection systems that keep pipelines from corroding, remote weather stations, mountain cabins, and eventually the first solar-powered water pumps and lights in villages that the grid had never reached — one by one, the off-grid niches lit up.

This is the first principle that bootstrapped the entire modern solar industry, and it is easy to miss because it runs against intuition. Solar did not begin its earthly career by competing with cheap grid power and winning. It began by competing with the absurd cost of extending the grid to places the grid did not want to go — and there, it did not need to be cheap in any absolute sense. It only needed to be cheaper than a wire. Each niche was small. But each niche generated revenue, and revenue funded the next factory, and every factory that ran taught the industry how to make cells a little cheaper still, which opened a slightly larger niche, which funded a slightly larger factory. The off-grid markets were the first rung of a ladder. Without them, there is no second rung, and no rooftop, and no vast desert solar farm. A technology that is too expensive for the mass market can still survive and improve if it can find a protected corner where it is the only option that makes sense. Berman found the corners.

There is a lovely, human footnote to all this, the kind of thing that never makes the textbooks. The people selling those early modules quickly learned that their competition was not other electricity — it was disappointment and distance. A telecom engineer facing a mountaintop repeater did the arithmetic on hauling diesel up the mountain twice a year for twenty years and found that the sun, which showed up for free every morning, simply won. The sales pitch was not idealism about clean energy. It was a spreadsheet. And that, in the end, is how solar first earned its living: not as a cause, but as the cheapest way to keep a light on where no wire ran.

Let me correct the misconception this chapter exists to demolish, because it is one of the most common stories people tell about solar and it is wrong in an instructive way. The popular version goes: solar cells were invented at Bell Labs in 1954, and then, once they got cheap enough, people started putting them on their roofs. That skips the most important chapter of the whole saga. Solar did not go from the laboratory to the rooftop. It went from the laboratory to outer space, and then — this is the part everyone forgets — from outer space to oil rigs, buoys, and mountaintop radio stations, and only after decades of that unglamorous off-grid apprenticeship did it finally arrive on anyone's roof. The rooftop is the last act, not the second. For its entire crucial adolescence, solar's home was not a suburban house but a navigation buoy pitching in the swell of a shipping lane.

A second misconception, subtler and worth naming: that cost declines in solar came from some single breakthrough, a better cell that suddenly made everything cheap. Berman's story shows the opposite. He did not invent a dramatically better cell — his cells were, by the physics, somewhat worse than the space-grade ones. He invented a cheaper way to make and package an acceptable cell, and he did it by attacking a dozen ordinary costs rather than one heroic one. Solar's affordability was, from the very start, a story of manufacturing and packaging and market-finding far more than a story of physics. Keep that in mind; it is the engine of nearly everything that follows.

And a third, gentler correction, one that guards the book's even hand. It is tempting, given Exxon's role, to spin a tale of oil companies secretly loving solar or secretly killing it. Neither is quite true. Oil companies funded early solar for ordinary business reasons — diversification, insurance, the plausible sense that energy was energy — and most of them later stepped back when the payoff looked distant, as companies routinely do with long-shot bets. There is no single villain and no single hero in the early economics of the cheaper cell. There is a chemist who asked a better question, a corporate parent that funded him for a while, and a scattering of lonely lights on the water that needed power and did not care where it came from.

So we can sum up where Elliott Berman left the world. Before him, solar electricity was a space-age jewel, priced beyond any earthly use at around a hundred dollars a watt. He, and the off-grid markets he chased, dragged it down toward twenty — not by perfecting the cell but by cheapening everything around it, and by finding the customers for whom "expensive" still beat "a wire." Oil rigs, buoys, remote telecom, railroad signals: the humble first rung of a very tall ladder. The industry now had revenue, factories, and a growing base of hard-won manufacturing experience — the ingredients, though no one yet used the phrase, of a learning curve. Solar had learned to sell. Its restless cousin, the wind, was about to learn the same lesson in a wilder and far more chaotic classroom, on the windswept mountain passes of California, where a very different force — not the price of a wire, but the temptation of a tax break — would summon thousands of turbines out of the ground almost overnight.

Chapter 32 — The California Wind Rush

Drive east out of the San Francisco Bay Area on Interstate 580, up over the brown folds of the Diablo Range, and you climb into a gap in the hills called Altamont Pass. On a summer afternoon the wind comes through it like water through a funnel, hot inland air rushing to fill the low pressure over the scorched Central Valley, and if you had stood in that pass in 1982 you would have watched something that had never existed anywhere on Earth: hundreds upon hundreds of small wind turbines, sprouting from the grass in ragged rows, blades of every size and design spinning at every speed, a mechanical forest thrown up almost overnight across a landscape that had held cattle and little else the year before. Some were elegant. Many were junk. All of them were spinning because of a document written in Washington and another written in Sacramento, and not, primarily, because anyone had carefully concluded that this particular ridge was the right place to catch the wind.

The California wind rush is one of the strangest and most instructive episodes in the whole history of renewable energy, and it teaches a lesson that has nothing to do with aerodynamics and everything to do with human incentives. It is the story of how policy can conjure an industry out of thin air in a matter of months — and how, if the policy is built carelessly, it can conjure an industry made largely of garbage. It is also, quietly, the story of how the modern wind turbine survived its own adolescence, because buried in all that garbage were a few honest machines, and the people who built them learned enough to build the giants that power the world today.

To see how the boom happened, you have to understand two pieces of paper. The first was federal. In the wake of the oil shocks of the 1970s — the 1973 embargo and the 1979 crisis that had, as we saw, put solar panels on Jimmy Carter's White House roof — the United States government decided to throw money at any energy source that was not imported oil. Among the tools it reached for was the tax credit. The details shifted over the years, but the essence was simple and potent: if you invested in a qualifying wind machine, the government would hand you back a large fraction of your investment as a reduction in the taxes you owed. Stack the federal credit on top of a generous California state credit, and in the early 1980s an investor could recover something close to half the cost of a wind turbine from the tax authorities before the blades had turned a single revolution. Half. Before generating a watt.

The second piece of paper was a federal law with a forgettable name and enormous consequences: the Public Utility Regulatory Policies Act of 1978, universally shortened to PURPA (rhymes, roughly, with "burp-uh"). Before PURPA, if you built a small power plant, the local electric utility had no obligation to buy your electricity, and could simply refuse, which killed most independent generation in its cradle. PURPA changed the rule. It compelled utilities to buy power from qualifying small producers, and — crucially in California — to buy it at prices pegged to what the utility would otherwise have had to pay to generate that power itself, prices that in the early 1980s, with everyone still terrified of expensive oil, were set high and guaranteed for years. So now the picture was complete. The government would pay you half the cost of building a turbine, and the utility was legally required to buy whatever it produced at an attractive, locked-in price. You did not even have to be sure the machine worked well. You just had to build it.

Here is the first-principles idea at the heart of this chapter, and it is one of the most important in all of energy policy, so let me state it carefully. Markets do not exist by nature; they are made. When a technology is new and expensive, there is no market for it, because no rational buyer will pay more for it than for the established alternative. A subsidy is a way of manufacturing a market that would not otherwise exist — of paying, out of the public purse, for the early, expensive units so that the industry can grow, gain experience, and eventually become cheap enough to stand on its own. This is a genuinely powerful and often necessary thing. But a subsidy is a blunt instrument, and it has a treacherous property: it rewards whatever behavior it literally specifies, not the behavior you actually wanted. If you reward electricity generated, you get people straining to generate electricity. If you reward money invested in turbines, you get people straining to invest money in turbines — whether or not those turbines ever generate much of anything.

Consider an analogy. Suppose a city, wanting more trees, announces that it will pay citizens for every sapling they plant, the payment made at the moment of planting. What will it get? A great many saplings jammed into the ground — and no particular reason for anyone to water them, since the reward came at planting, not at harvest. Some conscientious people will tend their trees into a real forest. Others will plant sticks in dust, collect the check, and walk away. The city wanted a forest; it specified planting; it got planting. California, in its wind credits of the early 1980s, largely rewarded the act of investing in a turbine rather than the act of producing electricity over a long lifetime — and it got, predictably, a landscape jammed with turbines, some of them barely functional, many of them planted more for the check than for the wind.

The people who chased that check earned a nickname that

has stuck to the era ever since: the "tax farmers," and their creations were "tax farms." The phrase is exact and damning. For a certain kind of investor — a dentist, a lawyer, a movie producer with a large tax bill and an accountant with an eye for shelters — a wind turbine in Altamont Pass was not really an energy project at all. It was a device for reducing taxes. The electricity was almost beside the point; what mattered was that buying the machine generated a paper loss and a tax credit that sheltered income from other, more lucrative endeavors. Whole partnerships were assembled to funnel wealthy investors' money into turbines whose chief output was not kilowatt-hours but deductions. And because the reward flowed from the purchase rather than from performance, the incentive to build a good, durable, well-sited machine was weak, while the incentive to build a cheap machine fast and in bulk was overwhelming.

The results were often dismal, and here the surprising facts begin. Thousands of turbines went up across three great California passes — Altamont east of the Bay, Tehachapi in the mountains north of Los Angeles, and San Gorgonio near Palm Springs, where the wind pours through the gap between two of Southern California's highest peaks. Many of the machines were small, underpowered, and poorly engineered; some were rushed to market by companies that had never built a turbine before and would never build another. Gearboxes disintegrated. Blades cracked and flew. Whole rows stood motionless, broken, their owners having already banked the only return they cared about. And yet — and this is the fact that stops people short — for a brief window in the mid-1980s, California contained something like ninety percent of the entire world's installed wind-generating capacity. One American state, thanks to two pieces of paper, briefly hosted almost all the wind power on the planet. It was the global capital of an industry that barely existed anywhere else, and much of what it hosted was mechanical rubbish.

A third surprising fact, and a darker one, would take years

to surface. The tax farmers who blanketed Altamont Pass with lattice towers did not know, and mostly did not think to ask, that the pass sat in the middle of superb raptor habitat — a corridor rich in golden eagles, red-tailed hawks, and burrowing owls hunting the ground squirrels in the grass below. The dense rows of small, fast-spinning turbines on open-lattice towers, which birds could perch on and which put many rotors at exactly the height where raptors hunted, turned Altamont into a killing field for exactly the most majestic and legally protected birds in the region. Altamont would become, for decades, the most infamous single example of wind power's toll on wildlife — a reputation the whole industry still labors under, and a problem we will return to in depth later in this book. For now, note only the bitter irony: the very haste and density that the tax incentives rewarded were what made Altamont so lethal to the eagles.

The boom, being built on paper, ended the way paper booms end: abruptly, when the paper changed. The lavish federal wind credits were allowed to lapse after 1985, and California's were wound down around the same time. Overnight, the arithmetic that had summoned the mechanical forest reversed. With no tax credit to harvest, a marginal, half-broken turbine was simply a marginal, half-broken turbine, worth nothing to a tax farmer and not much to anyone else. New installations collapsed. Companies that had existed only to feed the shelter market folded and vanished. The great fields of Altamont, Tehachapi, and San Gorgonio, so recently the frontier of a new energy age, stood half-derelict, many turbines never to spin reliably again, monuments to an incentive that had summoned quantity without ever demanding quality. To a casual observer at the end of the 1980s, the verdict looked clear: wind power had been tried in America, at enormous public expense, and it had failed.

But that verdict was wrong, and the reason it was wrong is the redemptive half of this story. Look closely at the wreckage and you find that not everything in it was junk. A minority of the machines had been built by people who actually meant it — engineers who cared whether the turbine survived the wind rather than merely qualified for the credit. And among the companies feeding the California boom were a handful of Danish manufacturers, working in a very different tradition. Denmark, as we saw much earlier, had a lineage of serious wind engineering running back to Poul la Cour in the 1890s and forward through Johannes Juul's sturdy Gedser turbine of the 1950s. Danish makers — Vestas above all, along with others of that generation — approached the California gold rush not as a tax game but as a market for real machines, and they built to a philosophy that had already earned a name: the "Danish concept," a three-bladed, upwind, unfashionably heavy and conservative design that valued surviving twenty years over squeezing out the last theoretical percent of efficiency.

Those machines, by and large, kept running. While flimsy turbines shed their blades and seized their gearboxes, the sturdy Danish designs stood in the passes and worked, season after season, proving in the harshest possible field trial which engineering choices endured and which did not. And this is the quiet miracle of the California wind rush: it functioned, for the surviving builders, as an immense, brutal, publicly subsidized learning laboratory. The Danes sold thousands of turbines into California, watched them succeed or fail under real load in real wind, carried the lessons home, and used them to refine designs that would, a decade later, seed the modern wind industry across Europe and then the world. The boom that looked like a fiasco was, for the companies that took it seriously, the classroom in which the modern turbine grew up. Vestas would go on to become one of the largest wind-turbine manufacturers on the planet. The genes of today's four-hundred-foot offshore giants trace, in part, straight back to the sturdy little machines that outlasted the tax farmers on the Altamont hills.

So we come to the two misconceptions this chapter must

gently correct, and they are mirror images of each other. The first is the boosters' misconception: that the California boom proved wind power worked. It did not, or at least not in the way the cheerleaders claimed at the time. A great deal of what was built was shoddy, poorly sited, and driven by tax arithmetic rather than energy engineering, and when the subsidies vanished, most of it stopped. Anyone who pointed to Altamont in 1984 as proof that wind was ready to power a nation was pointing at a bubble. The second is the opposite, the skeptics' misconception, and it is if anything more common and more lasting: that the collapse of the boom proved wind power failed. That is equally wrong. The boom's collapse proved that a badly designed subsidy produces badly designed machines — a lesson about policy, not about physics. The good turbines, the serious ones, endured the wind and the withdrawal of the credits alike, and it was those survivors, and the engineers who built them, who carried wind power forward. The California rush did not prove wind worked, and it did not prove wind failed. It proved something more subtle and more useful: that you get exactly what your incentives ask for, and that the durable machines and the durable companies were the ones that had been building for the wind all along, not for the tax code.

There is a worked example hiding in all this that is worth making explicit, because it captures why the durable machines mattered and the flimsy ones did not. Recall the iron law of wind that governs everything in this half of the book: the power in the wind rises with the cube of its speed. Double the wind speed and you get eight times the power; but equally, a turbine that is shut down or broken during the strongest winds forfeits a wildly disproportionate share of the energy it might have captured. A machine that runs reliably through the fierce afternoon gusts of Altamont Pass — precisely when the cube law is delivering its richest returns — earns many times more, over a year, than a flimsier machine of the same nameplate rating that keeps breaking down and sitting idle when the wind blows hardest. The tax farmers, rewarded at the moment of purchase, had no reason to care about this. The Danes, selling machines meant to earn their keep for two decades, cared about almost nothing else. The cube law quietly sorted the serious from the frivolous, and it was the serious that survived.

Where do you meet the California wind rush today? You meet it, remarkably, in the same three passes. Drive over Altamont now and the ragged mechanical forest of the 1980s is largely gone — the small, dense, bird-killing lattice turbines have been steadily torn out and replaced, over the past two decades, by a far smaller number of vastly larger, slower-turning, more efficient machines, sited more carefully and spaced more widely, in a deliberate effort to make peace with the eagles. Tehachapi and San Gorgonio, too, have been repowered with modern giants. The passes that were once the emblem of wind power's chaotic, subsidy-fueled infancy are now working examples of its disciplined maturity — the same wind, the same funnel of hills, but harvested at last by machines built to last. And every time an American politician debates a production tax credit for wind — a credit, note, that pays per kilowatt-hour actually generated, precisely the design lesson learned from the tax-farm fiasco — the ghost of Altamont is in the room, reminding lawmakers that what you reward is what you get.

To recap: in the early 1980s, generous federal and California tax credits stacked on top of PURPA's guaranteed purchases summoned thousands of small wind turbines onto three mountain passes almost overnight, briefly making one American state home to some ninety percent of the world's wind capacity — but much of what was built was shoddy "tax farm" junk, and when the credits lapsed after 1985 the boom collapsed, leaving Altamont with a wrecked reputation and a terrible toll on eagles. Yet the serious machines, above all the sturdy Danish designs, endured the wind and outlasted the subsidies, and the lessons their builders carried home seeded the modern wind industry. The rush proved neither that wind worked nor that it failed, but that incentives shape what gets built — the enduring lesson of a policy that asked for turbines and got them, good and bad alike. Solar, meanwhile, was undergoing a transformation of a wholly different character — not a boom-and-bust summoned by tax law, but a slow, relentless, decades-long slide in price so regular that one engineer would eventually notice it followed a law as reliable as the one that governed the computer chip.

Chapter 33 — Swanson's Law

Richard Swanson was standing in front of a room of investors, or perhaps a room of engineers — the setting varies in the retellings — trying to explain why anyone should believe that solar electricity, then a boutique curiosity costing many times the price of ordinary power, would ever become cheap. He was an electrical engineer by training, a Stanford professor who in 1985 had left the security of academia to found a company called SunPower and chase the dream of a genuinely efficient, genuinely affordable silicon cell. He had spent years listening to skeptics tell him that solar was a permanent luxury, forever too dear for the real grid, forever dependent on the kindness of subsidies. And he had a chart that said they were wrong. On it, the price of a solar module marched downward with a regularity that was almost eerie — not in fits and starts, not in dramatic breakthroughs, but in a steady, disciplined descent that had held for decades and showed no sign of stopping. Swanson had noticed something. The price of solar was not random. It obeyed a law.

The law now bears his name, and it is one of the quiet load-bearing beams of the modern world, so let me state it in its simplest form before we unpack it. Every time the cumulative amount of solar modules ever manufactured doubles, the price of those modules falls by roughly twenty percent. Not the price this year versus last year — the world can pump out solar faster in some years than others — but the price measured against total production ever. Double the stack of everything ever made, and the price drops about a fifth. That is Swanson's Law. It is not a law of physics, like Betz's limit for the wind or the band gap of silicon. It is a law of learning: a description of how good human beings get at making a thing when they make an enormous number of it.

Swanson did not invent the underlying idea, and he would be the first to say so. He was consciously echoing the most famous prediction in the history of technology: Moore's Law, the observation by Intel's Gordon Moore that the number of transistors that engineers could pack onto a computer chip kept doubling every couple of years, which is why the phone in your pocket outpowers the room-sized computers of the 1960s. Moore's Law had become a kind of secular prophecy in Silicon Valley, a self-fulfilling promise that the future would arrive on schedule and on a graph. Swanson, working just down the road in the same valley, on chips made of the very same silicon but tuned to catch light rather than to compute, saw that solar was following a curve of its own — cruder, slower, but every bit as relentless. If Moore's Law made computers cheap, Swanson's Law would make sunlight cheap. That was the bet, and it has paid off more spectacularly than almost anyone in power believed possible.

To feel the force of it, we need to understand the general idea behind the specific law, and here the human story reaches back long before Swanson. The phenomenon is called an experience curve, or a learning curve, and it was first noticed not in solar cells or computer chips but in airplanes. In the 1930s an aeronautical engineer named Theodore Wright, studying the assembly of aircraft, observed something that has since been confirmed across an astonishing range of manufactured goods: every time the total number of a thing produced doubles, the cost of making each one falls by a roughly constant percentage. The idea became known as Wright's Law, and Swanson's Law is simply Wright's Law wearing a solar panel. The constant percentage — the "learning rate" — differs from product to product. For solar modules it settles around twenty percent per doubling. For some products it is less; for a few it is more. But the shape is always the same: a steady, compounding decline, driven not by any single invention but by the accumulated grind of making the next one better than the last.

Why should making a thing get cheaper simply because you have made a lot of it? The reasons are wonderfully mundane, and this is the part that most people miss when they imagine technological progress as a series of eureka moments. Think of the first time you cooked an unfamiliar dish. It took an hour, you burned a pan, you wasted ingredients, and the result was mediocre. Cook it fifty times and you do it in twenty minutes, waste nothing, and it comes out perfectly, because you have learned where the shortcuts are, which steps matter, and which tools to keep within reach. Nothing about the recipe changed; you changed. A factory is a kitchen that never stops cooking, staffed by thousands of cooks who each notice small improvements — a slightly thinner wafer that saves silicon, a faster furnace, a jig that shaves a second off assembly, a supplier who has himself gotten cheaper by the same process, a defect finally traced to its cause and eliminated. No one of these is a breakthrough. Their sum, compounded over millions of units, is the collapse of a price.

Here is a second analogy that captures the compounding character of it. Imagine a language you are learning. The jump from knowing nothing to holding a halting conversation feels enormous and takes forever. But each doubling of your vocabulary and practice yields a similar-feeling leap in fluency, even though in absolute terms you are adding ever more words. Progress measured against effort feels steady, because you are climbing a curve where each step forward requires roughly doubling what came before. Solar's price decline works the same way: the first doubling of production, from a tiny base, was quick and cut the price by a fifth; but as the industry grew to gigantic size, each further doubling required manufacturing more panels than existed in all prior history — and still delivered its twenty percent. That is why the descent has been so long and so smooth. The doublings got harder to achieve, but the world kept achieving them, and each one paid the same dividend.

Now let us put numbers on the fall, because the numbers are frankly difficult to believe and they are the beating heart of this chapter. In 1977, near the very start of the terrestrial industry that Elliott Berman had helped to bootstrap, a solar module cost around seventy-six dollars for every watt of generating capacity. Seventy-six dollars a watt. To assemble a few hundred watts — enough for a modest set of household appliances — would have cost as much as a house extension. By the 2020s, the price of a solar module at the factory gate had fallen below thirty cents a watt, and in the great glut of Chinese overproduction it dipped, at moments, toward a dime a watt. Set those two numbers side by side: from about seventy-six dollars to under thirty cents. That is a decline of more than ninety-nine percent — in some accountings, better than ninety-nine point nine percent — in the price of the same fundamental product. It is one of the steepest and, more importantly, one of the most sustained cost declines in the entire history of industrial manufacturing. Very few technologies in human history have fallen so far, so consistently, for so long. Solar is one of the great deflationary marvels of the modern age, and it happened quietly, a few percent per doubling, decade after decade, while most of the world was looking the other way.

And "looking the other way" is not a figure of speech. It brings us to the most jaw-dropping fact in this chapter, and to a genuine intellectual scandal. The experts whose job it was to forecast the future of energy did not see this coming — not once, not vaguely, but repeatedly and badly, year after year. Consider the International Energy Agency, the world's most authoritative energy forecaster, whose projections governments and investors treat as gospel. For years running, the IEA published forecasts of how much solar the world would install, and for years running, reality blew past those forecasts almost immediately. The agency would draw a line into the future showing solar deployment leveling off, plateauing, growing only modestly — and the actual numbers would shoot past the projection within a year or two, again and again, so consistently that analysts began charting the successive forecasts as a fan of hopelessly pessimistic lines all sitting far below the soaring reality. The serial under-prediction of solar became a running embarrassment, a case study in how even the best experts, anchored to the assumption that the recent past would continue, systematically failed to believe a curve that had been running for decades right in front of them.

Why did they miss it so badly? Partly because a learning curve is genuinely counterintuitive: it asks you to believe that a thing will keep getting cheaper indefinitely, with no dramatic cause you can point to, simply because people keep making more of it — and the human mind, which craves a mechanism, distrusts progress it cannot attribute to a specific invention. Partly because forecasters, sensibly cautious, assume that any fast trend must slow down soon, and they had assumed exactly that about solar for so long that they kept being wrong the same way. And partly, one suspects, because solar was so small for so long that it was easy to dismiss as a rounding error, right up until the moment it wasn't. Swanson, standing in front of his skeptics in the 1980s and 1990s with his stubborn downward-sloping chart, had understood something the professional forecasters would take another twenty years to concede: that a learning curve, once established over decades, is a very hard thing to stop, and a foolish thing to bet against.

There is a lovely human texture to this that deserves telling, because Swanson himself was not a triumphant prophet but a working engineer whose own company rode the curve's cruel logic. A learning curve is a wonderful thing for the world and a terrifying thing for any single firm caught on it, because prices are always falling, which means yesterday's cutting-edge factory is tomorrow's money-loser, and a company must run flat out just to stay in place. SunPower built some of the most efficient silicon cells in the industry — Swanson's engineers were genuinely brilliant at wringing performance out of the physics — and the company survived and mattered for decades. But it also endured the relentless margin-crushing pressure that the very law bearing its founder's name imposed on everyone in the business. The curve does not care who discovered it. The name "Swanson's Law" itself was not coined by Swanson; it was bestowed, in 2012, by the British magazine The Economist, which noticed the pattern, saw the parallel to Moore's Law, and named it after the SunPower founder who had done so much to publicize it. Swanson got a law named after him and, like many who ride a great curve, a hard commercial road as the reward.

Let me now correct two misconceptions, because both are widespread and both matter. The first is the belief, still astonishingly common in casual conversation, that solar is "still too expensive" — a niche indulgence for the environmentally minded rich, propped up by subsidies, not a serious way to power an economy. This was true once. It has not been true for years. As of the mid-2020s, in a great many markets around the world, newly built utility-scale solar is the cheapest source of new electricity ever available to humankind, cheaper per unit than new coal, new gas, new nuclear, cheaper on an honest accounting even without subsidy. The global average cost of solar electricity fell by roughly ninety percent in the decade from 2010 alone. The person who says solar is too expensive is quoting a price list from an era that Swanson's Law has long since erased. There are real and serious questions about solar — about when it produces, about storing it, about where the panels are made, questions this book takes seriously — but raw cost per unit of energy is no longer among them.

The second misconception is subtler and cuts the other way, and I raise it to keep us honest. It is the retrospective illusion that this decline was foreseeable, even inevitable — that of course solar got cheap, anyone could have seen it. The historical record flatly contradicts this. The people paid to see it did not see it. The world's premier energy institutions under-predicted solar for the better part of two decades, revising their forecasts upward again and again and still falling short. The decline was not obvious; it shocked the experts, and it kept shocking them long after a reasonable person, looking at Swanson's chart, might have started to trust the trend. There is a real humility to be learned here about the limits of expert forecasting when a technology is riding an experience curve: the curve tends to be more reliable than the professionals betting against it. And there is a corresponding caution, to keep the ledger balanced — a learning curve describes the past with great authority and promises the future with none. It has run for decades; it may run for decades more; but no law of nature guarantees the next twenty percent, and honest analysis treats the extrapolation as a well-founded expectation, not a certainty.

Where do you meet Swanson's Law today? You meet it, most concretely, on your own utility bill and in the news, every time you read that some country or company has signed a contract to buy solar electricity at a price that would have been science fiction a generation ago — well under a few cents per kilowatt-hour, and in record-setting desert auctions, as we will see, at barely a cent. You meet it on the roofs of ordinary suburban houses, where an array that would have cost a fortune in Swanson's founding year now pays for itself in a handful of years. You meet it, invisibly, in the fact that solar has quietly become the fastest-growing source of new electricity on the planet, adding more capacity each year than any other technology — a growth that is not a policy miracle but the arithmetic consequence of a product that keeps getting cheaper by a fifth every time the world doubles its output. And you meet it whenever a forecaster, chastened at last, draws a line into the future that finally curves upward as steeply as the past did — trying, belatedly, to believe the law that Richard Swanson read off his chart forty years ago.

To recap: Richard Swanson, founder of SunPower, noticed that the price of solar modules falls by roughly twenty percent every time cumulative production doubles — a learning curve, in the tradition of Wright's Law and an echo of Moore's Law for computer chips — and this steady descent carried the price of a module from around seventy-six dollars a watt in 1977 to under thirty cents in the 2020s, a fall of more than ninety-nine percent that ranks among the greatest and most sustained cost declines in industrial history, and that the world's expert forecasters missed, badly and repeatedly, for decades. The law is not physics but the compounding of ten thousand small manufacturing improvements, and it makes nonsense of the still-common claim that solar is too expensive, while rebuking the hindsight that calls its rise inevitable. But a learning curve only tells us that making the thing got cheaper; it does not tell us how the thing is made, or why making it was ever hard in the first place — and to understand that, we must go down to the raw material at the very heart of the solar cell, to silicon itself, and to the strange, energy-hungry alchemy that turns humble sand into a substance of almost unimaginable purity.

Chapter 34 — The Price of Purity

Pick up a handful of sand on a beach and let it run through your fingers. Most of what falls is silicon dioxide — quartz — one of the commonest substances on the face of the Earth. Now imagine that somewhere in a factory, engineers are taking a material chemically related to that sand and purifying it to a standard so extreme that, in the finished product, fewer than one atom in a billion is anything other than silicon. Fewer than one in a billion. If the population of the entire planet were silicon atoms, the total allowable impurity would amount to a handful of people. That is the material at the heart of a solar cell, and the gulf between the sand on the beach and the mirror-bright ingot in the factory — a gulf measured not in chemistry so much as in energy, patience, and money — is the subject of this chapter. It is the story of why the second-most-abundant element in the Earth's crust was, for decades, the single most stubborn cost in all of solar power.

Let us begin with the abundance, because it is genuinely astonishing and it frames everything that follows. Oxygen is the most abundant element in the crust; silicon is second, making up roughly a quarter of it by weight. The rocks under your feet, the mountains, the sand of every desert and shore — a vast fraction of all of it is silicon locked up with oxygen as silica and silicate minerals. If solar cells were limited by the availability of raw silicon, we could pave continents in panels and never come close to running out. The Sun's cheapest fuel and its cheapest hardware are, at bottom, made of the same thing the beach is made of. So when someone tells you that solar power is limited by scarce, exotic materials, they cannot be talking about the silicon, which is as common as dirt because it very nearly is dirt.

And yet, for most of solar's history, silicon was the bottleneck — not because it was scarce, but because the silicon a solar cell needs bears almost no resemblance to the silicon in sand. A grain of quartz is silicon bonded to oxygen, riddled with traces of iron, aluminum, and a dozen other elements. A solar cell needs pure elemental silicon — the oxygen stripped away — and it needs that silicon almost inconceivably clean, and it needs it arranged, ideally, into a single flawless crystal. Getting from the one to the other is a journey through fire, poison gas, and repeated distillation, and every step of it costs energy. This is the central truth of the chapter, so let me state it plainly: silicon is cheap, but purity is expensive. What you pay for in a solar wafer is not the material. It is the extraordinary process of cleansing and ordering that material, and the mountain of energy that process consumes.

Why does purity matter so much? Here we need a little first-principles physics, and I will keep it in plain language. Recall from earlier in this book how a solar cell works. A photon of sunlight, striking the silicon, knocks an electron loose, freeing it to move and carry current. For that current to reach the wires and do useful work, the freed electron must travel through the silicon crystal without getting trapped or lost along the way. Now, a perfect silicon crystal is a smooth, orderly lattice, and a loose electron glides through it easily, like a marble rolling across a clean glass table. But every stray impurity atom — every atom of iron or copper or oxygen sitting where a silicon atom should be — is a defect in that lattice, a snag, a pothole. Impurities create traps that capture wandering electrons and quench them before they can escape to the circuit, converting the energy that photon delivered into useless heat. A cell built from dirty silicon is a cell full of potholes, and its electrons never reach the wire.

Here is an analogy that makes the sensitivity vivid. Imagine a vast concert hall in which a single whispered message must be passed from the front row to the back, person to person, without error. If every person in the hall is attentive and sober, the message arrives intact. But scatter through the crowd a few hundred people who mishear, or mumble, or deliberately garble what they pass along, and the message dissolves into noise long before it reaches the rear. In a solar cell, the electrons are the message, and impurity atoms are the garblers in the crowd — and because a single crystal contains a truly enormous number of atoms, even a "few hundred garblers per billion" is enough to wreck the signal. That is why solar-grade silicon must be purified to the neighborhood of 99.9999 percent — "six nines," in the trade's shorthand — and why the even more demanding silicon of computer chips is purified further still. The performance of the finished cell is astonishingly sensitive to contamination that, in almost any other industry, would count as spotlessly clean.

So how do you clean silicon to six nines? The workhorse method is a piece of chemical engineering called the Siemens process, developed by the German company of that name in the 1950s and still, in refined form, the backbone of the industry. The elegant idea at its core is distillation — the same principle that separates alcohol from a fermented mash, or fresh water from salt. You cannot easily pick impurity atoms out of solid silicon one by one. But you can convert the silicon into a gas, purify the gas, and then turn it back into a solid, leaving the impurities behind. In the Siemens process, crude metallurgical silicon — already smelted from quartz in a furnace, and perhaps 98 or 99 percent pure, nowhere near good enough — is reacted with hydrogen chloride to form a volatile compound called trichlorosilane, a liquid that boils at a comfortably low temperature. That liquid is then distilled, over and over, with each pass shedding a little more of its contamination, exactly as a moonshiner redistills to strengthen and clean the spirit. Finally the ultra-pure trichlorosilane gas is passed over hot silicon rods, where it decomposes and deposits fresh, gleaming, hyper-pure silicon atom by atom, growing the rods thicker until you have great fingers of polycrystalline silicon — polysilicon — clean to six nines and beyond.

The analogy to distilling water again and again is exact and worth dwelling on, because it captures both the method and the cost. If you distill water once, you remove most of the salt; distill the result again, and you remove most of what remained; a third time, most of what was left after that. Each pass is a multiplication of purity — but each pass also costs energy to boil and condense, and the last few nines of purity are wrung out at brutal expense, because you are heating and cooling enormous quantities of material to remove ever-tinier traces of contamination. The Siemens process is a distillation of silicon compounds, and it is voracious in its appetite for energy, running hot furnaces and hot rods for long hours to grow each batch. This is the deep reason polysilicon has always been costly: not because the atoms are rare, but because coaxing them to this purity means paying, again and again, the energy toll of distillation and deposition.

But purity is only half the battle, because a solar cell wants more than clean silicon — it wants ordered silicon, ideally a single unbroken crystal. And here we meet a second beautiful process, named for the Polish scientist Jan Czochralski, who discovered the underlying technique by happy accident in 1916, the story goes, when he absent-mindedly dipped his pen into a crucible of molten metal instead of his inkwell and drew out a thin thread of solidified metal that proved to be a single crystal. The Czochralski method, adapted for silicon, works like this. You melt the pure polysilicon in a crucible until it glows white-hot and liquid.

Then you touch the surface of the melt with a small "seed" crystal of perfect silicon, and you slowly, slowly withdraw the seed while rotating it. As the seed rises, molten silicon clings to it, cools, and freezes onto the seed's crystal structure, atom aligning with atom, so that the growing solid inherits the seed's perfect order. Pull steadily for many hours and you draw a single, flawless, cylindrical crystal — an ingot — sometimes as long as an arm and gleaming like polished chrome, every atom in ranks, grown from a pool of light-swallowing melt. It is one of the quietest marvels in all of manufacturing: a mirror-bright column of pure order, pulled slowly out of fire.

From that ingot you must finally make wafers — the paper-thin slices, each a couple of tenths of a millimeter thick, that become individual cells. And here is a fact that surprises almost everyone the first time they hear it, and that quietly haunted the industry's economics for years: a great deal of that painstakingly purified, painstakingly crystallized silicon is thrown away as dust in the very act of slicing it. To cut an ingot into wafers you must saw it, and a saw has a width; every cut turns a slice of silicon as thick as the blade into worthless powder. This wasted material is called kerf, and the "kerf loss" in traditional wafer sawing was staggering — for a long time, cutting an ingot into wafers turned something like half of the precious crystal into dust on the workshop floor. Picture spending days growing a flawless crystal at enormous energy cost, then feeding it to a saw that grinds half of it into sawdust. That is what wafering long meant. The industry has fought this waste hard, moving to thinner wire saws studded with diamond that cut a narrower kerf, and researchers keep chasing "kerfless" methods that would peel wafers off an ingot with no sawing at all — but for decades, the sawdust on the floor was a silent tax on every panel ever made.

Let me now surface three facts that together dispel the intuition most people carry about silicon and cost. First, the purity itself: solar-grade silicon, at six nines, is one of the purer bulk materials human industry produces, and the semiconductor silicon in a computer chip is purer still — cleaner, atom for atom, than almost anything else made at scale on Earth. Second, the waste: as we just saw, a large fraction of that hard-won crystal has historically ended up as sawdust, an extravagance that would be scandalous if the material were rare rather than merely expensive to purify. Third, and most counterintuitive, the abundance-versus-cost paradox: the raw silicon in a solar panel costs almost nothing, while the energy and processing to purify and crystallize it long dominated the price — which means that solar's cost problem was never a materials problem at all. It was an energy-and-process problem, exactly the kind of problem that a learning curve, grinding away year after year as we saw in the last chapter, is superbly equipped to solve.

The economics of all this were dramatized, in the 2000s, by one of the wildest boom-and-bust cycles in the whole industry — the great polysilicon price spike. As solar demand suddenly surged in the mid-2000s, driven by European subsidies, the world's small handful of polysilicon plants could not keep up. Building a new Siemens-process plant takes years and vast capital, and for a period the entire fast-growing solar industry was throttled by a shortage of purified silicon. Prices for polysilicon rocketed — by some accounts as much as tenfold above their earlier level — and the material that had always been the quiet bottleneck became a screaming one, with panel makers scrambling and signing desperate long-term supply contracts. Then, as always happens, the high prices summoned a wave of new plants, mostly in China, and by the time all that capacity came online the shortage had flipped to a glut, prices crashed, and many of the companies that had rushed in at the top were ruined. The polysilicon spike and crash of the 2000s is a perfect miniature of a commodity cycle, and it left a lasting mark: it was during that scramble that China built the enormous, cheap, energy-intensive polysilicon capacity that would help it come to dominate the entire solar supply chain, a story we will follow later.

Now to the misconception this chapter exists to correct, because it is one of the most common and most misleading things people believe about solar. The misconception is that silicon is scarce, or exotic, or intrinsically expensive — that solar power is somehow limited by a shortage of its key material, the way it might be limited by a rare metal. This is simply false, and the falsehood matters, because it feeds a broader and lazier claim that renewable energy is forever hostage to material scarcity. Silicon is sand. It is the second-most-common element in the crust, effectively inexhaustible for any plausible human purpose. What is expensive is not the silicon but the purity — the energy-hungry distillation of the Siemens process, the slow fiery patience of the Czochralski pull, the sawdust of wafering. And costs of that kind, unlike genuine scarcity, are exactly what human ingenuity and scale drive relentlessly downward. To say "silicon is expensive" is to confuse the price of the effort with the price of the stuff. The stuff is nearly free. The effort is what we have spent fifty years learning to make cheap.

A second, gentler misconception worth naming is that purifying silicon is so energy-intensive that a solar panel can never "pay back" the energy used to make it. It is true that turning sand into a wafer costs a great deal of energy — that is the honest heart of this chapter. But a panel then generates clean electricity for twenty-five or thirty years and more, and the accounting has long since tipped decisively in solar's favor: a modern panel repays all the energy invested in making it, silicon purification included, in something like one to two years of operation, leaving decades of net energy gain. The purity is expensive, but it is an investment that the Sun repays many times over. We will return to this energy-payback question in its own right; for now, note only that "energy-intensive to make" is not the same as "never worth making," and the difference is measured in the twenty-plus years a panel spends quietly paying the world back.

Where do you meet the price of purity today? You meet it every time you see a solar panel's deep, dark, faintly iridescent blue-black surface, so uniform and mirror-smooth — that flawless finish is the visible signature of a single crystal or a field of large crystals, grown at heat and cost from a pool of melt. You meet it in the geography of the supply chain: the reason so much of the world's polysilicon is made in a few specific places, some of them chosen precisely because electricity there is cheap, is that purification is so hungry for energy that the cost of power dominates where you can afford to do it. And you meet it, quietly, in the falling price on every solar quote, because the long descent of solar's cost has been, in large part, the story of humanity slowly learning to distill, crystallize, and slice silicon for less and less energy and less and less waste — turning the price of purity, decade by decade, from a fortune into an afterthought.

To recap: silicon is the second-most-abundant element in the Earth's crust, essentially sand, so its scarcity is a myth — but a solar cell needs it purified to around six nines, because a single stray impurity atom traps the very electrons the cell exists to collect, and it needs it grown into flawless crystals; achieving this means the energy-hungry Siemens distillation of trichlorosilane and the slow Czochralski pull of mirror-bright ingots, followed by a wafering step that long ground away much of the crystal as kerf dust. The cost of solar was never the material but the purity and the energy, which is why polysilicon spiked and crashed so violently in the 2000s and why the whole thing yielded, over time, to the learning curve. But purified, crystallized silicon is only one way to catch the Sun; there is another whole family of solar technology that uses a hundred times less material by spreading it in films thinner than a hair — and the contest between these two approaches, crystalline and thin film, is where we turn next.

Chapter 35 — Crystalline and Thin Film

Out in the Mojave Desert, where the sun hammers the ground for more hours of the year than almost anywhere in the United States, there are solar farms whose panels look subtly wrong to an eye trained on rooftops. Instead of the familiar grid of dark blue squares — the little cells with their fine silver lines, wired together like windowpanes — these panels are smooth, uniform sheets, a deep uninterrupted charcoal with no visible cells at all, as though someone had tinted a great pane of glass. They are made by a company called First Solar, and they contain almost no silicon. The active layer that catches the sunlight is a film of a compound called cadmium telluride, laid down thinner than a sheet of household plastic wrap — a few millionths of a meter thick, a hundred times thinner than the silicon wafer in an ordinary panel. Standing between a field of these and a field of conventional blue panels, you are looking at the two great rival families of solar technology, the two answers the industry gave to a single question: how little material can you get away with, and how much does cheapness cost you in performance?

That question defines this chapter, and the tension in it is a genuine engineering drama, not a settled matter with an obvious winner. On one side stands crystalline silicon — the technology of the wafer, the child of the fiery Czochralski pull we followed in the last chapter — which dominates the world market so thoroughly that it accounts for something like ninety-five percent or more of all solar panels made. On the other side stands thin film — cadmium telluride, amorphous silicon, and a copper-based compound with the unlovely acronym CIGS — which uses a tiny fraction of the material, can be flexible, and is cheaper to make in some respects, but which has never managed to unseat its crystalline rival from the mass market and instead holds a set of hard-won, valuable niches. To understand why the market went the way it did, and why thin film nonetheless refuses to die, we need to understand what each family actually is, in plain terms.

Start with crystalline silicon, and start by distinguishing its two forms, because the difference shows up in the very look of a panel and it is worth being able to read. Recall the flawless single crystal pulled from the melt: slice that into wafers and you get monocrystalline cells, each cut from one continuous crystal, its atoms in perfect ranks. Because the order is perfect, electrons travel through it beautifully, so monocrystalline cells are the most efficient — but growing that perfect crystal is slow and costly. There is a cheaper way: instead of pulling one flawless crystal, you can simply pour molten silicon into a mold and let it cool into a solid block.. As it freezes, it forms not one crystal but a jumble of many crystal grains packed together — like a mosaic rather than a single pane. Slice that block and you get multicrystalline, or polycrystalline, cells, recognizable by the shimmering, shattered-ice pattern of their surface, where each glinting facet is a separate grain. They are cheaper to make but slightly less efficient, because every boundary between grains is a snag where electrons can be lost — the concert-hall garblers of the last chapter, only now arranged along the seams between crystals rather than scattered through them.

For years the industry ran on both, with polycrystalline holding the low-cost end and monocrystalline the premium. But as the price of purifying and pulling silicon fell along the learning curve, the efficiency advantage of monocrystalline came to outweigh its cost penalty, and the market swung decisively toward mono. Today most new panels are monocrystalline, and they carry an alphabet soup of cell architectures that reward a moment's plain-language translation, because these acronyms are where a great deal of quiet progress hides. The workhorse design of the 2010s was called PERC — passivated emitter and rear cell — which was, in essence, a clever trick of adding a reflective, electrically insulating layer to the back of the cell so that light which would otherwise have passed straight through gets bounced back for a second chance at being absorbed, and so that electrons near the back surface are less likely to be trapped and lost. Think of PERC as putting a mirror behind a window and sealing the window's edges: more light caught, fewer electrons leaking away. PERC squeezed a meaningful extra slice of efficiency out of an ordinary cell for very little added cost, and it swept the industry.

Then came its successors. TOPCon — the name stands for a mouthful about a tunnel-oxide passivating contact, and you can safely forget the words — refined the same idea further, adding an ultra-thin protective layer that lets the useful electrons through while blocking the leaks even better, wringing out another percentage point or two of efficiency. And heterojunction cells, often shortened to HJT, take a different tack entirely: they sandwich the crystalline silicon wafer between whisper-thin layers of a different, amorphous form of silicon, which does a superb job of coddling the electrons at the surface and preventing losses, yielding some of the highest efficiencies of any commercial silicon cell. The details differ, but the through-line is simple, and it is the same story as the whole book: each architecture is a new trick for catching a little more light and losing a few fewer electrons, and each one, once perfected, ripples out across hundreds of gigawatts of manufacturing. A further refinement worth naming, because you can see it, is the bifacial panel — a cell built to catch light on both faces, so that sunlight reflected off the ground beneath the array adds a bonus to the light striking the front. Mount bifacial panels over pale desert sand or a white rooftop and they harvest the glow bouncing back up, a free few percent for the cost of a clear backsheet.

Now cross to the other family. Thin film abandons the wafer entirely. Instead of growing a crystal, slicing it, and wiring the slices together, you take a sheet of glass (or metal, or even flexible plastic) and deposit onto it an extremely thin layer of a light-absorbing semiconductor — vapor-coating it on, almost like spray-painting, in a continuous process. The great virtue is obvious in the numbers: where a silicon cell needs a wafer well over a hundred microns thick to absorb enough light, a thin-film absorber does the job in a layer perhaps one or two microns thick, because these particular materials swallow light far more greedily than silicon does. A hundredfold less active material; no ingots, no sawing, no kerf dust on the floor; a manufacturing line that runs more like printing a newspaper than like a jeweler cutting gems. Three chemistries have mattered. Amorphous silicon — silicon with no crystal order at all — was the earliest, the material in the little strips that powered pocket calculators for a generation, cheap and flexible but low in efficiency and prone to fading. CIGS — copper indium gallium selenide — offers higher efficiency and flexibility but has proven fiddly and expensive to manufacture consistently at scale. And cadmium telluride, the First Solar material, which turned out to be the great commercial success of thin film.

Here is the essential trade-off, and it is best understood as a three-way tug-of-war between efficiency, cost, and material. Crystalline silicon wins on efficiency: a good commercial silicon panel converts perhaps twenty to twenty-three percent of the sunlight striking it into electricity, comfortably ahead of most thin film. Thin film wins on material and, in some respects, on manufacturing simplicity: it uses a sliver of the semiconductor and skips the whole costly business of crystal-growing and wafering. So which matters more? The answer depends entirely on what you are trying to build, and this is the crux. If space is tight — a suburban roof of fixed size — you want the highest efficiency you can get, because every extra percent means more power from the same limited area, and there crystalline silicon's superior efficiency is decisive. But if space is cheap and abundant — a vast tract of empty desert where you can simply lay down more panels — then efficiency matters less and cost-per-watt and behavior in heat matter more, and there thin film can compete or even win.

Let me make that concrete with a worked example, because it captures why both families survive. Imagine you need to generate a fixed amount of power, and you have two panels to choose from: a silicon panel at twenty-one percent efficiency and a cadmium telluride panel at, say, eighteen percent. The silicon panel produces more power per square meter, so to hit your target you need fewer of them and less land. On a rooftop, where area is the binding constraint, silicon wins hands down — you simply cannot fit enough of the lower-efficiency panels up there. But now move to a utility-scale project on cheap desert land, where you can spread panels across a square mile without anyone caring about the extra acre. Now the binding constraint is not area but dollars per watt and how the panel behaves in brutal heat — and here comes the twist that has kept First Solar thriving. Silicon cells lose efficiency as they get hot; on a scorching desert afternoon, a silicon panel's output sags noticeably as its temperature climbs. Cadmium telluride degrades far less with heat. So in the hottest deserts, the thin-film panel's real-world, all-day, all-summer energy yield can close much of the gap with silicon's laboratory efficiency — and at a competitive cost per watt, on land that is nearly free, that has been enough to make First Solar one of the largest and most successful solar manufacturers in the world, and the dominant force in American utility-scale solar. Thin film did not conquer the rooftop. It conquered a corner of the desert, and it is a very large and lucrative corner.

So why did crystalline silicon win the mass market so overwhelmingly, despite thin film's real advantages? Several forces converged. Silicon rode the learning curve harder and longer, because so many more of it were made, so its cost fell faster than thin film's could — the very dominance became self-reinforcing, each doubling of silicon production driving silicon cheaper and widening its lead. Silicon's higher efficiency made it the natural choice for the rooftops and space-constrained installations that make up a huge share of demand. Silicon panels proved extraordinarily durable, routinely warrantied for twenty-five or thirty years with slow, predictable, gentle decline. And silicon had no dependence on relatively scarce elements: cadmium telluride relies on tellurium, one of the rarer elements in the crust, and CIGS on indium, both of which raise real questions about how far those technologies could scale even if everything else were equal, whereas silicon, as we saw, is effectively inexhaustible sand. The combination — cheaper through scale, more efficient, more durable, made of an unlimited material — is why silicon holds its ninety-five-percent-plus grip, and why the rooftops of the world are blue-black grids of wired wafers rather than smooth sheets of tinted glass.

This history sets up two misconceptions that deserve careful, even-handed correction, because they pull in opposite directions and both are common. The first, popular in the 1990s and 2000s and still lingering, is that thin film is "the future" — the sleek, cheap, flexible technology that would inevitably sweep silicon aside, if only it were given time. For decades, enthusiasts predicted that thin film's material savings and manufacturing simplicity must eventually triumph; venture capital poured into CIGS startups; the coming thin-film revolution was perpetually just around the corner. It never arrived in the form promised. Silicon's relentless cost decline outran thin film's advantages, and most of those startups failed. But — and this is the balancing correction — it would be equally wrong to conclude that thin film failed. It found real, durable, valuable niches: First Solar's cadmium telluride is a genuine commercial giant at utility scale in hot climates, precisely where its heat tolerance and land-tolerance shine, and flexible thin film serves applications where a rigid glass panel simply cannot go — curved surfaces, portable gear, lightweight installations. Thin film is not the future that swallowed everything, and it is not a failure; it is a successful specialist that lost the general contest but won its home ground. The honest verdict is "niche champion," not "coming revolution" and not "dead end."

The second misconception is simpler and more everyday: that all solar panels are basically the same — that a panel is a panel, an interchangeable blue rectangle. By now you can see how wrong this is. Panels differ in their fundamental material (crystalline silicon versus thin-film compounds), in their crystal structure (mono versus poly), in their cell architecture (PERC, TOPCon, heterojunction), in whether they catch light on one face or two (bifacial), in their efficiency, their behavior in heat, their cost per watt, their longevity, and the elements they depend on. These are not cosmetic distinctions; they determine which panel belongs on a cramped northern rooftop and which belongs in a blazing desert, which will still be producing power in thirty years and which will fade sooner, which can bend around a curved surface and which must lie flat and rigid behind glass. To choose a solar panel well is to weigh exactly these trade-offs — and the fact that the choice is real is precisely why both great families, and all their sub-varieties, continue to exist.

Where do you meet crystalline and thin film today? You meet crystalline silicon almost everywhere you look up: on the roof of nearly every solar-equipped house, on the carports and warehouses and the great majority of solar farms, in that familiar dark-blue grid of wafer cells laced with fine silver lines. You meet thin film mostly at scale and out of sight — in the vast First Solar arrays feeding American utilities from the sun-blasted Southwest, and in the flexible strips laminated onto backpacks, boat decks, and building facades where rigid glass would never fit. And you meet the whole restless family of cell architectures, quietly, in the steadily rising efficiency printed on the spec sheet of each new panel year over year — PERC giving way to TOPCon giving way to heterojunction — the visible face of an industry that never stops finding one more trick to catch one more photon. Watch, too, for the next fork in this road: researchers are now stacking an entirely different light-absorbing material, a crystal called perovskite, on top of silicon to catch the part of the spectrum silicon wastes, a tandem approach that has already broken efficiency records once thought impossible and that we will meet in full later in this book.

To recap: solar technology splits into two great families — crystalline silicon, made from wafers cut from grown crystals, which dominates over ninety-five percent of the market because it is efficient, durable, cheap through sheer scale, and made of inexhaustible sand; and thin film, which spreads a hundredfold less light-absorbing material, such as cadmium telluride, in a wafer-thin layer, sacrificing some efficiency for lower material use and better heat tolerance, and which lost the general market to silicon but reigns in real niches, above all First Solar's utility-scale arrays in hot deserts. Silicon's own progress runs through a ladder of ever-cleverer cell designs — mono over poly, PERC to TOPCon to heterojunction, bifacial catching light front and back — and the lesson is that panels are emphatically not all alike; the right one depends on whether area, cost, heat, or longevity is your binding constraint. Having seen why solar got cheaper, what it is made of, and how its two great families differ, we are ready at last to look under the hood of the cost decline itself — to take apart the learning curve, mechanism by mechanism, and see exactly how making more of a thing makes each one cheaper, which is where the deep economics of this whole revolution begins.

Chapter 36 — The Learning Curve

In 2011, an engineer at a solar module factory in eastern China stood at the end of a production line and watched a finished panel slide off the rollers, its blue cells glinting under the fluorescent lights. He had been in the business long enough to remember when a panel like this one might have cost the better part of two dollars for every watt it could produce. This one had cost him closer to one dollar a watt to make, and the price was still falling under his feet, month after month, faster than his accountants could redraw their spreadsheets. He told a visiting reporter that he no longer bothered to predict where the bottom was. Every time his team thought they had wrung the last cent out of the process, someone on the factory floor found another. A decade later, that same panel would cost less than a dime a watt to manufacture. Nothing else in the modern energy economy has behaved this way. To understand why, we have to look past the headline that solar keeps getting cheaper and ask the harder question underneath it: cheaper because of what, exactly?

We met the shape of this story earlier, when we followed the falling price of a solar cell from the Bell Labs marvel of 1954 down through the decades to the rooftops of today, and when we named the pattern after Richard Swanson, the

SunPower founder whose observation The Economist christened Swanson's law in 2012. That law says that every time the world doubles the total number of solar modules it has ever shipped, the price of a module falls by roughly twenty percent. It is a beautifully simple rule of thumb, and over four decades it has held with uncanny persistence, carrying module prices down by something like ninety-nine point nine percent. But a rule of thumb is not an explanation. Knowing that prices fall twenty percent per doubling tells you the rhythm of the drum without telling you who is beating it. This chapter is about the drummers. It is about the specific, physical, unglamorous reasons that a solar panel gets cheaper every year, and about why those reasons operate for solar and refuse, stubbornly, to operate for coal.

Start from first principles, with a distinction that economists make and the rest of us usually blur. The cost of generating electricity from any source divides into two buckets. There are the fixed costs, the money you spend once, up front, to build the thing: the steel, the silicon, the land, the labor of construction, the interest on the loan while it is all going up. And there are the variable costs, the money you keep spending for as long as the thing runs, chiefly fuel and the maintenance that fuel-burning demands. A coal plant is mostly the second kind of cost. You build it once, but then you must feed it a train of coal every few days for thirty years, and the price of that coal is set by a global commodity market you do not control. A solar farm is almost entirely the first kind. You build it once, and then it runs on sunlight, which arrives free and untaxed and will keep arriving long after the humans arguing about it are gone. This single structural difference is the key that unlocks everything that follows. A technology whose cost lives in the up-front manufacturing can be improved by manufacturing better. A technology whose cost lives in the fuel is chained to the price of that fuel, and no amount of cleverness at the plant will free it.

That is the deep reason solar behaves like a technology and coal behaves like a commodity, and the two words are doing real work. A technology is a recipe for making something, and recipes improve. Every microchip, every flat-screen television, every LED bulb has followed the same downward glide: as we make more of them, we learn to make them better and cheaper, and the learning compounds. A commodity is a raw material dug or pumped out of the ground, and while the machinery for extracting it improves, the resource itself does not get more generous. If anything it gets stingier, because we mine the easy seams first and leave the deep, thin, awkward ones for later. So the fascinating truth at the heart of this book is that we have quietly converted energy — for all of history a commodity you burned — into a technology you manufacture. And technologies ride a learning curve.

Let me open up that learning curve and show you the gears inside, because "we get better at it" is exactly the kind of hand-wave the style of honest writing forbids. When engineers and economists track the cost of a manufactured good against the total number ever produced, they find, over and over, a straight line on a particular kind of graph — the log-log plot, where each step along an axis is a multiplication rather than an addition. This is Wright's law, first noticed by an aeronautical engineer named Theodore Wright studying airplane production in 1936, and Swanson's law is simply Wright's law wearing a solar hat. The line is straight because the underlying process is multiplicative. You do not shave costs by a fixed number of pennies each year; you shave them by a fixed percentage each time cumulative production doubles. And a percentage of a shrinking number keeps shrinking, which is why the curve looks so dramatic when you plot price against time but so orderly when you plot it against volume.

So what, concretely, happens during each doubling? Several things at once, which is why the effect is so robust. The first is economies of scale, the plainest of them. A factory that makes a gigawatt of panels a year has certain fixed overheads — the building, the senior engineers, the research, the sales force. Spread those over ten gigawatts and the cost per panel falls before anything on the line has changed at all. The Chinese gigafactories that came to dominate the industry are, at bottom, an exercise in taking scale to a scale nobody had tried. When a single facility produces more modules in a year than the entire world installed a decade earlier, the per-unit overhead approaches a rounding error.

The second driver is manufacturing improvement, the accumulated thousand small refinements that a workforce discovers only by doing the work. This is the part that pure laboratory research can never supply, and it is the reason deployment matters so much more than most people assume. Somewhere on a production line, a technician notices that the furnace runs hotter than it needs to and trims the energy bill. An engineer works out how to cut the breakage rate on the wafer saws from four percent to two. A supplier finds a cheaper source of the silver paste that forms the fine metal fingers on a cell's face. None of these show up in a physics journal. All of them show up in the price. The learning is embodied in people and in tooling, and it can only be earned by making the next million panels, and then the million after that.

The third and fourth drivers are the ones an engineer can point to with a caliper. Consider the silicon wafer itself, the thin blue-grey square at the heart of every crystalline cell. In the early years those wafers were cut around three hundred microns thick — three-tenths of a millimeter — because sawing them thinner shattered too many. Over the years the industry learned to slice them to a hundred and eighty microns, then a hundred and fifty, and it is pushing thinner still. Silicon is one of the most expensive ingredients in a panel, so halving the wafer thickness quietly halves the silicon in every cell without changing the electricity it makes. In parallel, efficiency climbed. A cell that was fifteen percent efficient a couple of decades ago is now commonly twenty-two or twenty-three percent, meaning the same square of silicon, the same aluminum frame, the same pane of glass, the same truck to ship it, the same crew to install it, all now deliver half again as much power. Efficiency is a cost-killer in disguise, because so much of a solar system's price is not the cell at all but everything wrapped around it — and a more efficient cell dilutes all those fixed per-panel costs across more watts.

The fifth driver is the price of polysilicon, the ultra-pure feedstock from which wafers are grown. Purifying silicon to the required nine-nines cleanliness was, for years, a specialized and costly chemical business dominated by a handful of Western and Japanese firms. As demand exploded, Chinese producers built enormous new capacity, drove down the energy and chemical intensity of the process, and turned a once-scarce input into a glut. When polysilicon is cheap, every wafer downstream is cheap. The sixth driver is automation: lines that once needed rows of workers hand-placing cells now run with robotic stringers and automated inspection, cutting labor cost and, just as importantly, cutting the defect rate. And the seventh, threaded through all the others, is global competition. Dozens of manufacturers fighting for the same buyers, each terrified of being undercut, is a machine purpose-built for squeezing out cost. The oversupply that crushed manufacturer profits in 2024 was agony for the companies and a windfall for the world, because every one of them was forced to pass its learning straight through to the buyer.

Now stand this next to the record of fossil fuel and nuclear power, and the contrast is almost violent. Here is a genuinely surprising fact, the kind worth pausing on: over the long run, the real cost of electricity from coal and gas has been essentially flat, wandering up and down with the price of the fuel but going nowhere in particular, while the cost of solar fell by orders of magnitude. And nuclear, astonishingly, has often gone the wrong way. In several Western countries the cost of building a nuclear plant rose across the decades — an inverse learning curve that scholars have named, half in jest, "negative learning." Each new plant, hedged with fresh regulation and built by a workforce that had lost the muscle memory of the last one, cost more than the one before. The reasons are debated and partly particular to nuclear's safety burden, and a fair account notes that a more standardized, repeatedly-built reactor might have learned like anything else. But the plain historical fact stands: the fuel-and-megaproject model did not ride the curve, and the manufactured-module model did.

The reason returns us to first principles. There is no learning curve for a fuel. You can get better at digging coal, but the coal you dig is the same coal, and once it is burned the money is gone and you must buy more. A solar panel, by contrast, burns nothing. All of its cost is manufacturing cost, and manufacturing cost is exactly the kind of cost that learning devours. This is why the two technologies were destined to cross, and why, having crossed, they will not cross back.

From all this falls the single most important and most misunderstood policy lesson in the field, so let me state it plainly. Deploying solar, not merely researching it, is what drove the price down. For years the intuition in energy policy was that you fund laboratories, wait for a breakthrough, and deploy once the technology is cheap. But the learning curve runs the other way. Much of the cost reduction lived on the factory floor and in the installer's van, in knowledge that could only be won by building the next gigawatt. The countries that chose to deploy at scale before it was economically comfortable — Germany above all, with its feed-in tariffs paying above-market rates for solar power in the 2000s — were effectively buying down the cost curve, and not only for themselves. German households, through their electricity bills, paid a premium that pushed cumulative global production up another doubling, and every doubling lowered the price for the Texan and the Indian and the Nigerian who bought panels a decade later. It was, whether the payers intended it or not, one of the more consequential acts of involuntary generosity in economic history.

Which brings us to the misconception that most deserves

gentle correction: the charge that subsidies for solar "wasted" taxpayer money on an uneconomic technology. On a narrow accounting the early subsidies did pay for expensive electricity — those German rooftops of 2005 produced some of the priciest power on the grid. But that framing misses what the money actually bought. It did not buy kilowatt-hours; it bought movement along the learning curve, a cost reduction that then became permanent and global and free to everyone who came after. The subsidy was not a consumption expense; it was an investment in the price of the future, and the return on it — solar power now cheaper than anything humans have ever generated electricity from — has been extraordinary. One can argue, fairly, about whether the money was spent efficiently, whether tariffs were set too high or lingered too long. Those are real debates. But "the subsidies were wasted" mistakes the down payment for the whole cost of the house.

Let me leave you with the panel itself, decomposed, so the abstraction has weight in the hand. Take a modern crystalline module and follow the money. A meaningful slice of the cost is the polysilicon and the wafer cut from it — the silicon story, thinner and cheaper each year. Another slice is the cell processing, the doping and coating and printing of those fine silver lines, where automation and efficiency gains have bitten deep. Another is the module assembly: the tempered glass front, the polymer encapsulant that seals the cells against weather, the aluminum frame, the junction box, the labor of lamination. And then, once the panel leaves the factory, a whole second economy of cost that has nothing to do with the panel at all — the racking that holds it to the roof or the ground, the wiring, the inverter that turns its direct current into grid-ready alternating current, the permits, and the crew on the ladder. Here is the twist that surprises even people in the trade: as the module itself has grown so cheap, these balance-of-system costs, the mounting and wiring and soft costs of doing business, now often exceed the price of the panels they hold. The learning curve has done its work on the silicon so thoroughly that the frontier of cost has moved to the ladder and the paperwork.

So the story of why solar keeps getting cheaper is not one story but seven, braided together: scale, factory learning, thinner wafers, higher efficiency, cheaper polysilicon, automation, and merciless competition, each one squeezed a little harder with every doubling of the world's deployed capacity. It is the behavior of a technology, not a commodity, and it is the reason a manufactured panel could go where no lump of coal ever could. But a falling manufacturing cost, however dramatic, is not yet an answer to the question a utility actually asks, which is not "what does the panel cost" but "what does the electricity cost, over the whole life of the plant, against every rival on the grid." To answer that, the industry needed a single number that could set a wind farm and a gas plant side by side and let them fight fairly — and it is to that number, and the argument it settled, that we turn next.

Chapter 37 — LCOE

There is a moment in every argument about energy where someone reaches for a number, and for most of the twentieth century that number was a trap. A coal man would tell you his fuel cost a few cents a kilowatt-hour and rest his case. A solar advocate would counter that sunlight was free, which was true and useless, because the panels were not. The two were talking past each other, comparing an ongoing bill to a one-time purchase, and no honest verdict could come out of it. What finally broke the deadlock was not a breakthrough in physics or a fall in price but a change in bookkeeping — a single formula, unglamorous as a mortgage calculation, that at last let a wind farm and a gas plant stand in the same ring and be judged by the same rules. The formula is called the levelized cost of electricity, LCOE for short, and once the world learned to think in its terms, the entire argument about renewable energy quietly changed shape.

The idea behind LCOE is so simple that its power is easy to miss. Take a power plant of any kind. Add up every dollar it will ever cost you, from the first survey of the land to the last bolt removed at decommissioning forty years later: the construction, the fuel, the maintenance, the insurance, the interest on the money you borrowed to build it. Then add up every kilowatt-hour of electricity it will ever produce across that same lifetime. Divide the first number by the second. What comes out is the true, all-in cost of one kilowatt-hour from that plant — a cost that already contains the fuel and the steel and the loan and the labor, folded together into a single figure you can print on a card and compare against any other plant on Earth. That is LCOE. It is the great equalizer, and its genius is that it refuses to let anyone hide half their costs off the page.

Let me give you the analogy I find most useful, because LCOE is really a household idea dressed in utility clothes. Suppose you are choosing between two cars. One is cheap to buy but drinks fuel and needs repairs. The other costs far more on the forecourt but sips electricity and rarely visits a mechanic. How do you compare them honestly? Not by the sticker price, which flatters the first car, and not by the running cost, which flatters the second. You work out the true cost per mile over the whole life of the car — purchase plus fuel plus maintenance, divided by all the miles you will ever drive. That per-mile number is the LCOE of a car, and everyone who has ever agonized over a car purchase has done this arithmetic in their head. LCOE simply does it for power plants, where the stakes are billions and the "fuel efficiency" gap between a gas turbine and a solar farm is not a matter of a few miles per gallon but of one machine that must be fed forever and another that runs on weather.

Here is a second way to feel it. Think of two houses. The first is cheap to buy but has no insulation, so the heating bills bleed you every winter for as long as you live there. The second costs more up front but is so well built that it barely needs heating at all. The mistake is to compare the purchase prices, because the cheap house is only cheap until you move in. LCOE is the discipline of adding the winters to the price of the house. A gas plant is the cheap, draughty house: lower to build, but you pay for gas every year it runs. A wind farm is the expensive, well-insulated one: costly to raise, but the wind, like good insulation, asks for nothing after that.

Now watch what happens when you actually turn the handle, because the abstraction only bites when you carry a worked example through. Picture two hypothetical plants, each meant to serve the same town. The first is a gas-fired plant. It is relatively cheap to build, so its up-front cost is modest, and it can run whenever you ask it to. But it burns fuel, and the price of that fuel — call it a market price that drifts with the world's appetite for gas — must be paid for every kilowatt-hour it makes, every year, for thirty years. Most of its lifetime cost, when you finally tally it, turns out to be fuel. The second plant is a wind farm of the same output. It is expensive to build; nearly all its cost lands in the first year, in the towers and blades and the loan that paid for them. But after that its fuel bill is zero, because wind arrives free, and its maintenance is modest. When you divide each plant's total lifetime cost by its total lifetime output, something striking emerges. The gas plant's number is dominated by an ongoing fuel bill you cannot control; the wind farm's number is dominated by a one-time capital cost you fixed on the day you signed the loan. And as of 2024, when you run those sums with real figures, the wind farm's number comes out lower. Not lower with a subsidy bolted on. Lower on the raw arithmetic.

The figures themselves are worth stating carefully, because this is fast-moving ground and honest writing dates its numbers. As of 2024, the International Renewable Energy Agency put the global average levelized cost of utility-scale solar at roughly four point three cents a kilowatt-hour, and onshore wind at roughly three point four cents — the cheapest of any source of electricity on the planet. To feel how far that is, remember that solar's LCOE stood near forty-two cents a kilowatt-hour in 2010; it has fallen something like ninety percent in a decade and a half. The independent analysts at Lazard, whose annual survey has become a kind of industry scripture, tell a broadly similar story with their own methods, and the two do not agree to the last decimal — estimates vary, as they always will when you must guess future fuel prices and interest rates. But the headline they share is the one that reordered the debate: in most markets, new solar and new onshore wind are now cheaper than new coal or new gas, before any subsidy is counted at all. The cheapest new bulk power on Earth is now sunlight and wind. That sentence would have been laughed out of a boardroom in 2005.

And the auctions push even further, into territory that still startles the people who run them. When developers bid competitively for the right to build a solar farm and sell its power under a long contract, the winning prices have at times dropped below two cents a kilowatt-hour. In April 2021, a tender in Saudi Arabia — the Al Shuaibah project, blessed with relentless desert sun and cheap capital — set a world record with a bid of about one point zero four cents a kilowatt-hour. One cent. For a moment, sunlight in the Arabian desert became the cheapest electricity ever contracted by human beings, cheaper than the running cost alone of many a fossil plant already built and paid for. Such records are exceptional, stitched from perfect sun and near-free money, and they should not be mistaken for the every-day price. But they mark the frontier, and the frontier tells you which way the whole landscape is tilting.

This is the place to correct the misconception that outlived its truth. For years the reflexive objection was that renew-ables are only cheap because of subsidies — strip away the tax credits and the mandates, the argument ran, and the whole edifice collapses. That was a fair point once. It is no longer. The LCOE figures I have quoted are unsubsidized; they describe the cost of the electricity itself, before any gov-ernment has added a credit or a guarantee. Subsidies cer-tainly built the industry — we saw in the last chapter how they bought down the learning curve — but the industry no longer needs them to undercut its rivals. The cheapness is now baked into the hardware. To keep insisting that renew-ables are only cheap with subsidies is to describe the world of a decade ago as though it were still here.

But — and this is the caveat that separates honest advocacy from a sales brochure — LCOE does not tell the whole story, and anyone who waves it around as a final verdict is misusing it. The formula has a blind spot, and it is a serious one: it counts how much energy a plant produces but not when. A kilowatt-hour of solar generated at noon, when the grid is already awash in sunshine and prices sag toward zero, is worth far less than a kilowatt-hour delivered at seven on a still winter evening when demand peaks and the sun is long gone. LCOE treats both the same. It prices a wind farm's midnight gale that nobody needs identically to a gas plant's power delivered precisely at the moment of scarcity. In the language of the trade, LCOE captures the cost of energy but not its value, and value depends on timing.

There is more that LCOE quietly omits. It does not, in its plain form, count the cost of integrating a variable source into the grid — the transmission lines to carry wind from empty plains to distant cities, the batteries or gas plants held in reserve for when the weather fails, the extra reserves an operator must keep spinning because a cloud bank can dim a solar fleet in minutes. These integration and backup costs are real, and they grow as the share of variable power rises. A critic who says "your cheap solar needs expensive backup, so it is not really cheap" has put a finger on something true, and a fair-minded reader should grant it. The answer is not to pretend the costs away but to recognize that the industry has developed richer measures — a system-level or value-adjusted LCOE that tries to account for timing and integration — precisely because the simple number, for all its power, was never meant to carry the whole weight of a grid decision. The plain LCOE won the argument that renewables are cheap to generate. It cannot, by itself, settle the harder argument about how to run a grid on them. Those are two different questions, and confusing them is the most common error in the entire debate.

There is a subtlety buried inside the formula that deserves daylight, because it explains a fact that puzzles newcomers: how can solar's cost tick upward in a year when the panels keep getting cheaper? The answer is that the LCOE of a renewable plant is dominated not by fuel — it has none —

but by the cost of the money used to build it. Because nearly all of a wind or solar farm's expense lands up front, the interest rate on the loan matters enormously; the whole point of levelizing is to spread that capital cost, plus its financing, across decades of output, and the arithmetic of doing so leans hard on the assumed cost of capital. When interest rates rose sharply in 2022 and after, the cost of borrowing to build a solar farm rose with them, and so, remarkably, the levelized cost of solar edged up by a small fraction even as the panels themselves fell in price — by IRENA's reckoning solar's global LCOE nudged up by something like half a percent in 2024. For a fuel-burning plant, financing is a smaller share of the whole and rising rates bite less. This is the hidden pivot of renewable economics: a technology that has traded a fuel bill for a loan is a technology whose price answers less to the commodity markets and more to the central bank. It is one more way that sunlight behaves unlike coal, and one more reason a single headline number can never carry the full weight of the decision.

Hold both truths at once, then, because the honest position needs both. The first truth is genuinely transformative: on the metric that compares the cost of generating a kilowatt-hour, renewables have won, decisively and without subsidy, and they keep winning by wider margins each year. That is not propaganda; it is what the International Renewable Energy Agency, Lazard, and the world's electricity auctions all independently report. The second truth is that this metric, powerful as it is, measures cost and not value, and the value of power depends on delivering it at the right moment — which is exactly where variable sources are weakest and where the real engineering challenge of the transition now lives. A book that gave you only the first truth would be a brochure. A book that gave you only the second would be a brief for the other side. The reality holds both, and the interesting story of the next decades is precisely the tension between them.

Where do you meet LCOE in your own life? You meet it, invisibly, every time a utility decides what to build next, and increasingly the answer comes back the same: a solar farm, a wind farm, a battery, because the spreadsheet says so. You meet it when a corporation signs a power purchase agreement to buy a wind farm's output at a fixed price for twenty years, locking in a rate a gas plant cannot promise because no one can promise the price of gas two decades out. And you meet it, in a sense, in the very stability of that promise. Part of what makes renewables attractive on an LCOE basis is not just that they are cheap but that their cost is knowable. When almost all of a plant's cost is fixed on the day it is built, you can tell your customers what their power will cost in 2044. The gas plant cannot, because it is hostage to a fuel market no one commands. There is a quiet value in certainty that even the formula struggles to capture. It is the reason a technology company siting a data center, or a factory planning for the 2040s, increasingly signs directly with a wind or solar developer: not only because the power is cheap, but because its price is fixed, immune to the geopolitics of a gas pipeline or the shock of a war that sends fuel markets lurching. When Russia's invasion of Ukraine sent European gas prices to terrifying heights in 2022, the wind farms already built kept selling power at the same low, contracted rate they always had, indifferent to the panic, because their fuel had been free all along. That episode taught utilities a lesson no LCOE table quite spells out: a plant with no fuel bill is not merely cheap on average but stable in a crisis, and stability, in an anxious world, has a worth of its own.

Let me recap what this number did. Before LCOE, the debate about renewable energy was a shouting match between people comparing incomparable things — a fuel bill against a purchase price, a running cost against a capital cost. LCOE folded every cost a plant will ever incur, and every kilowatt-hour it will ever make, into one figure that let the contestants fight fairly. Run through that fair comparison, the result as of 2024 is that new solar and onshore wind are the cheapest new electricity humanity has ever had, subsidies or no. But the number has a blind spot — it counts how much power a plant makes, not when, and it stays quiet about the cost of stitching a weather-driven source into a grid that must be balanced second by second. That blind spot is the real frontier, and we will spend later chapters inside it. For now, one question presses harder than the rest. If the arithmetic has been this favorable for this long, someone had to actually build the factories that made the hardware this cheap — and the story of who did, and how a country went from bit player to near-monopolist of the world's solar supply, is where we turn next.

Chapter 38 — The China Story

In the summer of 2005, a solar company almost nobody in the West had heard of listed its shares on the New York Stock Exchange, and by the closing bell its founder had become, on paper, the richest man in mainland China. His name was Shi Zhengrong, and his company was Suntech, based in the city of Wuxi on the Yangtze delta. Shi was a physicist who had trained in Australia under Martin Green, one of the fathers of modern silicon cell design, and he had come home in 2001 with a doctorate, a fistful of patents, and an idea that struck the local officials of Wuxi as either visionary or reckless: that China, then a rounding error in global solar manufacturing, could become its factory floor. The city government put up much of the early money. Within four years Shi was a billionaire, and within a decade the company that made him one had gone bankrupt, crushed by the very avalanche of cheap panels it had helped set loose. Both halves of that story — the meteoric rise and the ruinous glut — are the story of how China came to make almost all the world's solar power, and neither half makes sense without the other.

The scale of what happened is difficult to convey without simply reciting the numbers, so let me recite them, because they are genuinely staggering. The manufacture of a silicon solar panel runs through a chain of four main stages: purified polysilicon is grown into crystalline ingots and sliced into wafers, the wafers are processed into light-catching cells, and the cells are wired and laminated into modules. As of 2024, China's share of world production at each of those four stages was, respectively, about ninety-three percent of polysilicon, roughly ninety-six percent of wafers, about ninety-two percent of cells, and around eighty-six percent of finished modules. Read that again. At the wafer step — the delicate business of slicing crystalline silicon into paper-thin squares — the rest of the planet combined makes barely one panel's worth in twenty-five. This is not dominance in the ordinary sense of a leading exporter with a healthy lead. It is something closer to a single-country monopoly over the physical supply of one of the technologies on which the world's energy future is being staked. And here is the fact that unsettles people most when they first meet it: it happened in about fifteen years, from a standing start.

How? The honest answer braids together several threads, and a fair account resists the temptation to reduce it to a single villain or hero. The first thread is the one Shi Zhengrong embodies: entrepreneurial talent, much of it Western-trained, returning home at exactly the moment Chinese local governments were hungry to seed new industries. Wuxi was not unique; across the country, provincial and municipal authorities competed to attract solar firms with cheap land, cheap loans, and cheap power, treating a gigawatt factory the way an American town might treat a new car plant. This is the second thread, and it is the decisive one: industrial policy at a scale and with a patience that no market economy attempted. The state, through its banks and its local governments, poured capital into solar manufacturing for years, tolerating losses that would have shut a Western firm on the theory that market share tomorrow was worth red ink today. When demand from Europe collapsed after 2011 and prices cratered, Western manufacturers — including once-mighty names in Germany and the

United States — went bankrupt by the dozen. Many Chinese firms went bankrupt too, Suntech among them. But the factories did not vanish. They were recapitalized, consolidated, and restarted, and the industrial capacity stayed inside China's borders, hardened by the very shakeout that killed the weaker players.

The third thread is relentless cost-cutting, the learning curve of the earlier chapters made flesh in a particular place. Chinese manufacturers did not merely build big; they built big and then attacked cost with a ferocity born of brutal domestic competition. They drove polysilicon prices down by mastering the energy-hungry purification process at scale. They thinned wafers, automated lines, squeezed suppliers, and undercut one another without mercy, and every cut in their cost became a cut in the world price. The gigafactory — a single plant producing more solar capacity in a year than the entire world installed in the early 2000s — became the standard unit of the industry, and China built them faster than anyone thought possible. By 2024, Chinese factories were capable of churning out something like six hundred and thirty gigawatts of modules a year, nearly double what the whole world could absorb, and the resulting glut pushed module spot prices down toward a dime a watt or below. For the companies this oversupply was a bloodbath of thin and vanishing margins. For the planet's decarbonization, it was a gift: solar panels had never been so cheap, and they were cheap precisely because Chinese manufacturers were, in effect, selling the world its energy transition at a loss.

It is worth dwelling on the speed of it, because speed is what makes this history feel almost unreal to anyone who lived through the earlier era. In the early 2000s, when Shi Zhengrong was raising his first money, the centers of solar manufacturing were Japan, Germany, and the United States, and China made a negligible sliver of the world's cells. By around 2010 China had become the largest producer. By the late 2010s it was not merely the largest but the overwhelming majority of every step. A shift that in most industries unfolds across a lifetime — the migration of a technology's manufacturing heartland from one side of the world to the other — happened here inside a single decade, faster than automobiles moved to Japan or textiles to Asia. And the wafer step is the sharpest illustration of all. The slicing of crystalline silicon into thin squares, a delicate and capital-intensive art, became so concentrated in Chinese hands that the rest of the planet combined came to make only a few percent of the world's wafers. There is essentially no other critical global industry in which one country holds so near to a total monopoly over an indispensable middle step. To build a solar panel anywhere on Earth without a Chinese wafer became, by the mid-2020s, almost impossible.

The human cost of that rise fell heavily on the West's own solar factories, and their stories are a sobering counterpoint. Germany, which had led the world into solar through its feed-in tariffs, watched its once-proud manufacturers collapse one after another as Chinese panels undercut them — Q-Cells, briefly the largest cell maker on the planet, filed for insolvency in 2012. In the United States, the failure of a federally backed manufacturer became a political scandal that shadowed clean-energy policy for years. These firms were not simply outcompeted on quality; they were outlasted on capital, unable to survive the years of losses that Chinese firms absorbed with state support and then compounded by a global price collapse. The factories that closed in Saxony and California did not reopen. The capacity that survived the shakeout survived in China. That is the flesh-and-blood meaning of an industrial-policy contest fought with patience and a tolerance for red ink, and it is why the outcome, once settled, proved so hard to reverse.

That double edge is the heart of an honest China story, and it deserves to be held steadily rather than resolved too quickly in either direction. On one side, China's scale slashed the global price of solar to a level no other country's industrial system was willing to reach, and in doing so accelerated the world's shift off fossil fuels by years, perhaps by a decade.

Every cheap panel on a rooftop in India or Nigeria or Texas is, in part, a dividend of Chinese overinvestment. Anyone who cares about cutting carbon has to reckon with the fact that the country most responsible for making clean energy affordable is China. On the other side, the same concentration is a strategic vulnerability of a kind the modern energy system has rarely faced. To stake your electricity supply — the foundation of everything else — on hardware that flows overwhelmingly from a single country's factories is to accept a dependence that could, in a crisis or a trade war, become leverage. The world spent the twentieth century worrying about dependence on a handful of oil exporters. The twenty-first may find itself worrying about dependence on one manufacturing power. Both things are true at once, and a reader who feels the pull of both is feeling the situation correctly.

The trade wars followed as night follows day. As Chinese panels flooded Western markets in the early 2010s, the United States and the European Union brought anti-dumping and countervailing-duty cases, accusing Chinese firms of selling below cost with state backing — which, in a sense, was precisely the strategy. Tariffs went up; Chinese firms routed production through third countries in Southeast Asia to slip around them; new tariffs chased them there. The pattern has recurred for over a decade and shows no sign of ending, because it reflects a genuine and unresolved tension: Western governments want both cheap clean energy and a domestic manufacturing base, and Chinese dominance lets them have the first only by surrendering the second. The Inflation Reduction Act of 2022 in the United States, with its lavish incentives for domestic clean-energy manufacturing, was in large part an attempt to buy back a foothold — to pay American factories enough to exist alongside Chinese ones. Whether such efforts can build a durable alternative supply chain, or merely a small subsidized enclave beside the Chinese giant, is one of the open questions of the decade, and its answer is still being written. There is an irony here that policymakers feel keenly. The very cheapness that makes Chinese panels so useful for cutting carbon is precisely what makes it so hard for any Western factory to survive without permanent subsidy, because no rival can match a supply chain that spent fifteen years and countless billions driving cost to the floor. A government that wants both cheap clean power and a home-grown industry finds the two goals quietly at war, and every tariff that protects the second raises the price of the first. There is no clean resolution to that tension, only a choice about which to weigh more heavily, and different countries are weighing it differently.

There is a harder and more painful thread that an honest book cannot skip, and it must be stated with care. A large share of the world's polysilicon — roughly forty percent — is produced in China's Xinjiang region, where the government's treatment of the Uyghur population has drawn credible and widely reported concerns about forced labor. In response, the United States passed the Uyghur Forced Labor Prevention Act, which presumes that goods with inputs from Xinjiang are made with forced labor and bars their import unless a company can prove otherwise. This has forced solar importers to trace their supply chains back to the quartz, and it has snarled shipments and reshaped sourcing. The factual situation is contested in its details, the Chinese government rejects the allegations, and I will not pretend to adjudicate here what courts, journalists, and human-rights investigators continue to examine. But the concern is serious and well-documented enough that no responsible account of solar's supply chain can leave it out. That the cleanest energy source humanity has ever built may carry, in part of its supply chain, one of the gravest labor concerns of the age is an uncomfortable fact, and discomfort is the right response to it. The industry's answer has been to develop polysilicon sources outside Xinjiang and to trace supply more rigorously; how fully that succeeds remains to be seen.

Now let me correct two misconceptions that recur whenever

this subject comes up, because both distort the history in revealing ways. The first is the wounded lament that the West invented solar power and therefore ought, by some natural right, to dominate its manufacture — that China somehow stole an industry that belonged to others. It is true that the foundational science was largely Western: Becquerel in France, the photoelectric theory from Einstein, the first practical silicon cell at Bell Labs in 1954. But invention and manufacture are different acts, and the West did not lose the manufacturing so much as decline to pursue it. Through the 2000s, Western firms and governments could have chosen to build the gigafactories and absorb the years of losses that scaling demanded, and by and large they chose not to, judging solar a niche and the capital better spent elsewhere. China made the opposite bet. To call the result theft is to misread a decision as a robbery. The West held the patents and let the factories go; China wanted the factories and built them. The lesson, if there is one, is that in a manufactured technology it is scaling, not inventing, that confers dominance — a lesson the earlier chapters on the learning curve should make feel almost inevitable.

The second misconception is the quiet snobbery that Chinese panels are somehow inferior — cheap in the pejorative sense, shoddy goods to be endured only for their price. This was perhaps arguable in the industry's rough early years; it is simply false now. Chinese manufacturers include the world's technology leaders, holding efficiency records and driving the frontier of cell design. The advanced cell architectures that dominate today's market, and the perovskite-on-silicon tandem that set the certified efficiency record of 34.85 percent in late 2024, came from Chinese firms and laboratories at least as much as from anywhere else. The panels going onto rooftops and into solar farms across the world are not a discount compromise; they are the mainstream product, made by companies that out-innovate as well as out-produce their rivals. To imagine otherwise is to comfort oneself with a story about quality that the market decided against years ago.

Where do you meet the China story today? You meet it, most likely, on your own roof or your neighbor's, or in the vast blue fields of a utility solar farm off the highway, because the odds are overwhelming that the panels came, in whole or in part, from a Chinese supply chain — the wafer almost certainly, the polysilicon very probably, the assembly quite likely. You meet it in the price, which is low because of a fifteen-year industrial gamble made a hemisphere away. And you meet it in the anxious speeches of Western energy ministers who want the cheap panels and fear the dependence, and who have not yet found a way to have the one without the other. Every solar installation is, in this sense, a small monument to a strategic bargain the world has half-consciously struck.

Let me draw the threads together. In a decade and a half, China went from a bit player to the maker of the overwhelming majority of the world's solar hardware — better than nine in ten wafers, roughly nine in ten cells and polysilicon, most of the modules. It did so through a combination of returning talent like Shi Zhengrong, patient state-backed capital, gigafactories built at unmatched scale, and cost-cutting so relentless it drove much of its own industry into loss. The result cut the global price of clean energy dramatically, a genuine boon to decarbonization, while concentrating the supply of a foundational technology in a single country, a genuine vulnerability — and it carries, in its Xinjiang polysilicon, a labor concern no honest account can wave away. But panels and wafers are only the visible end of a longer story. Behind every module and every turbine stands a hidden inventory of raw materials — the silver and copper and glass of solar, the steel and rare-earth magnets of wind — and the security of the whole transition rests on whether the world can dig, refine, and supply them. That bill of materials is where we go next.

Chapter 39 — The Supply Chain

On a windswept quay in the north of England, a single blade for an offshore wind turbine lay on a transporter, and a crowd had gathered to watch it leave. It was longer than a football pitch, more than a hundred meters of curved white composite tapering from a root as thick as a car to a tip you could close your hand around, and it took a specialized convoy, a police escort, and a carefully surveyed route with every roundabout measured in advance to carry just one of them to the harbor. Three of these would make a single rotor. A wind farm would need dozens of rotors. Standing beside that blade, you understand something that a headline about clean energy never conveys: that the transition off fossil fuels is, before it is anything else, a colossal act of material handling — a re-plumbing of the world's mines, smelters, and factories to produce hardware on a scale the planet has rarely attempted. Sunlight and wind may be weightless and free, but the machines that catch them are anything but.

This is the fact that gets lost in the poetry of renewable energy, and it deserves to be met head-on. Every source of electricity is, at bottom, a way of turning materials into power. A coal plant turns coal into power and keeps needing more coal forever. A solar farm and a wind turbine turn a fixed quantity of materials — steel, silicon, copper, glass, concrete, rare metals — into decades of power without further fuel. The renewable trade is to move the burden from an endless stream of fuel to a large one-time inventory of stuff. That is a good trade, an excellent one, but it is not a free lunch, and pretending clean energy is somehow immaterial is a fantasy that helps no one. The right way to think about it is the way an engineer thinks about any machine: draw up the bill of materials, the complete list of what goes into the thing, and follow each item back to the ground it came from.

Start with solar, and start with the surprise, because solar's material story has a genuinely counterintuitive character in it. The bulk of a solar farm is unremarkable and abundant — silicon refined from ordinary quartz sand, aluminum for the frames, steel for the mounting racks, tempered glass for the panel fronts, and polymer encapsulants, the tough transparent plastics that seal the cells against thirty years of weather. None of these is scarce. But threaded through every crystalline cell is a fine tracery of silver, painted on as a paste to form the thin metal fingers that collect the electric current the moment sunlight knocks it loose. Silver is the best electrical conductor there is, which is why it is used, and it is genuinely scarce, which is why it matters. The solar industry has grown to consume a remarkable share of the entire world's annual silver supply — a slice large enough that analysts seriously ask whether silver could become a brake on how fast solar can grow. Manufacturers know this and have spent years thinning the silver lines, substituting copper where they can, and squeezing the metal per cell down and down. But here is the tension in miniature: the very growth that makes solar a climate solution presses against the supply of a metal that has been precious since antiquity. The panel on the roof contains a little of the same element as the coin in a museum case.

Let me make the silver concrete, because a number you can hold changes how you see a solar field. Each crystalline cell carries only a whisper of silver — a fraction of a gram, painted in lines finer than a hair. But a single panel holds dozens of cells, a rooftop array holds dozens of panels, and a utility solar farm holds millions of cells, and when you multiply a whisper by the billions of cells the world now makes each year, the whisper becomes a roar. The industry has grown from a trivial consumer of silver into one that swallows a large share of the metal mined annually across the entire planet — a share substantial enough that some analysts model silver, of all the ancient precious things, as a possible ceiling on how fast solar can scale. This is why the fine art of "silver thrifting" — shaving the grams per cell, replacing silver fingers with electroplated copper, printing the lines ever thinner — is not a footnote but a strategic priority, quietly pursued in laboratories precisely so that a metal humans have prized for five thousand years does not become the thing that slows the sun.

Copper runs through the whole story, solar and wind alike, because copper is how electricity travels. Every panel's wiring, every turbine's generator windings, every cable snaking from a wind farm to the shore and on to the grid is copper, and the electrification of the world — not just its generation but its cars and heat pumps and transmission lines — is, in metal terms, largely a story about copper demand climbing steeply. Copper is not rare, but it must be mined and refined in quantities that will test the mining industry's ability to keep pace, and new mines take a decade or more to open. Aluminum, lighter and cheaper, does much of the structural work in solar frames and some of the conducting; glass, endlessly recyclable but heavy and energy-intensive to make, forms the protective face of every module. None of these is exotic. Their challenge is one of sheer volume: to build solar at the scale the transition demands is to demand more copper, more glass, more aluminum, more steel than the world has ever routed toward electricity before.

Now cross to wind, where the numbers turn from grams to tons and the imagination strains to hold them. A modern utility wind turbine is, by mass, mostly two of the oldest industrial materials we have: steel and concrete. The tower is steel, hundreds of tons of it, rolled into a tapering tube that can stand a hundred meters and more against the wind. The foundation, especially onshore, is a great buried raft of reinforced concrete, more hundreds of tons of it, anchoring the whole structure against the enormous leverage of the wind pushing on the rotor far above. Follow one turbine through its bill of materials and the figure that lands hardest is the steel: a single large turbine can contain on the order of hundreds of tons of it in tower and machinery, and a wind farm multiplies that by scores of machines. This is why the health of the wind industry is tied to the price of steel, and why a spike in steel and shipping costs in the early 2020s pushed the cost of offshore wind up and stalled projects that had penciled out fine a few years before. Wind power, for all its futurism, rests on the two materials that built the nineteenth century. And that dependence runs the other way too, in a fact that unsettles the tidy image of clean energy: making steel and cement, the twin pillars of a turbine, is itself among the most carbon-intensive activities on the planet, together responsible for a large share of global industrial emissions. A turbine therefore begins its life in carbon debt, having caused emissions in the forging of its tower and the pouring of its foundation, and it must run for a period — typically well under a year of generation — before the clean power it makes has repaid that debt and every kilowatt-hour after is pure gain. Over a twenty-five-year life the payback is overwhelming, the emissions a small fraction of a fuel-burning plant's. But the debt is real at the start, and honesty requires naming it: the machines that free us from burning carbon are, in their very construction, temporary borrowers of it.

The blades are the exception, and they are a marvel worth pausing on. A turbine blade is not steel and not concrete; it is a composite, a sandwich of fiberglass — sometimes carbon fiber in the most demanding spots — bonded with resin and often stiffened with a core of balsa wood or foam. In engineering terms a blade is a giant, precision-made wing, shaped to the same aerodynamic logic as an aircraft's, and manufactured to tolerances that must survive a hundred million flexing cycles over twenty-five years in rain, ice, and gale. Here is a fact that surprises people who have only seen turbines from a distance: those serene white blades are among the most sophisticated large composite structures made anywhere, closer in their engineering to an aerospace part than to a piece of civil infrastructure, and each one is essentially handmade, laid up in a mold longer than a blue whale. The same composite mastery that makes a blade strong and light also makes it stubborn at the end of its life — the thermoset resins that bind it do not melt down for reuse the way steel or aluminum do — and that is a problem we will come back to.

Inside the nacelle, the housing atop the tower, sits the piece of the wind story that has drawn the most geopolitical attention: the generator, and in many designs, its magnets. A turbine turns slowly — a large rotor sweeps around perhaps ten to twenty times a minute — and there are two ways to turn that stately rotation into grid-frequency electricity. One is a gearbox that steps the speed up to a conventional generator; the other is a direct-drive generator, a large ring studded with powerful permanent magnets that needs no gearbox at all, trading mechanical complexity for a heavy magnetic one. Those permanent magnets are made from an alloy of iron, boron, and rare-earth elements — chiefly neodymium, with a pinch of dysprosium or terbium to help them hold their magnetism when hot. A single large direct-drive machine can carry hundreds of kilograms of magnet material, a substantial fraction of it rare-earth metal. And rare earths bring us to the second great chokepoint of the transition, the one that mirrors the solar supply chain we met in the last chapter.

Here the misconception to correct is written into the name itself. Rare-earth elements are not, in fact, geologically rare. Neodymium is more abundant in the Earth's crust than lead; the whole family is scattered widely across the planet. What is rare is the willingness and capacity to process them, because separating these chemically similar metals from their ores is a difficult, dirty, and toxic business that most countries have been happy to let someone else do. That someone else is overwhelmingly China, which refines something on the order of ninety percent of the world's rare-earth elements regardless of where the raw ore was dug. The chokepoint is not a hole in the ground but a chemical plant, and it sits, like so much of the transition's supply chain, inside one country's borders. This is why "rare earths are running out" is exactly the wrong worry. They are not scarce in the crust; they are concentrated in the refinery, and a concentration of processing is a different and more fixable vulnerability than a genuine shortage of the element. The wind industry has responded by trimming the heaviest rare earths out of its magnet recipes, favoring geared designs that need far less magnet, and encouraging mining and refining outside China — Australia, the United States, and others are slowly building capacity. But as of the mid-2020s the dependence remains real.

Let me draw the two bills of materials side by side as a worked example, because seeing them together is clarifying. Build one large wind turbine and you must supply, roughly, hundreds of tons of steel for the tower and machinery, hundreds of tons of concrete for the foundation, three enormous composite blades each a handmade aerospace-grade wing, tons of copper for the generator and cabling, and — in a direct-drive machine — hundreds of kilograms of rare-earth magnet. Build one solar farm of comparable output and you must supply vast quantities of glass, aluminum, and steel racking, silicon refined from sand, copper wiring, the polymers that seal the panels, and a modest but genuinely constraining quantity of silver spread microscopically across millions of cells. Both are material-intensive; neither is material-free; and each leans on a different pressure point — silver and copper for solar, steel and rare earths for wind. This is the physical reality behind every clean-energy pledge: a pledge to mine, refine, and route these materials at a pace the world has not tried before.

There is also the land and the transport, easy to forget and impossible to skip. A solar farm is land-extensive, spread flat across acres to catch a diffuse resource, though the ground beneath and between panels can graze sheep or host pollinators or, increasingly, grow crops in the arrangement called agrivoltaics. A wind farm spreads its turbines across a landscape but leaves the land between them largely farmable, so a field can grow wheat and electricity at once. And moving the hardware is its own feat: the blade on the English quay, the transformer that takes a special rail car, the monopile foundations for offshore turbines that need purpose-built installation vessels which exist in scarce supply. The supply chain is not only mines and factories; it is roads measured for oversized loads and harbors dredged for turbine-installation ships. The scarcity of those ships is itself a bottleneck that surprises people: the specialized vessels that can hammer a monopile into the seabed or lift a nacelle to a hub height taller than a cathedral number only in the dozens worldwide, they cost a fortune, and their booking calendars are full for years. An offshore wind project can have every turbine, blade, and cable ready and still wait on a single ship. It is a reminder that the transition is limited not only by metals in the ground but by the mundane machinery of moving them — the cranes, the convoys, the dredged channels, the trained crews — an industrial logistics as demanding as the manufacturing itself.

All of this raises the question of the end, which the industry is only now being forced to confront and which we will meet properly in a later chapter: what happens to a solar panel or a turbine blade after its twenty-five or thirty years are up? A panel is largely recyclable — glass, aluminum, and silicon can be recovered, and the small quantities of silver and other metals are worth reclaiming — though the recycling infrastructure is still immature. A blade is the harder case, that stubborn thermoset composite that has too often ended its life buried in a landfill, though "recyclable blade" resins and new processes are emerging. For now I only plant the flag: the same material intensity that makes these machines is a material stream that will one day flow the other way, and a transition that thinks only about building and not about un-building has not finished its arithmetic.

So the machines that catch the weightless sun and wind are, themselves, some of the most material-hungry infrastructure of our time — a bill of materials running from ordinary sand and steel to precious silver and geopolitically fraught rare earths, none of it truly rare in the ground but much of it bottlenecked in the refining and concentrated in a few hands. To build the transition is to secure these materials, and to secure them is a task of mining, chemistry, diplomacy, and recycling that is only beginning. But suppose we solve all of it — suppose the silver and steel and magnets flow freely and the factories hum. We would still be left with the deepest problem of all, the one no amount of material can fix, because it is not about the machines but about the resource. The sun sets every evening and the wind comes and goes as it pleases, and the grid that carries their power must be balanced perfectly, second by second, whether they are cooperating or not. That is the trouble with intermittency, and it is where we turn next.

Chapter 40 — The Trouble with Intermittency

In a control room in Folsom, California, a wall of screens tracks the electricity flowing across the most solar-saturated grid in the United States, and on a clear spring afternoon the operators watch a curve that has become quietly famous in the trade. Around midday, as the state's millions of solar panels pour power onto the grid, one particular line — the demand the grid must meet from everything that is not solar — sags into a deep valley, so much sunshine arriving that the rest of the system has almost nothing left to do. Then, as the afternoon wears on and the sun sinks toward the Pacific, that same line begins to climb, gently at first and then with alarming steepness, rising by thousands upon thousands of megawatts in the space of a few hours as the solar fades to nothing and thirty million people come home, turn on lights, cook dinner, and charge their cars. Plotted across a day, the curve has a plump belly and a long neck rearing up at sunset, and someone at the California grid operator, staring at it years ago, said it looked like a duck. The name stuck. The duck curve is now the single most recognizable picture of the central technical problem of renewable energy, and this chapter is about what that problem actually is.

Let me be precise about the word, because "intermittency" gets thrown around as if it were a verdict, and it is really just a description. Solar and wind are what engineers call variable, or non-dispatchable, sources. Dispatchable means you can call for the power when you want it: you tell a gas plant to produce more and it burns more gas; you ask a hydro dam to open its gates and it does. Variable means the opposite — the power arrives on the resource's schedule, not yours. The sun stops every night, without fail, and dims under passing cloud. The wind rises and falls with the weather, strong one hour and slack the next, blowing hardest often at the times you least need it. These sources do not produce electricity when you ask; they produce it when the sky decides. And that would be a mere inconvenience, except for one iron fact about electricity that makes it the defining challenge of the entire field.

The fact is this: electricity, almost uniquely among the things we use, must be produced at the very instant it is consumed. The power lighting your room right now was generated a fraction of a second ago, hundreds of miles away, and delivered to you at the speed of an electric field down a wire. There is, in the grid itself, essentially no storage. The whole vast machine — every generator, every wire, every appliance — is locked together in a single synchronized dance, and the amount of power flowing in must match the amount flowing out at every moment, or the system's frequency drifts and, if the mismatch is large enough, the grid destabilizes and fails. This is why running a grid is less like filling a reservoir and more like a feat of continuous, split-second balancing. Supply must chase demand up and down through every minute of every day, matching it not approximately but exactly, forever.

Two analogies help make this vivid, and both are worth holding. The first is a bathtub that must be kept exactly full — not nearly full, exactly full — while people at one end keep dipping cups of water out at unpredictable rates and you, at the other end, must open and close the tap moment by moment to keep the level precisely at the brim. Let the level rise and it overflows; let it fall and it runs dry; your only job, all day, is to match the tap to the cups. That is a grid operator's life, and demand is the people with the cups: it rises in the morning, peaks in the evening, falls at night, and you must follow it. Now imagine that a large part of your tap is no longer under your control — that a solar tap gushes on its own all midday and shuts off entirely at dusk, and a wind tap opens and closes as the weather likes. You must still keep the tub exactly full using whatever controllable taps remain, working around the ones that do as they please. The second analogy is a seesaw that must be kept perfectly level, generation on one side and demand on the other, with weights being added and removed from the demand side constantly and now, on the generation side, some of your weights appearing and vanishing according to the sky. The operator's task is to keep the plank level regardless.

It is worth separating two different difficulties that variability creates, because they call for different answers. The first is simple variability: the sun follows a known daily arc, output rising through the morning and falling through the afternoon, a pattern as predictable as the calendar. That is manageable precisely because it is foreseeable — you know sunset is coming and can plan for it. The second and thornier difficulty is uncertainty: you do not know exactly when a cloud bank will roll over a solar fleet, or whether tomorrow's wind will blow as forecast. Variability you can schedule around; uncertainty you must hold reserves against. This is why weather forecasting has become, unexpectedly, a core discipline of the modern grid. The better an operator can predict the next hour's and next day's sun and wind, the less he must keep expensive backup idling just in case, and the cheaper and smoother the whole system runs. A great deal of the practical art of integrating renewables turns out to be the art of predicting the weather well enough to trust it.

Now return to the duck and read its anatomy, because the

picture repays close attention. The belly of the duck — that midday valley where net demand sags — is the sign of solar's success: so much sunshine is flowing that the grid needs little else, and on the sunniest days California and other solar-heavy regions have watched the wholesale price of electricity fall to zero and even below. Negative prices. Generators paying to offload power they cannot easily switch off, because at that moment the grid has more electricity than it knows what to do with. It is one of the genuinely strange facts of the modern grid that at solar noon in a sunny region, power can briefly become a thing you are paid to take away. But it is the duck's neck that keeps operators awake. As the sun sets, solar output collapses from its midday flood to nothing over a couple of hours, and it does so at exactly the moment human demand surges into its evening peak. The grid must therefore ramp up its other, dispatchable generation — gas plants, hydro, imports, batteries — with astonishing speed, summoning many thousands of megawatts in a few hours to fill the hole the sun leaves behind. That evening ramp, growing steeper every year as more solar is added, is the sharp end of the intermittency problem: not that the sun goes down, which we have always known, but that it goes down fast, together, and precisely when the day's demand is climbing.

It helps to put a rough number on the neck, because scale is what makes it daunting. On a spring evening, California's grid has at times had to replace something on the order of thirteen or fourteen thousand megawatts of vanishing solar within about three hours — the output of a dozen large power plants, summoned into existence between the appetizer and the dessert. Imagine what it would take to store your way through that ramp instead of burning gas. A shipping-container-sized battery of the kind now dotting substations holds a few megawatt-hours; to carry a grid through a steep multi-thousand-megawatt evening for several hours takes not one such box but tens of thousands of them, marshaled into installations that are, in energy terms, among the largest machines humans have ever assembled.

This is exactly why grid-scale storage has exploded — the world added a record of roughly two hundred gigawatt-hours of battery capacity in 2024 alone — and why the duck's neck, more than any other feature of the curve, has become the business case for the battery. The belly created a surplus; the neck created a market.

The stakes of getting the balance wrong are not abstract, and one episode made them vivid. In 2016, the entire state of South Australia — a grid rich in wind — went dark in a matter of seconds after a violent storm tore down transmission towers and knocked the system off balance. The blackout became a political football, blamed by some squarely on renewables, though later analysis found the immediate cause was the physical destruction of transmission lines and the way the system's protection settings responded, not wind power as such. What is telling is the sequel: South Australia answered by building, among other things, one of the world's first giant grid batteries, which proved so fast at steadying the system's frequency that it paid for itself far quicker than anyone expected and touched off a global rush to build more. A crisis that looked like an argument against renewables became, in the end, a proving ground for the tools that make them work — a pattern worth remembering whenever a single bad day is offered as a verdict on the whole enterprise.

This is the place to plant a flag on the concept that organizes so much thinking about grids, and which the rise of renewables has thrown into question. The old grid was built around baseload: large plants — coal, nuclear, big hydro — designed to run steadily around the clock, providing a constant floor of power, with more flexible plants layered on top to follow demand up and down. Baseload was cheap, steady, and dispatchable, and for a century it was the backbone of the system. Variable renewables do not fit that picture. They are not a steady floor; they are a fluctuating input that the rest of the system must accommodate. Some argue this means renewables cannot replace baseload and the whole system is therefore doomed to need conventional plants forever. Others argue that "baseload" was always just a description of the cheapest plants of a particular era, not a physical law, and that a grid built around flexibility — storage, transmission, demand that shifts to follow supply — can dispense with the old constant-floor model entirely. That larger argument is one we will take up in full later, when we look at the solutions and the public fight over them. Here I only want the concept clear, because you cannot understand the intermittency debate without knowing what the old system assumed.

Which brings me to the misconceptions, and these matter more than usual because they poison the public conversation. The first and most common is the claim that every watt of solar or wind must be matched, one for one, by a watt of conventional backup held in reserve for when the weather fails — that renewables are therefore a doubling of the system, a costly redundancy. This is wrong, and understanding why is worth the effort. The grid is not a collection of individual generators each needing its own understudy; it is a single system, and systems have properties that their parts do not. The wind is not still everywhere at once — a calm in one region is often a gale five hundred miles away, so a geographically spread fleet of wind farms produces far steadier power than any one of them. The sun does not set on the whole continent simultaneously. Wind and solar tend to complement each other across seasons, wind often blowing strongest in winter when the sun is weakest. Demand itself can flex, shifting to follow supply. And storage can carry midday's surplus into the evening's neck. Because of all this, the amount of firm backup a system truly needs is far less than a naive one-for-one accounting suggests. The right question is never "what backs up this one turbine" but "what does the whole system need to stay balanced across all conditions" — and the answer depends on diversity, flexibility, and storage, not on shadowing every panel with a gas plant.

The second misconception is the sweeping claim that intermittency makes renewables impossible — that a grid simply cannot run on sources that come and go, and that the whole enterprise is a fantasy destined to end in blackouts. This overstates the difficulty into an impossibility, and the plainest refutation is that it is already being done. As of the mid-2020s, whole regions run reliably on very high instantaneous shares of wind and solar — Denmark, South Australia, and parts of Germany and Spain routinely meet large fractions, sometimes nearly all, of their electricity demand from the wind and sun for hours at a stretch, and the lights stay on. Intermittency is not a wall; it is an engineering problem, and engineering problems have engineering answers — better forecasting, more transmission, more storage, more flexible demand, smarter markets. That does not make it easy or free. The costs and challenges rise as the renewable share climbs toward the very top, and honest people disagree about how far and how fast the system can be pushed and at what price. But "hard and costly at high penetration" is a completely different statement from "impossible," and the loudest voices often blur the two. The truth sits in between: intermittency is real, it is the central technical challenge of the transition, and it is being solved, imperfectly and expensively but genuinely, right now.

Let me leave you inside the control room one more time, watching the neck of the duck rear up at sunset, because that image holds the whole chapter. Everything about running a grid on sun and wind comes down to that: the resource arrives on nature's schedule, the grid must be balanced on humanity's, and the gap between the two is the problem the entire field exists to close. It is closed with forecasting that turns uncertainty into mere variability, with a geographic spread that smooths the individual gusts and clouds into a steadier whole, with flexible plants and shiftable demand, and above all with storage that lets the fat belly of midday feed the hungry neck of evening. None of it changes the underlying fact that solar stops at night and wind blows as it pleases. All of it is the art of building a system that keeps the bathtub exactly full anyway.

So here is the recap. Solar and wind are variable, producing power when the sky permits rather than when we ask, and electricity is uniquely unforgiving because it must be generated at the very instant it is used, the whole grid balanced second by second like a bathtub kept precisely at the brim. California's duck curve captures the sharpest version of the challenge: a midday glut of solar so large it can drive prices below zero, followed by a steep evening ramp as the sun vanishes and demand peaks together. It is a real and serious problem, but it is an engineering problem, not an impossibility, and it does not require shadowing every renewable watt with a fossil one. To reason about it properly, though, we need one more idea — a number that tells us how much of the time a wind turbine or a solar panel actually delivers, and why a machine rated for a hundred megawatts almost never produces a hundred megawatts. That number is the capacity factor, and it is where we go next.

Chapter 41 — Capacity Factor and the Myth of the Nameplate

On a bright rating bench in a factory outside Aurich, in the flat wind-country of northern Germany, a technician clips a metal plate to the base of a freshly built turbine. The plate is small, hardly bigger than a paperback, and stamped into it is a single decisive number: the machine's rated power, its maximum, the figure the marketing brochures will quote and the newspapers will repeat. Say it reads five megawatts — five million watts, enough in principle to light a small town. That number is true. It is also, in a way I want to spend this chapter unpacking, one of the most misleading true numbers in all of energy. Because the plate tells you what the turbine can do in a single perfect instant, when the wind blows at exactly the right speed and every bearing spins free. It says nothing at all about what the turbine will actually do across a year of real weather. And the gap between those two things — between the nameplate and the harvest — is where more confusion about renewable energy lives than almost anywhere else.

Engineers call that plate the nameplate capacity, and the honest number that lives underneath it they call the capacity factor. The capacity factor is simply the ratio of what a generator actually produces over some stretch of time to what it would have produced had it run flat out at its rated power the whole time. If our five-megawatt turbine ran at full tilt every hour of a year, it would generate a certain fixed amount of energy. It never does. The wind drops, the wind gusts too hard and the machine feathers its blades to protect itself, the turbine sits idle for maintenance. So the real annual output comes in at some fraction of that theoretical maximum — and that fraction, expressed as a percentage, is the capacity factor. For onshore wind today it typically runs somewhere between about 30 and 40 percent. For solar photovoltaic panels it is lower, roughly 15 to 25 percent depending on where you plant them. For offshore wind, out where the breeze is stronger and steadier, it climbs to something like 40 to 55 percent. And for a nuclear power station, which burns its fuel steadily around the clock, it sits up near 90 percent.

Now, the moment those numbers land, a certain reflex fires in a lot of readers, and it is worth naming and disarming right away, because it is the central myth of this chapter. The reflex says: if a solar farm only runs at 20 percent, then it's 80 percent useless — it's broken most of the time, it's wasteful, it's a swindle. I have heard versions of this at dinner tables and read them in serious opinion columns. It is completely wrong, and understanding why it is wrong is the key that unlocks everything else. A low capacity factor is not a measure of failure or waste. It is an expected, designed-in property of a machine that harvests a resource which comes and goes. The Sun sets every night; a solar panel producing nothing at two in the morning is not malfunctioning any more than a fishing net is malfunctioning when the tide is out. The capacity factor bakes that rhythm into a single honest number. It is not the machine's report card. It is a description of the weather.

To see this cleanly we need to separate two ideas that everyday speech smears together: power and energy. This is the first-principles distinction the whole chapter turns on, so let me draw it out with care. Power is a rate — how fast energy flows at a given instant. It is measured in watts, and its multiples climb the familiar ladder: a kilowatt is a thousand watts, a megawatt a million, a gigawatt a billion, a terawatt a trillion. Energy is the accumulated total — power added up over time — and it is measured in watt-hours and their multiples, the kilowatt-hours that appear on your electricity bill. A watt is like the speed on a car's speedometer; a watt-hour is like the distance the car has actually traveled. The nameplate on our turbine is a speedometer reading — the top speed the machine can hit. The capacity factor is what tells you the distance it will really cover, because it accounts for all the hours the machine spends going slower than flat out, or parked.

Here is the analogy I keep coming back to, because it makes the whole thing click. Think of a sprinter and a marathon runner. The sprinter can hit an astonishing top speed — call that his nameplate, his rated power. But he can only hold it for ten seconds. Over the course of a full day, the total ground he covers is modest. The marathoner never comes close to the sprinter's peak speed; her rated power is far lower. Yet she holds a steady pace for hours and, across a day, covers far more distance. Nuclear is the marathoner: an unspectacular but relentless pace, a capacity factor near 90 percent. Solar is a sprinter who is only allowed to run when the Sun is up: brilliant peaks at noon, nothing at night. If you judged the marathoner by her top speed you would call her slow, and you would be a fool. If you judge a solar farm by its capacity factor and call it broken, you are making exactly that mistake — confusing the peak with the total, the speedometer with the odometer.

Which brings us to the practical trap, the one that costs real money and produces real bad journalism: comparing generators by their nameplates alone. You will read that a country has built a one-gigawatt solar farm, and somewhere nearby that it has an old one-gigawatt nuclear plant, and the natural assumption is that the two are equivalent — a gigawatt is a gigawatt. They are not remotely equivalent in what actually matters, which is energy delivered over the year. The nuclear plant, purring along near 90 percent, turns its gigawatt of capacity into roughly nine times what the solar farm produces from the same nameplate at 20 percent. Put the other way around, to match the annual energy of that one nuclear plant you would need to build something like four or five gigawatts of solar nameplate, plus a way to store and shift the power to when it's needed. This is not an argument against solar — solar's per-unit cost has fallen so far that overbuilding the nameplate can still be the cheaper path, as later chapters will show. It is simply an insistence on comparing like with like. When someone tells you how many gigawatts of something were installed, the honest follow-up question is always: at what capacity factor? Multiply the two and only then do you know how much electricity actually arrived.

Let me make it concrete with a worked example, the kind of back-of-the-envelope sum I want you to be able to do yourself. Take a single wind turbine rated at three megawatts, running at a capacity factor of 35 percent — a fair figure for a good onshore site. A year contains 8,760 hours. If the turbine ran flat out the whole time it would produce three megawatts times 8,760 hours, which is 26,280 megawatt-hours of energy. But it doesn't run flat out; it runs, on average across the year, at 35 percent of that. So the real annual harvest is 26,280 times 0.35, which comes to about 9,200 megawatt-hours, or 9.2 million kilowatt-hours. A typical European home might use somewhere around three or four thousand kilowatt-hours a year. So one such turbine, spinning on its own out in a field, quietly supplies the annual electricity of roughly two to three thousand homes. That is the number the nameplate could never have told you. It only emerges when you fold the weather in.

A few things about capacity factors genuinely surprise people, so let me offer three. The first is that the capacity factor of wind turbines has been climbing steadily, and not because the wind got stronger. It's because the machines got taller and their blades got longer, reaching up into faster, steadier air and sweeping a wider circle of it. A turbine installed in the 1990s might have managed a capacity factor in the low 20s; a modern onshore machine on a good site can reach the high 30s or beyond, from the very same wind resource, simply by being built bigger and smarter. The capacity factor, in other words, is partly a measure of engineering, not just geography. The second surprise is how well the best offshore machines do: out in the North Sea, a modern turbine can post a capacity factor above 50 percent, edging into territory that would once have seemed the exclusive preserve of fossil and nuclear plants — because offshore the wind rarely stops. The third surprise cuts the other way, and it's important for honesty: even a fossil-fuel gas plant often runs at a capacity factor well below its nameplate, sometimes only 40 or 50 percent — not because it can't run more, but because the grid only calls on it when it's needed. A low capacity factor, it turns out, isn't unique to renewables at all. It can simply mean a machine is being used as a backup rather than a workhorse.

That last point exposes a subtler confusion worth correcting head-on, because even careful people fall into it: the idea that capacity factor is a kind of efficiency. It is not. Efficiency asks how much of the energy that arrives at a machine gets turned into useful output — how much of the sunlight hitting a panel becomes electricity, how much of the wind's kinetic energy a rotor captures, and there we run into hard physical ceilings we've met earlier in this book, the roughly 33 percent single-junction limit for a simple silicon cell, the Betz limit of about 59 percent for any wind rotor. Capacity factor asks something entirely different: not how well the machine converts the energy it receives, but how often the energy shows up in the first place. A solar panel can be a marvel of efficiency, converting a fifth of the sunlight that strikes it, and still have a capacity factor of 20 percent — because for two-thirds of every day there is simply no sunlight to convert. The panel is not failing at night. There is nothing there to fail at. Conflating the two numbers — treating a 20 percent capacity factor as though it meant the panel wastes 80 percent of the sunlight — is perhaps the single most common error in casual energy talk, and now you can spot it instantly.

There is a related myth that I want to dismantle plainly, because it comes dressed up as a knockout argument: "solar only works 20 percent of the time, so it's 80 percent useless." Two things are wrong with it. First, as we've seen, capacity factor is not uptime — it doesn't mean the panel is dead for 80 percent of the hours; it means that, averaged across all the hours including the productive noon peaks and the empty nights, output comes to a fifth of the theoretical maximum. On a clear day at midday the panel may be running at nearly its full nameplate; that peak just gets averaged in with the zeros of night. Second, and this is the part that critics of renewables and boosters of renewables both sometimes forget, the low capacity factor is already priced in. When engineers quote the cost of solar electricity — the levelized cost that has fallen some 90 percent since 2010, down to a few cents a kilowatt-hour — that figure is calculated by dividing the plant's lifetime cost by the actual kilowatt-hours it produces over its life, capacity factor and all. Nobody is pretending the panel runs day and night. The 20 percent is baked into the price you're quoted. So the fact that the panel sleeps at night has already been paid for, honestly, in the cost per unit. It is not a hidden catastrophe waiting to ambush you; it is an accounting input that was folded in from the start.

It helps to see where the capacity factor actually comes from, hour by hour, because the number that looks so abstract on a spreadsheet is really a story about a machine responding to the sky. Recall the operating regime of a wind turbine we met earlier. Below a certain cut-in speed — a gentle breeze of three or four meters a second — the blades won't turn at all, and the turbine produces nothing. As the wind picks up past that threshold, output climbs steeply, following that ferocious cube law by which doubling the wind speed multiplies the power eightfold. Then, at what engineers call the rated wind speed, somewhere around twelve to fifteen meters a second, the turbine hits its nameplate and holds there, deliberately shedding any extra wind by pitching its blades, because the generator can take no more. And if the wind grows violent, past a cut-out speed of around twenty-five meters a second, the machine shuts down entirely to save itself. Now lay a whole year of real wind against that curve. A great many hours fall in the gentle middle, where the turbine runs at a fraction of its rated power; a modest number of hours are calm enough that it produces nothing; and only a lucky slice of the year brings the steady strong wind that lets it run flat out. Average all those hours together — the zeros, the middling, the full-tilt — and you get the capacity factor. It is not a fudge factor or a discount; it is the true shape of the wind, folded into a single number.

There is one more misconception worth naming before we leave the subject, because it is the mirror image of the first and just as common among enthusiasts as the "80 percent useless" line is among skeptics. It is the idea that a higher capacity factor is always better, full stop — that we should prize the marathoner and scorn the sprinter. But capacity factor on its own tells you nothing about cost, and cost is what actually decides what gets built. A solar farm with a capacity factor of only 20 percent can still produce the cheapest electricity on the grid if the panels are cheap enough, and they now are. What matters in the end is not how high the capacity factor climbs but how few cents each delivered kilowatt-hour costs once you have divided the plant's whole lifetime bill by everything it actually produces. The capacity factor is an input to that calculation, not the verdict. A low number is not a mark of shame any more than a high one is a badge of virtue; both are simply facts about a resource, to be fed into the honest arithmetic of price.

Where do you meet all this in the wild? On any energy news story you read from now on. When a report says a nation added "50 gigawatts of solar last year," you now know to reach quietly for a capacity factor — call it 15 percent in a cloudy northern country, 25 in a sunny one — and multiply, to translate that headline power into real annual energy before you compare it to anything. When a politician holds up a single nuclear plant against a field of turbines and calls the comparison obvious, you know the missing number that would make it honest. And when you see the specification sheet for a rooftop system — "8 kilowatts," it will say, proudly — you know that the 8 kilowatts is the noon-in-June best case, and that what the roof actually banks over a year is that figure stretched across the hours and thinned by every cloud and every night, the capacity factor doing its quiet, honest arithmetic in the background.

So let me set the record straight in plain terms before we move on. The nameplate is a promise about a single perfect instant; the capacity factor is the truth about the year. Power is a rate, energy is the total, and only energy heats your home or lights your street. A low capacity factor is not waste, not inefficiency, and not failure — it is the expected signature of a machine that harvests a coming-and-going resource, and its cost is already counted in the price. The sprinter and the marathoner are simply built for different races. But naming all of this only sharpens the real question, the one that has been lurking under every sentence: if the Sun sets and the wind falls, and the machines fall silent right on schedule, how on earth does a grid built on them keep the lights on? That is where we turn next.

Chapter 42 — Keeping the Lights On

In a low-slung control room outside Copenhagen, a woman named — let us call her the shift operator, for the people who do this work rarely get named — sits before a wall of screens that glow with the whole electrical life of Denmark. One display shows a single number that never stops trembling: the frequency of the national grid, hovering around 50 cycles per second, flickering in the third decimal place like a heartbeat on a monitor. Another shows the wind pouring in from the turbines on the western coast and out in the North Sea; on a good gusty afternoon that figure alone can exceed everything the country is using, and Denmark has run whole days on which wind supplied more than half its electricity and moments at which it supplied it all. Her job, minute by minute, is to keep that trembling frequency number from drifting — to make sure that in every single second, the electricity being generated across the country exactly matches the electricity being drawn out of it. She is, in effect, balancing a spinning plate the size of a nation, and she is doing it with a fuel that answers to the weather rather than to her.

That instantaneous matching is the first principle beneath everything in this chapter, and it is stranger and stricter than most people realize. Electricity, at the scale of a grid, is almost impossible to store in the wires themselves; it must be made at the very instant it is used. If, second by second, generation and demand fall out of step, the grid's frequency begins to slide — too much generation and it rises, too little and it falls — and if it slides too far, protective systems trip, plants disconnect to save themselves, and the failure can cascade into darkness across a region in the space of a few breaths. So the frequency, that number trembling around 50 cycles a second in Europe or 60 in North America, is the grid's vital sign. Keeping it steady means keeping supply and demand married together instant by instant, forever, with no intermission. This was hard enough when the generators were steady coal and nuclear plants you could dial up and down at will. The puzzle this chapter is about is how you do it when a large and growing share of your supply is sunlight and wind, which arrive on their own schedule and cannot be dialed up on command.

The reflexive answer, the one you hear most often, is that you simply can't — that a grid leaning on weather-driven power is a grid courting blackout, and that beyond some modest share the whole thing must fall over. It is a serious-sounding objection and it deserves a serious answer, because the reality is that grid operators around the world are already running systems on very high shares of wind and solar, day in and day out, and keeping that frequency needle rock-steady while they do it. They manage it not with a single silver bullet but with a whole toolkit, half a dozen techniques that each shave a piece off the problem, and it's worth walking through them, because together they are the reason the lights stay on.

The first and in many ways deepest tool is geographic diversity — the simple, saving fact that the wind is nearly always blowing somewhere. A single turbine is at the mercy of a single patch of sky; if the air goes still over one hillside, that machine falls silent. But spread ten thousand turbines across a thousand miles of coastline, plains, and hills, and the stillnesses no longer coincide. A calm over Jutland may sit beside a gale over the North Sea; a windless afternoon in one province overlaps a blustery one three hundred miles east. The wider the area you draw your power from, the smoother the combined flow, because local lulls and local gusts average out against one another. This is why the size of the balancing area matters so much, and why connecting grids together with high-capacity transmission lines is one of the most powerful things you can do — a subject we'll return to in force. A bigger, better-connected grid is a calmer grid, because it can borrow from wherever the weather happens to be cooperating. And modern lines can move power astonishingly far; electricity harvested from wind on a coastal plain can be delivered to a city many hundreds of miles inland, so that the geographic averaging works across whole subcontinents rather than single countries.

The second tool is the happy accident that solar and wind are often complementary — that when one flags, the other frequently rises. Think about what makes a sunny day and what makes a windy one. Bright, still, high-pressure weather is the friend of the solar panel and the enemy of the turbine. Grey, blustery, low-pressure weather is the reverse: little sun, but plenty of wind. And across the year the pattern deepens, because in many temperate regions the winter, when the Sun is weak and low, is precisely the windy season, while the calm of high summer is when sunlight is at its most generous. The two resources are, to a useful degree, mirror images — and a grid that draws on both together sees a flatter, more forgiving supply than either could offer alone. It is like having two suppliers whose off-days rarely fall on the same afternoon.

The third tool is to stop treating demand as a fixed thing that supply must chase, and to let some of the demand move instead. This is called demand response, and the idea is that certain electricity uses don't actually care exactly when they happen. A cold-storage warehouse can be chilled a little harder when power is abundant and coast for an hour when it's scarce. A fleet of electric cars can be told to charge in the small hours, or to pause charging for twenty minutes when the grid is tight. An aluminum smelter can trim its draw for a stretch in exchange for a lower price. None of this dims a single home's lights; it simply shifts flexible loads toward the moments when the wind is blowing and the Sun is up, so that demand bends gently to meet supply instead of the other way round. For a weather-driven grid, a demand that can flex is nearly as valuable as a supply that can.

The fourth tool, for now, is flexible dispatchable generation — plants you can turn up and down quickly to fill the gaps: hydroelectric dams, which can go from nothing to full power in minutes and are the cleanest form of this flexibility, and, still, natural gas plants, which many grids lean on as the fast-responding backup that covers the moments when wind and sun fall short together. This is the part of the current system that renewable advocates are least comfortable with and that honesty demands we name plainly: at today's stage of the transition, most high-renewable grids still keep flexible gas or hydro in reserve to firm up the supply, and much of the story of the coming chapters — storage, long-duration storage, a rebuilt grid — is the story of gradually replacing that fossil backstop with something clean.

It is worth pausing on frequency itself, because it is the quantity that makes all of this so unforgiving, and a small analogy makes the danger vivid. Imagine the whole grid as a great spinning flywheel, kept turning by every generator connected to it, and think of demand as a hand pressing gently on the rim, trying to slow it. As long as the generators push exactly as hard as the hand presses, the wheel spins at a steady rate — the steady frequency. But let demand suddenly outstrip supply, and the hand presses harder than the generators can answer; the wheel slows, the frequency dips. Let supply outstrip demand, and the wheel speeds up, the frequency rises. The wheel is colossal and it cannot be allowed to slow or race by more than a hair, because the machines plugged into it — and the generators themselves — are built to run at one speed. Let the frequency wander too far and protective relays begin snapping generators off the wheel to save them, which removes push, which slows the wheel further, which snaps off more generators, a runaway unwinding that can black out a region in seconds. This is why grid operators speak of frequency with something close to reverence, and why every tool in this chapter is ultimately in service of that one trembling number.

The fifth tool is forecasting, and it has quietly become one of the most important. If you can predict the wind and the sunshine accurately a day or even a few hours ahead, then their variability stops being a series of ambushes and becomes a schedule you can plan around. Modern weather models fed into machine-learning systems can now forecast a wind farm's output for the next day with remarkable accuracy, so operators know in advance when to warm up a reserve, when to charge a battery, when to line up an interconnector. Variable is not the same as unpredictable, and the better we forecast, the more the "unpredictable" part shrinks.

There is a sixth tool, and in a way it is the one that ties all the others together: the size of the balancing area and the strength of the interconnections that stitch neighboring grids into one. A small grid, walled off from its neighbors, must balance every wobble entirely from its own resources; a large, well-connected grid can lean on a whole continent's worth of generation and demand at once. Europe's synchronized grid, spanning dozens of countries, behaves in this respect as a single enormous machine, so that a factory switching on in one nation is quietly absorbed by generators a thousand miles away. Enlarging the balancing area is, in a sense, the master tool, because it amplifies every other one — geographic diversity works better across a continent than a county, forecasting errors in one region cancel those in another, and flexible plants anywhere can rescue a shortage everywhere. It is also, not coincidentally, the tool that runs most directly into the hard limits of building new transmission lines, a bottleneck we will meet again.

Let me put a picture to how these tools work together, because the metaphor helps. A grid operator running a high-renewable system is less like an engineer flipping switches and more like the conductor of an orchestra. No single player holds the whole melody; the wind section swells and fades, the solar section comes in bright at midday and drops out at dusk, the hydro and gas players stand ready to fill any silence, and the demand-response players can be cued to soften their part on a moment's notice. The conductor's whole art is timing — keeping every entrance and exit exactly on the beat so that the sound, the frequency, never wavers. Or think of a juggler who has learned to keep more and more balls in the air not by gripping any one of them harder but by watching all of them at once and nudging each at the right instant. That distributed, anticipatory balancing — many small corrections rather than a few big levers — is the real skill of running a modern grid, and it is a skill that has grown, not shrunk, as the share of weather-driven power has climbed.

It is worth adding that voltage, not just frequency, must be held within tight bounds too — a second dial the operator watches alongside the first. Where frequency is a property of the whole grid at once, voltage is more local, sagging where demand is heavy and far from generation, and it must be propped up region by region with equipment that injects what engineers call reactive power. This used to come, once again, largely from the big spinning plants; in a renewable grid it increasingly falls to the inverters of solar and wind farms to supply it, another old free service that must now be deliberately provided. The point is that keeping the lights on is not one balancing act but several running at once — frequency across the whole system, voltage in every corner of it — all held steady in the same unforgiving, instant-by-instant present.

The evidence that it works is no longer theoretical, and a few facts here genuinely startle people. Denmark has for years drawn around half of its electricity from wind across a full year, and far more than that in windy stretches. South Australia, a grid at the end of a long thin wire, has run for days at a time with wind and solar supplying the overwhelming majority of its power, and has installed a big battery that stabilizes the whole system — more on that in the next chapter. There have been afternoons on which entire regions, and even whole countries for an hour or two, have run on nearly 100 percent renewable electricity, the fossil plants idled, the grid humming along on sun, wind, and water alone. A decade ago serious engineers doubted such moments were possible without the lights flickering; now they pass almost without comment. And the reach of modern transmission still surprises: on the largest high-voltage links, power can travel more than a thousand miles with only modest losses, which means a still evening in one place can quite literally be rescued by a gale a country away.

Which brings us to the misconception that has to be handled with real care, because it is the one people reach for as proof, and getting it right matters more than scoring a point. The claim is that high renewable penetration causes blackouts — and the exhibit most often produced is the catastrophic Texas outage of February 2021, when much of the state lost power in a deadly winter storm. It is worth being scrupulous here. That disaster was not, at root, a renewables story. A brutal cold snap froze equipment across every kind of generation — gas wellheads and pipelines froze, coal piles and instrumentation froze, a nuclear unit tripped, and yes, some wind turbines iced up too, though wind was a small part of the expected winter supply and much of the shortfall was in gas. The deeper causes were a grid that was lightly connected to its neighbors and so couldn't borrow much power from outside, generators that hadn't been weatherized for extreme cold, and market and planning choices made long before the storm. Blaming the blackout on wind is like blaming a multi-car pileup on the one car painted a bright color. The honest lesson of Texas is nearly the opposite of the popular one: it is a lesson about the value of the very things this chapter has described — strong interconnection to neighbors, geographic diversity, weatherized and reliable equipment across all sources, and careful planning. Well-run grids with high renewable shares handle variability every day; poorly connected, poorly prepared grids can fail with any fuel mix. The fuel is rarely the villain; the design and the preparation are what decide whether the lights stay on.

So let me gather the thread. Keeping the lights on means marrying supply to demand in every single second, and holding the grid's frequency steady as a heartbeat while you do it. With a weather-driven supply that job is harder, but it is being done, right now, all over the world, by leaning on geography that averages out the local lulls, on sun and wind that cover for each other, on demand that can flex, on hydro and, for now, gas that fill the gaps, and on forecasts that turn ambushes into schedules — the conductor keeping a hundred players on the beat. The blackouts that critics point to turn out, on inspection, to be failures of connection and preparation rather than of clean power itself. But every tool in this chapter has a limit. Geography and forecasting and flexible backup can smooth the hour-to-hour swings; none of them can, on their own, save the surplus of a windy noon to spend on a still midnight. For that you need to catch the energy and hold it — to store the sunshine itself. That is where we go next.

Chapter 43 — Storing Sunshine

In the flat scrubland near Jamestown, South Australia, in the last weeks of 2017, a row of white steel cabinets went live beside a wind farm and quietly rewrote what a lot of engineers believed a battery could do. It had been built on a dare. The state had suffered blackouts, the billionaire Elon Musk had bet on Twitter that his company could install the world's biggest lithium-ion battery in a hundred days or it would be free, and the bet was on. The machine that resulted — a hundred megawatts of power backed by well over a hundred megawatt-hours of energy, a facility the press nicknamed the Tesla big battery and the engineers called Hornsdale — began doing something no coal plant could. When a distant generator tripped and the grid's frequency lurched, Hornsdale answered in a fraction of a second, faster than any human hand or spinning turbine could react, injecting power to catch the fall before it spread. It made money, it stabilized the grid, and it demolished, almost overnight, the lazy assumption that there was simply "no way to store" electricity. That assumption is the myth this chapter exists to take apart, because storage is the great enabler — the thing that turns sunshine and wind from a resource that comes and goes into one you can spend when you choose.

To think clearly about storage we need two ideas kept firmly apart, because nearly every muddle about batteries comes from mashing them together. The first is power: how fast a store can deliver or absorb energy, measured in watts and their multiples — kilowatts, megawatts, gigawatts. The second is energy: how much it can hold in total, measured in watt-hours, the kilowatt-hours and megawatt-hours we've been counting all along. The analogy I want you to carry through this whole chapter is a water tank. The size of the tank — how many liters it holds — is the energy. The width of the pipe running in and out of it — how fast the water can flow — is the power. You can have a huge tank with a narrow pipe, which holds a great deal but dribbles it out slowly, or a small tank with a fat pipe, which empties in a rush but soon runs dry. When you read that a battery is "a hundred megawatts and a hundred and fifty megawatt-hours," you are being told the width of its pipe and the size of its tank, and both numbers matter, because they answer two different questions: how hard can it push, and for how long. A hundred-megawatt, hundred-and-fifty-megawatt-hour battery can shove a hundred megawatts into the grid for about an hour and a half before the tank runs dry. Miss this distinction and every claim about storage becomes confused; hold onto it and the whole field snaps into focus.

There is a third number that governs any store, and it is the one nature charges you for: round-trip efficiency. No storage is free. When you put energy in and take it back out, some of it is always lost along the way — to heat, to friction, to the chemistry itself. Round-trip efficiency is the fraction that survives the journey. Put a hundred kilowatt-hours into a good lithium-ion battery and you might get about ninety back; the missing ten were the toll for the round trip. This is why storage is not the same as generation and never quite free: every stored kilowatt-hour costs you a little more than a kilowatt-hour to make, because of what the tank swallows on the way through. The better the round-trip efficiency, the smaller that toll, and lithium-ion's high efficiency — comfortably above 85 or 90 percent — is a large part of why it has come to dominate.

And dominate it has, on the back of one of the steepest price collapses in the history of manufactured goods. The cost of lithium-ion batteries has fallen by something like 90 percent over the past decade — a decline driven first by the electric car, which pulled the whole industry to a colossal scale, and then spilling over into the grid. As the price fell, grid-scale battery farms went from curiosities to a boom. As of 2024 the world had installed on the order of 375 gigawatt-hours of grid batteries, with a record of roughly 200 gigawatt-hours added in that single year — the fleet essentially doubling in twelve months. What these installations do supremely well is two distinct jobs. The first is time-shifting over a handful of hours: soaking up the surplus of a sunny midday, when solar farms produce more than the grid can use and prices sag toward nothing, and releasing it into the ravenous early evening, when the Sun has set but everyone comes home, switches on, and demand spikes. A battery lets you buy cheap noon sunshine and sell it back at the expensive dusk — the daily arbitrage that is now the bread and butter of grid storage. The second job is the one Hornsdale made famous: near-instant grid services, responding in milliseconds to catch a frequency wobble, providing the split-second reflexes that keep the whole system stable. That millisecond speed genuinely surprises people; a battery can react to a fault faster than a spinning generator can even begin to change its output, which makes storage not just a bucket of energy but one of the fastest stabilizing tools a grid has ever had.

But — and this is where honesty about limits matters as much as enthusiasm — lithium-ion batteries are, for now, short-duration machines. They are all fat pipe and modest tank. Most grid batteries are built to deliver their rated power for something like one to four hours, and then they are empty. This is not a flaw so much as an economic fact: the pipe (the power electronics) is relatively cheap to enlarge, but the tank (the cells themselves) is not, and the cost of the tank scales straight up with how many hours of storage you want. Four hours of lithium storage is affordable and everywhere. Forty hours starts to strain the economics. Four hundred hours — the kind of storage you'd need to ride through a long, dark, windless winter spell — is, with today's lithium chemistry, wildly, prohibitively expensive. Which sets up a crucial correction to a hope people often voice: no, today's batteries cannot do seasonal storage. They cannot bank the surplus sunshine of July to spend in the gloom of January. That is a genuinely different and harder problem, and it has its own chapter waiting. Batteries are a superb answer to the daily rhythm and a poor answer to the seasonal one, and pretending otherwise helps no one.

Now let me correct the misconception this chapter opened against, because the deeper irony is that the technology which really dominates energy storage is not the fashionable lithium battery at all, and it is more than a century old. It is pumped hydro. The idea is almost comically simple: when you have surplus electricity, use it to run pumps that push a vast quantity of water from a lower reservoir up to a higher one. The water now sits at altitude, holding potential energy the way a wound spring holds tension. When you need the power back, you let the water fall down through turbines, and it generates electricity on the way down, exactly like a conventional hydroelectric dam. The upper reservoir is the tank; the flow of water is the pipe; and the round-trip efficiency of a good pumped-hydro scheme is excellent, often around 80 percent or better. This is the water-tank analogy made literal — the store really is a tank of water uphill. And here is the fact that upends the whole popular picture: pumped hydro accounts for roughly 90 percent or more of all the grid energy storage installed on Earth. The lithium batteries make the headlines; the water uphill quietly does most of the work, and has been doing it since long before anyone owned a mobile phone. "There's no way to store electricity" is not merely wrong; it has been wrong for a hundred years.

The scale of pumped hydro, once you notice it, is genuinely humbling, and a couple of figures bring it home. Some of these schemes hold not megawatt-hours but gigawatt-hours — thousands of times the energy of a single battery container — because the "battery" is a lake. Consider what it takes: to store a serious quantity of energy you must lift a truly staggering tonnage of water a few hundred meters into the air, and the reason it works is that there is so very much water. A large pumped-hydro station can pour out its gigawatt of power for the better part of a day before the upper reservoir runs dry, a feat of endurance no battery farm on Earth can yet match. This is why, when people speak of storage as though it were a brand-new frontier waiting on some breakthrough, engineers who know the field tend to smile. The frontier was crossed in the 1890s, when the first pumped-storage plants appeared in the Alps, and the technology has been quietly maturing ever since. What is new is not storage itself but cheap, fast, modular storage — the battery — that can be dropped anywhere without a mountain and a lake.

There is a human dimension to the battery revolution worth pausing on, because it explains why the price collapsed so fast. Grid storage did not, in the main, drive its own cost down. It rode down on the back of the electric car. For a decade the world's automakers and battery giants poured tens of billions into lithium-ion manufacturing to put batteries into vehicles, and that colossal investment drove the learning curve — the same kind of relentless cost decline we saw with solar panels — steeper and faster than anyone in the power industry could have managed alone. The grid simply reaped what the car industry sowed: as cells rolled off ever-larger production lines for cars, they became cheap enough to stack in shipping containers beside a solar farm. It is one of the transition's quiet ironies that the machine most responsible for making grid storage affordable was never a grid machine at all. It was the automobile.

Pumped hydro has its own limits, of course — it needs the right geography, two reservoirs at different heights within reach of each other, and building a new one is a slow, mountainous, expensive undertaking. And so alongside the two workhorses, a whole menagerie of other storage ideas is being tried, each aimed at a niche the incumbents don't fill. Flow batteries store their energy in tanks of liquid electrolyte, and because you can make the tanks as big as you like independently of the power electronics, they decouple the pipe from the tank in a way lithium can't — promising cheaper long-duration storage, if their cost can be brought down. Thermal storage banks energy as heat, in molten salt or hot rocks or even bricks, to be drawn back later as heat or electricity. Gravity storage revives the pumped-hydro principle without the water, hoisting massive weights up towers with surplus power and letting them descend to generate. Compressed-air storage pumps air into underground caverns under pressure and releases it through turbines. Most of these remain small or experimental, but together they represent a healthy hedge — a recognition that no single technology will fit every duration and every place.

Let me ground all this in a tangible, worked example of the kind I've tried to give throughout. Picture the most common building block of modern grid storage: a battery packed into a standard shipping container, the sort you see stacked on ports and freight trains. One such container typically holds a few megawatt-hours of energy — let's say three megawatt-hours, which is three thousand kilowatt-hours, as a round working figure. Now recall that a typical home might draw something on the order of one kilowatt on average through an evening. So that single container, fully charged, holds roughly three thousand kilowatt-hours — enough, in principle, to carry about a thousand homes through three hours of that evening peak, or a few hundred homes through a longer stretch, before it needs recharging from the next day's sun. Line up dozens of those containers in rows, as the big battery farms do, and you have a facility that can swallow a solar farm's midday glut whole and pour it back out across the evening for a small city. The mental picture is worth keeping: a battery farm is just a field of shipping containers, each one a tank of stored electricity, sized in the same kilowatt-hours that tick over on your own meter at home.

One more limit deserves an honest mention, because storage is sometimes sold as though it were pure magic: batteries wear out, and they lose a little energy every time they cycle. A lithium cell degrades gradually with use, holding slightly less each year, so that a battery farm rated at a certain energy today will hold measurably less a decade hence — a real cost that has to be counted into every honest reckoning of what stored power actually costs. And the round-trip toll we met earlier never goes away: every kilowatt-hour that passes through a store arrives a little diminished. None of this is disqualifying — the numbers still work handsomely for the daily job — but it is why storage is a genuine cost added to the system, not a free pass, and why the cheapest clean grid uses as little of it as it can get away with rather than as much as possible. Storage is a tool to be used sparingly and well, not a wand to be waved.

Where do you meet this today, beyond the news? In more places than you might guess. If you have solar panels and a home battery on your wall, you are running a miniature version of the noon-to-dusk time-shift in your own garage, storing the day's surplus to run your lights through the evening. Utility battery farms now sit quietly beside solar and wind installations on every inhabited continent, and in California and Texas they have grown large enough to reshape the evening electricity market, blunting the price spike that used to greet every sunset. And the pumped-hydro schemes have been there all along, their reservoirs tucked into hills from Wales to Virginia to the Alps, filling by night and emptying by day, the oldest and largest batteries in the world hiding in plain sight as lakes.

So let me gather the recap. Storage is the enabler that lets a weather-driven grid spend its energy on its own schedule rather than the sky's. It runs on two numbers kept firmly apart — power, the width of the pipe, and energy, the size of the tank — and it always pays a toll in round-trip efficiency for the round journey in and out. Lithium-ion batteries, 90 percent cheaper than a decade ago and able to respond in milliseconds, are superb at the daily job of shifting a few hours of power and stabilizing the grid, while century-old pumped hydro quietly holds most of the world's stored energy as water uphill, and a bestiary of flow, thermal, gravity, and compressed-air schemes stands ready for the niches between. The claim that we simply cannot store electricity is one of the most durable and most wrong ideas in all of energy. But every technology in this chapter shares one hard boundary. They are, at heart, machines for shifting energy across hours, or at best a day or two. None of them can yet reach across the seasons, to save the abundant sunshine of summer for the long, dark, windless week of deepest winter. That gap — the genuinely unsolved, genuinely hard part of the whole story — is where we turn next.

Chapter 44 — The Long-Duration Problem

There is a word the Germans use for the weather that keeps energy planners awake at night, and it is worth saying aloud: Dunkelflaute. It means, roughly, "dark doldrums" — a dull, windless, overcast spell that can settle over northern Europe in the depths of winter and refuse to budge for days. The Sun barely clears the horizon and hides behind cloud; the air goes slack and the turbines drift to a stop. Picture such a week descending on a continent that has staked its electricity on sun and wind. The solar farms produce a trickle. The wind farms produce almost nothing. And the battery fleet we spent the last chapter admiring — those fields of shipping containers that so gracefully shift a sunny noon into a busy dusk — runs flat in an afternoon and then sits there, empty, for six more days. This is the long-duration problem, and it is the genuinely hard part of the whole renewable story, the part where honest enthusiasts stop smiling and reach for their calculators. Everything up to now has been about shifting energy across hours. This chapter is about bridging days, and even seasons — a different beast entirely.

To feel why it is a different beast, return once more to the water tank, because the economics turn on a distinction we drew in the last chapter and must now push harder. A storage system has a pipe and a tank — power and energy — and crucially, they cost differently. For a lithium battery, the pipe is relatively cheap and the tank is expensive, and the tank's cost scales straight up with duration. Doubling the hours you want to store roughly doubles the cost of the cells. This is perfectly tolerable when you are buying four hours of storage. It becomes ruinous when you are buying a week. To carry a grid through a seven-day Dunkelflaute with lithium batteries, you would need to build a tank not four hours deep but well over a hundred hours deep — thirty or forty times as much stored energy per unit of power — and the bill scales up almost in lockstep. The very technology that is a bargain for the daily rhythm becomes economic madness for the seasonal one. That, in a sentence, is why "just add more batteries" is the wrong answer to the long-duration problem, and it's the first misconception this chapter has to retire. More batteries is exactly the right instinct for shifting hours and exactly the wrong one for bridging seasons, because the cost of the tank grows with every hour you demand of it.

And the seasonal gap is not a small version of the daily one; it is vastly larger, and this genuinely surprises people who have only thought about the day-night cycle. Consider the shape of the year in a temperate country. Sunlight is feeble in winter — the days are short, the Sun sits low and weak, and a solar panel that drenches you in power in June produces a fraction of that in December. Yet winter is precisely when electricity demand is highest, when the cold bites and the lights burn from mid-afternoon. So the seasonal mismatch runs exactly the wrong way: least sun when you need the most power. Wind helps, because in many places winter is the windy season, which is part of why the two resources pair so well — but wind has its own long lulls, and the Dunkelflaute is defined by the failure of both at once. The upshot is that the amount of energy you would need to store to move summer's surplus into winter's deficit is not a little bigger than the daily storage need. It is enormous — orders of magnitude larger, a reservoir of energy so vast that filling it with batteries is not even worth costing out. The daily problem asks you to save a few hours. The seasonal problem asks you to save a season. They are not the same problem wearing different hats.

So if batteries can't do it, what can? The honest answer is that there are several candidate paths, none of them yet mature, and the field has not decided among them — which is itself the answer to the second misconception, the gloomy one that says: the problem is unsolved, therefore renewables cannot scale, therefore the whole project is doomed. That does not follow. "Unsolved" is not "unsolvable," and here there is not one promising path but four, any of which, or any blend of which, could carry the day. The uncertainty is about which will win, not whether any can. Let me walk through them, because they are genuinely different bets.

The first path is almost embarrassingly blunt: overbuild and curtail. Instead of storing the energy for the lean times, simply build so much wind and solar capacity that even on a gloomy winter's day the vastly oversized fleet still produces enough to keep the lights on. On sunny, windy days such a fleet would produce far more than anyone could use, and you would deliberately throw the surplus away — curtail it, spilling the excess the way a full reservoir spills water over a dam. To the thrifty mind this sounds wasteful and even offensive: building three or four times the solar you "need" and then binning most of its output on good days. But here is the counterintuitive fact that makes it serious, and it flows straight from how cheap solar and wind have become. When the energy itself costs almost nothing to produce, wasting it is cheap; it is the storage that is expensive. Past a certain point it can genuinely be cheaper to build a great surplus of near-free generation and curtail the excess than to build the colossal store that would save it. Curtailment is not failure; at high renewable shares it is a rational design choice, a way of buying reliability with cheap steel and silicon instead of expensive chemistry. The overbuild path trades the hard problem of storage for the easier problem of building lots of panels.

The second path is transmission at continental scale — the geographic diversity of an earlier chapter, pushed to its logical extreme. A Dunkelflaute is a regional event; it settles over a patch of the continent, not the whole planet. Somewhere beyond its edges the wind is blowing or the Sun is shining. If you lace the continent together with enough high-capacity power lines, you can import your way through the dark spell, drawing on a sunny south while the cloudy north sits becalmed, or on a windy Atlantic seaboard while the interior goes still. A continent wired tightly enough behaves almost like a single vast machine in which local weather cancels out. The catch is that these are transmission lines of a scale and length that dwarf anything we routinely build today, crossing borders and jurisdictions, and building them is as much a political and permitting challenge as an engineering one — a theme that will dominate the next chapter.

The third path is long-duration storage of a fundamentally cheaper kind than lithium — technologies built from the ground up to make the tank cheap even when the pipe is modest, which is exactly the opposite of lithium's economics. Flow batteries, which store energy in tanks of liquid electrolyte you can simply make bigger, are one candidate. More striking is a chemistry that has drawn a great deal of attention: the iron-air battery, which stores energy by, in effect, rusting iron in a controlled way and then reversing the rust. Iron is one of the cheapest and most abundant materials on Earth, and an iron-air battery is slow and heavy and would make a hopeless car battery — but it doesn't need to be fast, it needs to be cheap and deep, holding power not for four hours but for a hundred. These technologies are aimed squarely at the multi-day gap, and several are moving from the laboratory toward first commercial plants, though it is too early to know which, if any, will win on cost.

The fourth path reaches back, remarkably, to an idea we met near the start of this book — to the Danish schoolmaster Poul la Cour, who more than a century ago used the surplus power of his experimental windmill to split water into hydrogen and oxygen, storing the wind as a bottled gas. That is green hydrogen, and it is the leading candidate for true seasonal storage. The logic is elegant: in summer, when solar and wind pour out more electricity than anyone needs, use the surplus to run electrolyzers that split water into hydrogen; store that hydrogen — in tanks, in salt caverns, in the vast underground spaces that once held natural gas; and in winter, burn it or run it through fuel cells to make electricity again. It is, quite literally, storing summer sunshine as a chemical to be spent in the dark of winter — power-to-gas, the modern realization of la Cour's hundred-year-old intuition. Hydrogen's great virtue is that the "tank" can be enormous and cheap — an underground cavern costs little per unit of energy stored — which is exactly what the seasonal problem demands. Its great vice, and honesty requires dwelling on it, is inefficiency: the full round trip from electricity to hydrogen and back to electricity loses well over half the energy, far worse than a battery's modest toll. And the whole chain — electrolyzers, storage, fuel cells — remains expensive and immature, with green hydrogen costing several times what planners hope it eventually might. It is a serious candidate for the seasonal niche precisely because nothing else stores such vast quantities so cheaply, and a doubtful one everywhere that a battery would do, because it wastes so much on the way through.

It is worth dwelling for a moment on how these paths might combine, because the real future will almost certainly not pick just one. Picture a grid a few decades hence that leans mainly on wind and solar. For the daily rhythm it uses lithium batteries, shifting each afternoon's surplus into the evening — the cheap, solved part. For the ordinary run of cloudy or calm days, a few days at a stretch, it draws on some cheap deep store, iron-air or a flow battery, sized in the tens of hours. It is wired to its neighbors by fat transmission lines, so that a regional lull is met by imports from wherever the weather is kinder. It is deliberately overbuilt, so that even a mediocre day yields enough, and on the good days it cheerfully spills the excess. And for the rare, brutal, continent-wide Dunkelflaute that outlasts even all of that — the once-a-winter siege of dark and calm — it falls back on a reserve of hydrogen made from summer's spilled sunshine, or, in the transition years, on the gas plants it has not yet retired. No single technology carries the whole load; each covers the timescale it is cheapest at, handing off to the next as the calm stretches longer. The art is in the layering, and the layering is why "which technology wins" is slightly the wrong question. Several will win, each at its own duration.

The economics of this layering hide a fact that surprises almost everyone the first time they meet it, and it is worth stating baldly: for the longest, rarest gaps, it is often cheaper to build capacity you will hardly ever use — or to waste energy you have already made — than to store energy for the moment you might need it. A store built to cover a once-in-three-years week of darkness sits idle almost all the time, earning nothing, its cost spread over a tiny sliver of use. Against that, spilling the surplus of a thousand sunny days is close to free, because the sunlight cost nothing to begin with. This inversion — waste being cheaper than thrift — runs against every instinct we have about energy, formed in an age when fuel was precious and every joule was dear. In a world where the fuel is free and falls from the sky, the calculus flips, and the cheapest reliable system is often a deliberately wasteful one. Learning to be comfortable with throwing away free energy may be one of the harder mental adjustments the transition asks of us.

Notice what these four paths have in common, and it is the deep first principle of the whole chapter: the seasonal problem is an energy problem, not a power problem. For the daily grid we mostly worry about power — can we push enough watts at the evening peak. For the seasonal grid we worry about sheer quantity — can we hold enough total energy to last a dark week or bridge a lean season. And the economics of storage split exactly along that seam. Technologies where the tank is cheap even if the pipe is modest and the efficiency mediocre — hydrogen in a cavern, iron slowly rusting, water uphill — are the ones that suit long duration, because when you need a hundred hours of energy, the size and cost of the tank is everything and the width of the pipe hardly matters. Lithium, brilliant for delivering great power over a few hours, is simply the wrong tool, not because it is bad but because it is optimized for the opposite job. Match the tool to the timescale and the confusion clears.

Where do you meet the long-duration problem today? You meet it in the news every time a cold, calm, cloudy week settles over Europe and electricity prices spike as grids fire up their remaining gas plants to fill the gap — that gas is currently the seasonal backstop, and replacing it is the whole point of everything in this chapter. You meet it in the pilot plants now being built: the first commercial iron-air installations, the green-hydrogen "hubs" being planned around industrial ports, the salt caverns being surveyed for hydrogen storage. And you meet it, quietly, every time an energy planner argues about whether to build more transmission, more overbuild, or more storage — because that argument, at bottom, is an argument about how to solve the long-duration problem, and no one yet knows the cheapest answer.

So let me set the record straight before we move on. The hours-long problem is essentially solved: batteries and pumped hydro handle the daily rhythm well and cheaply. The days-and-seasons problem is not solved, and it is genuinely hard, because the seasonal energy gap dwarfs the daily one and the cheap daily tool, lithium, becomes ruinous when stretched across a week. But "hard and unsolved" is not "impossible," and there is not one path forward but several viable ones — overbuilding cheap generation and cheerfully wasting the surplus, wiring a continent together so the weather cancels out, cheap deep storage like iron-air and flow batteries, and green hydrogen storing summer's sun as winter's fuel in a hundred-year-old echo of Poul la Cour. "Just add more batteries" is the wrong answer, and "it's unsolved so renewables can't scale" is a false conclusion; the truth sits in between, in a set of real options still competing to be cheapest. And note how many of those options — the continental transmission lines above all — depend not on some new chemistry but on rebuilding the electrical grid itself, the aging one-way network we inherited from the age of coal. That grid, and why it must change, is where we turn next.

Chapter 45 — The Grid Must Change

Somewhere in a utility office in the American Midwest sits a spreadsheet that ought to be famous, because it captures the real bottleneck of the energy transition better than any photograph of a turbine ever could. It is the interconnection queue — the list of wind farms, solar farms, and battery projects that have applied to plug into the grid and are waiting for permission. The list is staggeringly long. Across the United States, well over a terawatt of proposed generation and storage — more than the entire existing capacity of the country's power plants — sits in these queues, financed, engineered, ready to build, and stuck, sometimes for five years or more, waiting in line for the studies and the wires that would let it connect. Read that again: the machines are not the problem. The panels are cheap, the turbines are proven, the money is willing. What they are all waiting for is the grid — the aging web of wires that was never designed for them, and that must now be rebuilt on a scale and at a speed the industry has not attempted in generations. This chapter is about that rebuilding, and about the uncomfortable truth that the single biggest obstacle to renewable energy in the rich world today is not technology at all. It is the grid, and the permission to build it.

To understand why the grid must change, you have to understand what it was built to be. The electrical system we inherited is a creature of the fossil age, and its whole architecture reflects the machines it was born to serve. Picture a handful of enormous central power stations — coal plants, big hydro dams, later nuclear stations — each one a giant sitting at a fixed point, burning fuel or holding back a river, generating a colossal steady stream of power. From those giants, electricity flowed outward and downward: up onto high-voltage transmission lines that carried it long distances, then stepped down through substations onto the lower-voltage distribution lines that fanned out to homes and factories. Power moved one way, from the few big producers to the many small consumers, like water flowing downhill from a few great reservoirs through an ever-branching delta. The whole system was engineered around that assumption — one-directional, centralized, and dispatchable, the giants at the top turned up and down on command to match whatever the consumers below happened to draw.

A renewable grid violates every one of those assumptions, and that is why it cannot simply be poured into the old pipes. It is distributed rather than centralized: instead of a few giants, it has millions of sources — vast wind farms on remote plains, solar arrays in far deserts, and, remarkably, solar panels on the roofs of ordinary houses. It is two-way rather than one-directional: when your neighbor's rooftop panels produce more than the house is using at midday, the surplus flows backward, up the distribution line that was only ever designed to carry power down. The delta is being asked to run in reverse. And it is weather-driven rather than dispatchable: its output follows the sky rather than the operator's command. An electrical grid built for a dozen obedient giants is now being asked to knit together millions of small, scattered, fickle, two-way sources — and the wonder is not that it strains, but that it works as well as it does. Which brings us to the misconception that has to go first: the idea that renewables are "plug and play," that you can simply build the panels and turbines and slot them into the existing grid like appliances into a wall socket. They are not, and you cannot. The grid itself needs deep and expensive rebuilding, and pretending otherwise is why those interconnection queues have grown so monstrously long.

The most urgent piece of that rebuilding is transmission — new long-distance power lines, and lots of them. Here geography is the whole story. The best wind blows where few people live, on the open plains and out at sea; the best sunlight falls on deserts far from any city. But the electricity is needed in the cities, hundreds of miles away. So a renewable system depends, far more than the old fossil one, on moving enormous quantities of power across great distances — from the windy plains and the sunny deserts to the population centers that will burn it. That means building thousands of miles of new high-voltage lines, and it is here that a particular technology earns its keep: high-voltage direct current, or HVDC. Most of the grid runs on alternating current, which is ideal for local distribution, but over very long distances alternating current leaks energy. Direct current at high voltage leaks far less, which makes HVDC the tool of choice for the long hauls — the electrical superhighways that can carry a gale's worth of power from a distant coast to an inland metropolis a thousand miles away with only modest losses. A future built on remote wind and desert sun is a future laced with these HVDC superhighways, and building them is one of the great construction projects the transition demands.

Below the great highways, the humble distribution grid — the local wires on the poles outside your window — must also grow smarter. It was built to be dumb and one-way: send power down, and never think about it again. Now it must cope with power flowing both ways as rooftops feed in, with electric cars drawing heavily at unpredictable hours, and with the need to sense and respond in real time to all of it. This is the "smart grid" — distribution laced with sensors, controls, and communication so that it can manage two-way flows, balance loads street by street, and orchestrate the millions of small devices, from home batteries to car chargers, that now hang off its edges. The one-way road is becoming a two-way web, and a web needs traffic signals the old road never did.

Now we come to the most subtle piece of the rebuilding, and the one that most rewards a first-principles explanation, because it is a hidden service most people have never heard of and yet it holds the whole grid together: inertia. Recall from an earlier chapter that an alternating-current grid runs at a precise frequency — 50 or 60 cycles per second — and that keeping this frequency steady, instant to instant, is the essence of keeping the lights on. In the old grid, that steadiness came almost for free, as a happy side effect of physics. The great fossil and hydro plants generate power by spinning massive metal rotors, tons of steel whirling in perfect time with the grid's frequency. Those spinning masses carry enormous rotational momentum — inertia — and that inertia acts like a giant flywheel on the whole system. When something disturbs the balance — a plant trips off, a surge of demand hits — the spinning masses resist the change by their sheer momentum, releasing or absorbing energy for a few crucial seconds and slowing the frequency's drift long enough for operators and reserves to respond. This stabilizing inertia was never bought or planned; it simply came along, unbidden, with every spinning generator. It was the shock absorber nobody knew they were relying on.

The trouble is that solar panels and wind turbines do not provide it — at least not naturally. A solar panel has no spinning mass at all; it feeds the grid through an inverter, a piece of power electronics that converts its direct current into grid-matched alternating current. A wind turbine does spin, but it is decoupled from the grid's frequency through its own electronics. So as the great fossil flywheels are retired and replaced by inverter-based sources, the grid's natural inertia drains away, and with it the shock absorber that used to buy those critical seconds. A grid with too little inertia is a twitchy, brittle grid, where a single fault can send the frequency plunging faster than anything can catch it. This is one of the genuinely hard engineering challenges of a high-renewable system, and it is a large part of what the interconnection studies in that long queue are actually studying.

The answer is to make the electronics do deliberately what the spinning steel once did by accident. Modern inverters can be built to be "grid-forming" rather than merely "grid-following" — instead of passively syncing to a frequency the giants set, they can actively help set and hold the frequency themselves, standing the whole logic of the grid on its head. And they can be programmed to provide synthetic inertia: sensing a frequency wobble and injecting or absorbing power in milliseconds to mimic the steadying push of a physical flywheel, drawing on batteries or on the wind turbine's own spinning blades to do it. The grid is learning to manufacture, through clever software and fast electronics, a stabilizing service that it used to get for free from tons of whirling metal. It is a quiet revolution in how a grid holds itself together, invisible to everyone but the engineers, and it is essential to a system without the old giants at its heart.

To feel how strange the loss of inertia really is, consider a demonstration that has actually played out on real grids. On systems like Ireland's or South Australia's, which at times run on very high shares of wind and solar, operators once had to impose a hard rule: a minimum number of conventional spinning generators had to be kept running at all times, burning fuel they did not strictly need, purely to keep enough inertia on the wheel. The renewables could have supplied the power; what they could not supply was the shock absorber. Operators were, in effect, paying to keep flywheels spinning. It was only as grid-forming inverters and fast batteries matured that these grids could begin to relax the rule, letting synthetic inertia stand in for the real thing and pushing the fossil floor lower. South Australia's big battery, the one we met stabilizing frequency in an earlier chapter, was among the first machines to prove that a box of electronics could do a spinning generator's stabilizing job — and do it faster. Watching that happen, engineers realized that inertia, long treated as an immovable law of the grid, was really just a service, and services can be provided in more than one way.

There is a second hidden service the old giants supplied for free, and it deserves a mention because it too must now be manufactured deliberately: the ability to ride through a fault and to restart the grid after a total collapse. A large conventional plant can, in principle, help bring a dead grid back to life from a standing start — a capability engineers call black start — and can hold steady through the electrical jolt of a nearby fault without tripping off. Inverter-based sources had to be taught these tricks one by one, in software, and the teaching is still going on. It is a reminder that the grid we inherited was not just a set of wires but a bundle of quiet, unpriced services — inertia, fault ride-through, black start, voltage support — that came free with spinning steel and now must be re-created, deliberately and at cost, in a system without it. The panels and turbines were the easy part. Reproducing everything the old machines did without anyone noticing is the deep and unglamorous work.

The analogy I find most honest for all of this is that we are upgrading a one-way road network into a smart, two-way web. The old grid was a system of one-way streets: power poured down from the reservoirs of a few big plants to the consumers below, and traffic never ran the other way. The new grid must be a dense web in which power flows in every direction — up from rooftops, across from distant wind farms, back and forth as batteries charge and discharge — coordinated by sensors and signals at every junction, with a nervous system of controls that the old road never possessed. You do not build that web by widening the old streets. You rebuild the network, from the great highways down to the local lanes, and you give it a brain.

A single figure captures how far behind the wires have fallen. Study after study of deeply decarbonized grids concludes that the amount of transmission a country needs must roughly double, or more, over the coming decades — and yet the pace at which new high-voltage lines are actually being built has, in many rich countries, barely moved for years, or even slowed. Generation can be built in months; a major transmission line can take a decade or more from first proposal to first current, most of it consumed not by construction but by permitting, litigation, and the patient assembly of consent from every jurisdiction along the route. The result is the paradox of our moment: the cheapest electricity in history, ready to build, dammed up behind a shortage of wires. It is as if we had invented an abundant new crop and then discovered we had no roads to carry it to market.

Which brings us, finally, to the misconception that matters most for anyone trying to understand why the transition is slower than the falling price of panels would suggest. It is tempting to assume the remaining obstacle is technological — that we are waiting on some breakthrough in the machines. Mostly, we are not. The panels and turbines are cheap and proven; even the grid-forming inverters and the HVDC lines are engineering we know how to build. The real obstacle, in the rich world above all, is permitting and politics. Building a long transmission line means winning approval across every county, state, and landowner it crosses, running a gauntlet of environmental reviews and local objections that can stretch a project out for a decade or kill it outright. The interconnection queues are clogged not for want of technology but for want of process, wires, and permission. It is entirely possible to build a grid that runs on sun and wind; what is proving genuinely hard is getting permission to build the wires that would carry it. The bottleneck is not in the laboratory. It is in the planning office and the courtroom, and no amount of cheaper silicon will clear it. That is a sobering thought, but also, in its way, a hopeful one — because a problem of permission and politics is a problem we already know how to solve when we decide it matters, in a way we can never guarantee for a problem of physics.

So let me gather the whole thread. The grid we inherited was built for the fossil age — centralized, one-way, dispatchable, a delta carrying power downhill from a few great giants to the many. A renewable grid is the opposite in nearly every respect — distributed, two-way, weather-driven — and it cannot be poured into the old pipes. Making it work means a vast build-out of new transmission, HVDC superhighways to move remote wind and desert sun to the cities, a smarter two-way distribution grid at the local level, and grid-forming inverters that manufacture, in software and fast electronics, the stabilizing inertia the spinning giants used to supply for free. None of it is plug and play, and the deepest obstacle turns out not to be technology but the slow, contested business of getting permission to build. The road must become a web, and someone must be allowed to lay it. We have followed the electron now from the photon that freed it, through the cell and the blade, onto the grid and into storage, and up against the hard limits of season and inertia. It remains to see how far this can be pushed — out past the coastline to the giant machines rising from the sea, and up to the utility scale where the whole system is headed next.

Chapter 46 — Offshore Wind

In the autumn of 1991, a squat little tugboat nosed out from the Danish coast near the island of Lolland, trailing a barge, and set eleven white turbines into the shallow, grey water of a bay called Vindeby. Each one stood on a concrete base that had been floated out and sunk into the seabed like a giant plant pot. They were small by any standard — rotors about the width of a tennis court is long, generators rated at 450 kilowatts apiece, which is to say that all eleven together, running flat out, could produce not quite five megawatts, roughly the appetite of a large supermarket. Fishermen grumbled. Engineers held their breath. Nobody was entirely sure the machines would survive a North Sea winter, let alone twenty of them. They survived twenty-five. When Vindeby was finally dismantled in 2017, it had spent a quarter of a century proving a single, world-changing proposition: that you can plant a wind farm in the sea and it will pay you back.

I begin with Vindeby because it is easy to forget how radical the idea once seemed, and how completely it has been vindicated. Today a single offshore turbine can out-produce the entire Vindeby farm several times over, and the waters off northern Europe bristle with thousands of them. But the logic that sent those first eleven machines to sea is the same logic that drives the industry now, and it is worth stating plainly, because it explains everything that follows.

The reason to go offshore is the wind itself. Over open water there are no hills, no forests, no cities to snag and slow the moving air. The sea surface is smooth, so the wind blows faster close to the ground — or rather, close to the water — than it ever would over rough land. And it blows more steadily, with fewer of the lulls and gusts that make onshore wind so fickle. Recall the single most important fact about wind power, the one we met when we first stood beneath a turbine: the energy in moving air rises with the cube of its speed. Double the wind speed and you get eight times the power. So a stretch of sea where the wind averages, say, ten meters per second instead of the seven or eight you might find at a good inland site is not a little better. It is dramatically, disproportionately better. A modest gain in average wind speed translates into a large gain in energy harvested, and out at sea those gains are there for the taking.

There is a second reason, and it is about geometry as much as meteorology. Out on the water you have room. A turbine's swept area — the disk its blades trace as they turn — grows with the square of the blade length, so a rotor twice as long captures four times as much wind. Bigger is better, relentlessly, and the sea lets you build big in a way the land never quite does. There are no neighbors to shadow, no roads too narrow to truck a hundred-meter blade down, no ridgeline that runs out. You can float a component that would be impossible to move by highway. As a result, the largest turbines in the world are all offshore machines, and they have grown to a size that is genuinely difficult to picture until you are standing next to one.

Consider the scale. A modern offshore turbine's blade tip, at the top of its arc, can reach well over 250 meters above the waves. The Eiffel Tower's first observation level — the one with the restaurants, where the tourists stop — sits at about 57 meters. The whole tower, to the tip, is around 330 meters. A big offshore turbine reaches past the first level, past the second, and is closing on the summit. In October 2024, the Chinese manufacturer Dongfang rolled out a machine rated at 26 megawatts, with a rotor about 310 meters across — wider than the Eiffel Tower is tall. A single turbine of that class, the engineers claimed, could generate on the order of 100 gigawatt-hours of electricity in a year. Hold that number for a moment, because we are going to unpack it.

And there is a third reason to go to sea, the least glamorous but perhaps the most consequential: that is where the people are. Roughly forty percent of humanity lives within a hundred kilometers of a coast. The great load centers of the world — London, Tokyo, Shanghai, New York, the German industrial belt, the Chinese seaboard — press right up against the water. Onshore wind is often strongest in remote, empty places, far from the cities that need the power, so you must string long transmission lines across the countryside to deliver it, fighting landowners and permitting boards the whole way. Offshore wind sits just over the horizon from the demand. The electricity has a short journey home.

So the case is compelling: stronger wind, steadier wind, room for giants, and proximity to the load. Why, then, did it take until 1991 for anyone to try, and why is offshore power still, decades later, more expensive than its onshore cousin? The answer is a single unforgiving word. The sea.

Here is a first-principles way to feel the difference. Onshore, a turbine stands on a patch of solid ground; you pour a concrete footing, bolt the tower to it, and you are essentially done with the foundation. Offshore, the "ground" is thirty or forty meters of cold saltwater and then a seabed of sand, clay, or rock that you cannot see. To make a fixed-bottom turbine — the kind that has dominated the industry so far — you must reach down through all that water and anchor the machine to the earth beneath. The commonest solution is brutally simple in concept: the monopile, a single vast steel tube, sometimes ten meters in diameter and hundreds of tonnes in weight, hammered into the seabed by a hydraulic pile-driver that strikes it thousands of times. On top of the buried pile sits a "transition piece," and on top of that the tower and turbine. In deeper or trickier water, engineers switch to a jacket foundation — a lattice of steel legs, like a miniature oil-rig, spreading the load across three or four smaller piles. Either way, the foundation is often the single most expensive part of the whole enterprise, and it is invisible, sunk in the dark beneath the waves.

That is only the beginning of the sea's demands. Salt water is corrosive; it eats steel and electronics with a patience that never tires, so every surface must be coated, sealed, and protected. Waves pound the structure a thousand times an hour, flexing the steel, working the joints, so the whole thing must be engineered against fatigue for a design life of decades. Storms roll in with no mountains to break them. And when something goes wrong — a bearing fails, a sensor dies, a blade needs inspecting — you cannot simply drive a truck up. You send a crew out by boat, timing the trip to a calm-enough weather window, and they clamber from a heaving vessel onto a ladder bolted to the foundation, or you fly a technician in by helicopter and winch them down onto a platform the size of a small patio, high above the water. Maintenance that would take an afternoon on land can take days at sea, and the specialized installation vessels — the jack-up ships that plant their legs on the seabed and lift themselves clear of the waves to work — cost a fortune to charter and are in short supply.

Then there is the matter of getting the power ashore. Each turbine feeds its electricity through cables laid along the seabed to an offshore substation — itself a hulking platform standing on its own foundation — which gathers the output of the whole farm, steps up the voltage, and sends it home through a thick export cable buried under the seabed all the way to a connection point on the coast. Those subsea cables are a notorious weak point; a snagged anchor or a shifting sandbank can sever a link that takes months and a specialist repair ship to mend. Everything about the sea is harder, slower, and more costly than the equivalent job on land.

Which brings us to the first great misconception about this technology, and it is worth naming squarely. Offshore wind is not, as the casual phrase would have it, "just wind farms in water." It is a different discipline, closer in spirit to offshore oil-and-gas engineering than to a row of turbines in a field. The turbine on top may look familiar, but everything below the nacelle — the foundation, the marine cables, the substation, the installation fleet, the maintenance logistics — belongs to the harsh grammar of the sea. That is why offshore power has consistently cost more per unit than onshore, and why, for years, it needed generous support to get built at all.

And yet the costs fell, and they fell hard. The mechanism was competition. Rather than guaranteeing developers a fixed, comfortable price, governments in Europe began auctioning the right to build offshore farms, awarding the seabed lease to whoever would sell the power most cheaply. Developers, betting on ever-larger turbines and the falling costs that come with experience — the same learning curve we have watched drive down the price of solar and onshore wind — bid lower and lower. In the mid-2010s, some offshore projects in the North Sea were bid at prices that would have seemed fantastical a decade earlier; a few were even offered "subsidy-free," meaning the developers expected to earn their return from the wholesale market alone. The North Sea became the industry's proving ground: Britain, Germany, Denmark, the Netherlands, and Belgium ringed those shallow, windy waters with farm after farm, each bigger than the last, until the modern giants arrived. Britain's Hornsea projects and the sprawling Dogger Bank development — the latter built on a shallow sandbank far out in the North Sea, where the water is shallow enough to build but the coast is a distant smudge — count among the largest power stations of any kind ever constructed, with capacities measured in gigawatts.

But I promised you even-handedness, and the recent story is not one of smooth, endless decline. As of 2024, offshore costs had ticked back up. The reasons were mundane and stubborn: a burst of general inflation, higher interest rates that bite especially hard on capital-heavy projects, and tangled supply chains that left turbines, vessels, and cables scarce and dear. Some developers, having won auctions at prices set before the inflation hit, found they could no longer build profitably; a few walked away from contracts and forfeited their deposits rather than lose more. The industry's average cost of offshore power, having fallen for years, stopped falling and in places rose. Estimates vary and the picture is fluid, so treat any single figure with caution, but the honest summary is this: offshore wind grew cheaper for a long time, then hit a rough patch, and where its costs settle next is genuinely uncertain.

Let me make one turbine concrete, because the numbers are more vivid than the abstractions. Take that 26-megawatt machine — or, to use a figure more representative of what is being installed at scale today, imagine a turbine rated at 15 megawatts, a size now common in the North Sea. Rated power is the maximum it can produce when the wind is strong enough; a 15-megawatt turbine at full tilt makes 15 million watts, enough in that instant to supply thousands of homes. But it is rarely at full tilt. To know its real annual harvest we need the capacity factor — the fraction of its theoretical maximum output it actually delivers over a year, averaged across calms and gales. And here is one of the quiet wonders of going to sea. Onshore turbines typically run at a capacity factor somewhere in the mid-twenties to low-forties percent. Offshore machines, bathed in that stronger, steadier wind, routinely reach around fifty percent, and the best sites do better. That is a remarkable figure: a well-sited offshore turbine spends its year delivering, on average, roughly half of its maximum possible output — comparable, in that respect, to some conventional power plants.

So: 15 megawatts of rated power, times the 8,760 hours in a year, gives about 131 gigawatt-hours if it never stopped. Multiply by a capacity factor of, say, 0.5, and you get roughly 65 gigawatt-hours a year from a single turbine. If a typical home uses somewhere around 3,500 to 4,000 kilowatt-hours a year — the figure varies enormously by country — then one such turbine supplies on the order of sixteen or seventeen thousand homes, all by itself, spinning quietly over the horizon. Scale that to a farm of a hundred turbines and you have a genuine power station, feeding a city, built on the sea. Now you can see why the 26-megawatt Dongfang machine, in stronger Chinese waters, could plausibly reach that headline 100 gigawatt-hours a year — one machine, the annual electricity of tens of thousands of households.

The geography of who builds these things has shifted, too, and holds a second misconception worth correcting. It is tempting to assume offshore wind is a mature, universal technology, humming away off every coast. It is not. For most of its history it was an almost entirely European affair, with Britain long the world leader by installed capacity, closely joined by Germany and the pioneering Danes. Then China, having conquered onshore wind and solar manufacturing, turned to the sea and built at a pace that soon carried it to the front of the pack. Meanwhile, other coastlines that seem obvious candidates have barely begun. The United States, with thousands of kilometers of windy Atlantic shelf, had almost no offshore wind at all until the 2020s; its industry is young, its first commercial-scale farms only recently switched on, and its progress has been buffeted by exactly the cost inflation and supply-chain troubles I described. Offshore wind is not everywhere, and it is certainly not everywhere mature. In much of the world it is just being born.

Where do you meet it today, if you are not a North Sea fisherman? You meet it on the electricity bill of a British household, a meaningful slice of whose power now comes from turbines its owner will never see. You meet it in the skyline off the Danish and Dutch coasts, where day-trippers photograph the distant white forest on the horizon. You meet it, increasingly, in the news from Massachusetts and New York and California, where the American industry is fighting to establish itself. And you meet it in a number that would have astonished the crew of that little tugboat at Vindeby: the world now has on the order of eighty gigawatts of offshore wind installed, and adds several more every year — thousands of times the capacity of those first eleven machines.

There remains, though, a hard ceiling built into everything I have described, and it points us straight to the next chapter. All of it — the monopiles, the jackets, the whole fixed-bottom industry — depends on being able to reach the seabed and anchor to it. That works in water up to roughly sixty meters deep, out on the shallow continental shelves that ring places like the North Sea. But most of the world's ocean is far deeper than that, and most of the world's best offshore wind blows over water where no pile could ever reach the bottom. To harvest it, you cannot stand on the seabed at all. You must float.

Which is exactly what a handful of engineers, off the coast of Scotland, had already begun to do.

Chapter 47 — Floating Wind and Floating Solar

Twenty-five kilometers off Peterhead, on the north-east shoulder of Scotland, where the North Sea deepens toward a hundred meters and no pile could ever find the bottom, five turbines stand in a loose row in the open water. Look closely and something is wrong with them — or rather, something is missing. There is no visible foundation, no fat steel tube rising out of the waves. Each turbine sits atop a long, slender cylinder that bobs, ever so slightly, with the swell. The whole machine is floating, held in place not by a pile driven into the seabed but by three thick chains slanting down into the dark. This is Hywind Scotland, switched on in 2017, the first floating wind farm the world had ever seen, and standing on the deck of a survey boat beneath one of its towers you feel you are watching a magic trick: a structure taller than most skyscrapers, weighing thousands of tonnes, riding the ocean like a buoy and calmly making electricity.

To understand why anyone would attempt something so counter-intuitive, we need to return to the ceiling that ended the last chapter. Fixed-bottom offshore wind — the monopiles and jackets of the North Sea — works only where the water is shallow enough to reach, up to roughly sixty meters. That sounds like a lot until you look at a chart of the world's coasts. Off much of Japan, off California and Oregon, off the deep-plunging edges of the Mediterranean, off large parts of the Atlantic and the whole Pacific rim, the seabed drops away almost immediately into water hundreds or thousands of meters deep. And here is the fact that reorganizes the entire map of offshore wind: most of the world's offshore wind resource lies over water too deep for a fixed foundation. The best winds, the strongest and steadiest, blow over exactly the seas where you cannot plant a pile. If offshore wind is to be more than a North Sea specialty, it has to learn to float.

The physics of how it does so is, at its heart, the physics of a boat, and it comes down to a single principle discovered by Archimedes more than two thousand years ago. A floating object is held up by buoyancy: it displaces a volume of water, and the weight of that displaced water pushes back up with a force exactly equal to the object's own weight. A steel turbine is far denser than water and would sink like a stone on its own, so the trick is to attach it to a hollow structure — a great sealed float — large enough that the water it displaces weighs as much as the whole assembly. Build the float big enough and the turbine rides high and stable, just as a steel ship, mostly empty air inside its hull, floats despite being made of metal that would sink if melted into a bar.

But floating is not the same as staying put, and a turbine that drifts is useless — it would wander off station and tear its power cable loose. So the second half of the trick is mooring. Chains or synthetic ropes run from the floating platform down to anchors fixed on the seabed, holding the machine within a tight watch-circle while still letting it heave and sway gently with the waves. And the third challenge, the subtlest, is stability against tipping. A wind turbine is grievously top-heavy — a massive nacelle and long blades perched at the summit of a tall tower — and the wind pushing on that rotor exerts a colossal overturning force. On land or on a pile, the rigid foundation simply resists it. A float must resist it while bobbing on the water, and that is genuinely hard.

Engineers have converged on a few designs, each a different answer to the tipping problem. The spar, which is what holds up the Hywind turbines, is essentially a very long, slender cylinder, weighted heavily at the bottom so that its center of gravity sits far below the waterline. Like a fishing float or a child's weighted toy that always rights itself, a spar resists tipping because any lean lifts the heavy keel and gravity pulls it back upright. Its drawback is that the ballast hangs so deep it needs deep water and a sheltered fjord to be assembled upright before towing out. The semisubmersible takes the opposite approach: a wide, raft-like structure of several buoyant columns linked together, floating shallowly like a broad-beamed barge, gaining its stability from sheer width — it is hard to tip a wide platform, just as a broad-hulled catamaran is steadier than a narrow canoe. And the tension-leg platform is tethered by cables pulled bar-taut straight down, so the buoyant hull strains upward against its moorings and the tension holds it rigidly in place. Each has its trade-offs of cost, water depth, and ease of assembly, and the industry has not yet settled on a winner.

It is worth pausing on just how much resource this unlocks, because the numbers reorder one's sense of the whole industry. Off a country like Japan, an archipelago ringed by deep water, the shallow shelf suitable for fixed foundations is a thin fringe, while the vast bulk of the wind blowing over its seas rides above water hundreds of meters deep. The same is true off California and Oregon, where the continental shelf plunges almost from the beach, and off much of the Mediterranean, and off the storm-rich Atlantic coasts. Study after study has concluded that the majority of the world's technical offshore wind potential — in many maritime countries, the great majority — lies in water too deep for a monopile. Fixed-bottom offshore wind, for all its North Sea triumphs, has only ever been fishing in the shallows at the edge of a far larger sea of energy. Floating wind is how you reach the rest of it, and that is why so serious an effort now goes into a technology that is still, by any honest measure, young and dear.

The human story here runs, once again, through Scandinavia and through a handful of stubborn engineers who kept insisting the sea could be tamed. The Norwegian energy company that built Hywind had cut its teeth for decades on floating oil platforms in the brutal North Sea, and floating a wind turbine was, in a sense, a graceful pivot of that hard-won expertise from fossil fuel to wind. A single prototype spar-mounted turbine had been bobbing off Norway since 2009, enduring the winters, teaching its makers how a floating machine actually behaves when the swell and the wind and the rotor's own gyrations all conspire at once. The lessons of that lonely prototype fed directly into the five-turbine Hywind Scotland farm, which in its early years posted capacity factors that startled even its builders — regularly around half, and in some winter months considerably more, because the deep-water winds it tapped were so strong and so relentless. The magic trick worked, and it worked better than the skeptics had allowed.

Which lands us on the first great misconception about all of this. Floating wind sounds like science fiction — a turbine on a raft, tethered to the abyss — and it is easy to file it away as a someday technology, forever ten years off. It is not. It is operating now, has been for years, generating real electricity onto real grids off Scotland and Portugal and in scattered pilots elsewhere, with far larger commercial arrays under development off Japan, California, and the Mediterranean. The engineering is proven; what remains is to drive down the cost and scale it up, the same grinding, unglamorous work of repetition and learning that took fixed-bottom offshore from the eleven toy turbines of Vindeby to the gigawatt giants of Dogger Bank. Floating wind is early, and it is expensive, and it is absolutely not imaginary.

If the deep sea is one frontier, there is a gentler one much closer to home, and it turns the same principle of buoyancy to a completely different purpose. Picture not the open ocean but a calm inland reservoir behind a hydroelectric dam, its surface flat and glinting, and floating on it, moored in neat rows, thousands of solar panels. This is floating solar — "floatovoltaics," in the industry's cheerful coinage — and it is one of the fastest-growing niches in the whole renewable story, booming above all across Asia, where land is scarce and reservoirs are plentiful.

The idea is disarmingly simple. Instead of covering fields with panels, you mount them on buoyant plastic pontoons and set them afloat on water you already own — the reservoir behind a dam, an irrigation pond, a flooded quarry, a lake beside a water-treatment works. The panels themselves are ordinary; only the mounting is different, exchanging steel racks driven into soil for floats that ride the surface, anchored to the bank or the bed so they cannot drift. And the appeal is that it solves several problems at once, which is rare enough in engineering to be worth savoring.

The first benefit is land, or rather the saving of it. A large solar farm needs a large area of ground, and in crowded, mountainous, or intensively farmed places that ground is precious and contested. A reservoir, by contrast, is a flat, empty, sun-drenched expanse that is doing nothing but sitting there. Cover part of it with panels and you generate serious power without displacing a single field or forest. For a country like Japan, or the packed river deltas of southern China, that arithmetic is compelling.

The second benefit is water, and it is a lovely piece of physical serendipity. An open reservoir loses a great deal of water to evaporation — the sun beats down on the surface all day and lifts it invisibly into the air, a loss that matters enormously in hot, dry regions where every drop is fought over. Float a canopy of solar panels over part of that surface and you shade the water beneath, cutting the evaporation substantially. So the panels do double duty: they make electricity and they conserve water, a genuine two-for-one that pairs especially well with hydroelectric dams, where the reservoir is the fuel and losing it to the sky is losing money. Better still, a floating solar array atop a hydro dam can share the same power lines and even the same connection to the grid, and the two sources complement each other neatly — the sun shines by day, and the stored water can be released to generate through the evening and the cloudy spells, a kind of natural battery sitting right beneath the panels.

The third benefit is subtler and repays a moment of first-principles thinking, because it corrects a fact that surprises almost everyone: solar panels do not love heat. It seems paradoxical — surely the hotter and sunnier, the better? — but a photovoltaic cell actually loses efficiency as its temperature climbs. The reason lies in the physics we met long ago: a panel converts light to electricity by knocking electrons free across the material's band gap, and as the silicon heats up, that band gap narrows slightly and the freed charges lose a little of their orderly push, so the voltage sags and the cell delivers less power for the same sunlight. A panel baking on a black rooftop in still summer air can run twenty or thirty degrees hotter than the air around it, and pay for it in lost output. Now float that same panel a few inches above cool water, with the breeze skating across the surface, and it runs cooler — and a cooler panel is a more efficient panel. Floating solar arrays, cooled from below by the water they shade, often produce a useful few percent more electricity than the identical panels would on hot ground nearby. The water they protect from evaporation returns the favor by keeping them at their best.

That brings us to the second misconception, and it is the mirror image of the one about floating wind. Where floating wind is dismissed as too futuristic, floating solar is dismissed as too flimsy — surely panels on plastic floats will be smashed by the first storm, corroded by the water, tangled by the waves? In fact these systems are engineered precisely for the wet, moving environment they live in: the floats are made of tough, ultraviolet-resistant plastics, the arrays are moored with slack that lets them rise and fall with changing water levels, and the electrical connections are sealed and rated for the conditions. They are not ordinary garden panels dropped onto pool toys. They are purpose-built, and the largest of them have already ridden out years of weather on reservoirs across India, China, and Southeast Asia, some of them rated at over a hundred megawatts-peak — enough, at full sun, to supply a substantial town — floating serenely where nothing but water was before. India's Omkareshwar project, commissioned on a reservoir in 2024, is one of a growing fleet of such installations, and the global market has been expanding at roughly twenty percent a year.

Let me draw the two halves of this chapter together with a single worked comparison, because the buoyancy principle links them more tightly than it first appears. In both cases, the fundamental move is the same: displace enough water to hold up your equipment, then moor it so it cannot wander. But the demands could hardly be more different. A floating solar panel weighs a few kilograms, sits inches above a calm reservoir, and needs only enough float to stay dry and enough anchor to stay put — a problem an engineer can solve with cheap plastic and a length of rope. A floating wind turbine weighs thousands of tonnes, towers hundreds of meters into gale-force winds over open ocean, and must resist an overturning force great enough to flip a lesser structure clean over — a problem that draws on generations of offshore-platform engineering and costs a fortune to get right. Same principle, opposite ends of the spectrum of difficulty. That is why floatovoltaics is already a mass-market product spreading across the reservoirs of Asia, while floating wind is still a premium, pioneering technology feeling its way toward scale. Both, though, are answers to the same underlying scarcity: we are running out of easy, empty, sunlit and windswept land, and the water — deep ocean and calm reservoir alike — is where the next great expanses lie.

Where do you meet all this today? You meet floating wind

in the electricity that flows ashore from Hywind's five turbines into the Scottish grid, and increasingly in the plans of coastal nations — Japan, South Korea, the American West Coast, France, Norway — whose best wind blows over water too deep for anything else, and who are lining up the auctions and the port upgrades that a floating industry will need. And you meet floating solar every time you see a photograph of a reservoir quilted with dark panels, a sight that was almost unknown fifteen years ago and is now commonplace from Singapore to the hydro dams of China, where the panels shade the water, cool themselves in the process, and feed their power out along the lines the dam already built.

What unites the two frontiers, and what makes them fitting bookends to a single chapter, is a lesson this whole book keeps circling back to. There is far more solar and wind resource in the world than we could ever use — the sun and the wind are effectively limitless — and the real constraints are always practical: where to put the machines, how to reach the resource, how cheaply we can build. Floating wind and floating solar are two ingenious answers to the "where," reaching for energy in places we once thought unreachable, using nothing more exotic than a principle Archimedes worked out in his bath.

But there is one more way to catch the sun that we have so far set aside, older than the photovoltaic panel and stranger to modern eyes — a way that uses no electron, no band gap, no silicon at all. It works with mirrors and heat, it descends directly from the desert dreamers we met near the start of this story, and it can do one thing that no ordinary solar panel can: keep making power hours after the sun has gone down.

Chapter 48 — Concentrated Solar Power

In the Mojave Desert, near the California–Nevada border, three slender towers rise out of a sea of mirrors. There are more than three hundred thousand of the mirrors, each the size of a garage door, spread across five square miles of scrubland and tilted with computer-controlled precision so that every one of them throws its patch of reflected sunlight at the same spot: the top of the nearest tower, where a boiler glows so fiercely that the desert air around it shimmers and, at certain hours, the whole tower-top burns too bright to look at directly. This is Ivanpah, and when it switched on in 2014 it was the largest solar power station of its kind in the world. Yet if you drove past it expecting to see solar panels, you would find none. Not one. Ivanpah makes electricity the way a coal plant does — by boiling water into steam and spinning a turbine — except that its furnace is the sun, gathered and hurled by a third of a million mirrors onto a single burning point.

This is the other solar, the one most people forget exists, and it deserves a chapter of its own because it works on an entirely different principle from everything we have discussed since the photovoltaic effect. A solar panel, as we have seen at length, is a quantum device: a photon strikes silicon, knocks an electron loose across the band gap, and a current flows, directly, silently, with no moving parts and no heat engine. Concentrated solar power — CSP for short, and I will use the short form once we have met it — does something far older and, in a sense, far cruder. It does not care about photons and electrons at all. It simply gathers sunlight into a small, blazing spot to make things very hot, and then uses that heat exactly as we have used the heat of burning coal or gas for two centuries: to boil water, drive a steam turbine, and turn a generator. It is, at its heart, a heat engine with the sun for a fire.

To feel why concentration matters, hold a magnifying glass over a scrap of paper on a bright day. The sunlight falling on the whole lens is spread thin — a thousand watts on a square meter is warm but harmless. But focus that same light down to a pinpoint and the paper smokes and catches. You have not added any energy; you have merely collected the energy from a wide area and delivered it to a tiny one, and the concentration is everything. The temperature a surface can reach depends on how intensely light is piled onto it, and a lens or a mirror that gathers the sun from a broad aperture and squeezes it to a point can drive that spot to hundreds or even thousands of degrees. Every CSP plant is a magnifying glass writ enormous, trading a great field of cheap mirrors for a small, ferociously hot receiver.

This is a very old dream, and the reader who has come this far will recognize its ancestors. We met, near the beginning of this story, the French inventor Augustin Mouchot, who built solar-concentrating steam engines in the 1870s and showed one off at the Paris Exhibition of 1878, and his assistant Abel Pifre, who ran a printing press on focused sunlight. And we met, most memorably, Frank Shuman, the American engineer who in 1912 and 1913 built the world's first genuine solar thermal power station at Maadi, on the outskirts of Cairo, where long parabolic troughs of mirror concentrated the Egyptian sun onto pipes of water, raising steam to drive an engine that pumped the Nile onto the fields. Shuman's plant worked. It was killed not by any technical failure but by the First World War and, soon after, by the discovery of cheap oil, which made his elegant desert machine look quaint. But the lineage never died, and modern CSP is the direct heir of Shuman's troughs — the same idea, a century on, built at a scale he could only have dreamed of.

The engineers of the modern revival converged on a handful of ways to do the concentrating, and it is worth knowing them apart, because they look very different in the landscape. The oldest and, for a long time, the most common is the parabolic trough: long, curved mirrors shaped like a gutter, running in parallel rows across the desert, each focusing sunlight onto a black pipe suspended along its focus. A fluid — often a special heat-transfer oil — runs through the pipe, is heated to a few hundred degrees, and carries that heat away to a boiler. California's SEGS plants, built through the 1980s and 1990s in the Mojave, were troughs, and they ran for decades. The second design is the one at Ivanpah: the power tower, or central receiver, in which a whole field of flat mirrors called heliostats — each independently steered to track the sun and bounce its light at a fixed target — surrounds a tall tower with a receiver at its summit. Because a tower gathers the light of thousands of mirrors onto a single point, it can reach much higher temperatures than a trough. There are two further variants you will meet less often: the linear Fresnel design, which uses flat or gently curved mirror strips as a cheaper approximation of a trough, and the dish, a mirrored satellite-dish shape that focuses onto a small engine at its focal point and reaches the highest concentrations of all, though it has struggled to scale.

Now we come to the feature that makes CSP genuinely special, the one thing it can do that an ordinary solar panel cannot, and it follows directly from the fact that CSP traffics in heat rather than electricity. Heat is easy to store. Electricity is hard to store — that is the whole vexed problem of batteries that haunts the rest of this book. But heat you can simply keep in a well-insulated tank, the way a thermos keeps coffee hot for hours. And so the best CSP plants add a component that transforms them: thermal energy storage, usually in the form of molten salt.

The idea is beautiful in its simplicity. Ordinary salts, the cousins of table salt, melt at a few hundred degrees into a clear, water-thin liquid that holds heat superbly and stays stable at high temperature. A CSP plant with molten-salt storage uses its mirror field to heat a great volume of this salt — pumping it from a "cold" tank at perhaps 290 degrees to a "hot" tank at nearly 570 — and stores the hot salt in an insulated tank. During the day, when the sun is up, the plant can run its turbine directly and bank the surplus heat in the hot tank. Then, when the sun sets and every photovoltaic panel for a thousand miles goes dark, the CSP plant keeps running: it draws the hot salt back out of the tank, uses it to boil water and spin the turbine, and generates electricity into the night. Spain's Gemasolar plant demonstrated this dramatically, running around the clock through midsummer, its turbine humming under the stars on heat harvested hours before. A power plant that runs on sunlight after dark sounds like a contradiction, and it is one of the quiet marvels of energy engineering that it is simply, routinely true.

That single ability — to shift the sun's energy across time, from the glut of noon to the demand of the evening — is CSP's answer to the great weakness of solar power, its intermittency. A photovoltaic farm stops dead at sunset; to shift its power into the evening you must pair it with a separate battery. A molten-salt CSP plant carries its storage inside itself, in cheap tanks of hot salt, and can dispatch power on command for hours after the sun is gone. For a grid operator who needs electricity most in the evening peak, exactly when solar panels are fading, that is a valuable trick.

There is an elegant thermodynamic bonus buried in this arrangement, worth spelling out because it explains why the power-tower design in particular has drawn so much effort. The efficiency with which any heat engine can turn heat into work depends, at bottom, on how hot the heat is — the greater the gap between the hot source and the cool surroundings, the more of the heat can be converted, a principle the French engineer Sadi Carnot worked out in the 1820s and which governs every steam turbine ever built. A parabolic trough, with its long focal line, can only concentrate the sun so much, and its heat-transfer oil tops out at a few hundred degrees. A power tower, gathering thousands of mirrors onto a single point, concentrates far more intensely and can drive its molten salt to nearly six hundred degrees — hotter salt, a bigger temperature gap, a more efficient turbine, and, not incidentally, more energy packed into each tank of stored salt. That is the prize the tower designers were chasing: not just storage, but hotter, denser, more valuable storage. It is also, as we will see, part of why the towers proved so much harder to build than the humble troughs.

So why, given all this, did CSP lose? Because it did lose — decisively. For a while in the 2000s it looked as though CSP and photovoltaics were running neck and neck to become the solar technology of the future, and serious people bet on the mirrors. Then the price of ordinary silicon panels collapsed, driven by the manufacturing juggernaut and the relentless learning curve we have traced elsewhere in this book, falling so far and so fast that photovoltaic electricity became the cheapest power humanity has ever produced. CSP simply could not keep up. It is a fundamentally more complicated machine — a field of precision-steered mirrors, a high-temperature receiver, a fluid loop, a boiler, a steam turbine, tanks of molten salt kept perpetually liquid — where a photovoltaic panel is a slab of glass with no moving parts that can be mass-produced by the shipload. Complexity is expensive, and mass production is cheap, and the market chose the cheap slab of glass. By the late 2010s, new CSP construction had slowed to a trickle while photovoltaics were being installed by the hundreds of gigawatts.

CSP had been out-competed, and the honest verdict is that on raw cost per kilowatt-hour it never really recovered.

But — and this is the second thing an even-handed account must say — CSP is not dead. It is a niche, but a living one, and it survives precisely where its special abilities are worth paying for. Where storage is prized, where the grid needs power that can be summoned after sunset, where the sun is strong and direct and land is cheap, CSP still gets built. Spain remains a stronghold. The Gulf states have raised CSP towers as part of vast solar complexes. China has built several. And the most spectacular modern example stands in the Moroccan desert outside the town of Ouarzazate, on the edge of the Sahara: the Noor complex, one of the largest concentrated solar installations on Earth, combining parabolic troughs and a towering central receiver, with molten-salt storage that lets it feed Moroccan homes into the evening — a plant that turns the Saharan sun into dispatchable power on a scale that would have left Frank Shuman speechless.

There is a first-principles reason CSP lives where it does, and it is worth being precise about, because it also explains where CSP can never work. CSP needs strong direct sunlight — the sharp, unscattered beam of a clear desert sky. A magnifying glass will not focus a hazy, overcast sky to a point; it needs the sun as a distinct disc casting hard-edged shadows, because only that direct beam can be reflected and concentrated onto a receiver. Diffuse light, scattered across a cloudy dome, cannot be focused at all. An ordinary photovoltaic panel, by contrast, will make some power even under cloud, from the diffuse skylight. This is why CSP belongs to the deserts — the Mojave, the Sahara, the Arabian sands, the high plateaus of China and Spain — and why it makes no sense at all in cloudy northern Europe, where a mirror would spend half the year staring at grey. The technology is chained to the sunbelt by the very optics that make it work.

Two misconceptions are worth naming plainly. The first is the assumption, so natural now that it is almost invisible, that a "solar power plant" means a field of photovoltaic panels. After this chapter you know that is not always so: some of the world's largest solar plants contain no panels at all, only mirrors and boilers and turbines, and they belong to a completely different branch of the family tree — the heat branch, descended from Mouchot and Shuman, not the electron branch descended from Bell Labs. The second misconception is the obituary: that CSP failed and is therefore finished. It failed to win the mainstream, yes, out-competed by cheap silicon. But it survives as a specialist, valued for storage and heat in the places that need those things, and there is even renewed interest in its very high temperatures for industrial process heat — for driving the furnaces and reactors of heavy industry, which need heat, not electricity, and which cannot easily be run off a solar panel.

Honesty requires that I tell one cautionary tale, because CSP's history is not all triumph. In the Nevada desert stood a plant called Crescent Dunes, a single tall tower ringed by ten thousand heliostats, built in the mid-2010s as a showcase for molten-salt storage. On paper it was the future: a power tower that could store the day's heat and generate deep into the night. In practice it struggled. A leak in the molten-salt tank forced a long shutdown; the plant never reliably hit its promised output; costs overran and its finances collapsed, and for a time it sat idle, a very expensive monument to the gap between a beautiful concept and a working machine. It is fair to state the reasons without either gloating or glossing: Crescent Dunes was a pioneering design pushed to a difficult scale, and pioneering at scale is where energy projects most often stumble. It is also fair to add that other molten-salt towers, Gemasolar and Noor among them, have run far more successfully, so the technology itself is not condemned by one troubled plant. The lesson is the one this book keeps relearning — that the distance between a demonstration and a dependable power station is measured in hard, expensive years.

Where do you meet CSP today? You meet it, if you drive Interstate 15 between Los Angeles and Las Vegas, in the three glowing towers of Ivanpah rising from the Mojave, visible for miles. You meet it in the evening electricity of Moroccan and Spanish towns, drawn from tanks of salt that were hot at noon. You meet it, most of all, as a road not taken — a reminder that the solar future could have looked very different, all mirrors and steam, and that it turned out otherwise only because a slab of silicon glass got improbably, world-changingly cheap.

And that cheapness is exactly where we turn next. For the same collapse in the price of the photovoltaic panel that doomed CSP to a niche did something else at the other end of the scale. It made it possible to carpet the desert not with mirrors but with panels — millions upon millions of them — and to build solar power stations bigger than any Ivanpah, bigger than any coal plant, the largest power installations the world has ever assembled.

Chapter 49 — The Utility-Scale Solar Farm

In the Thar Desert of Rajasthan, in the far northwest of India, near a dusty crossroads village called Bhadla, the ground has been paved with silicon. Stand at the edge of the Bhadla Solar Park and the panels run to the horizon in every direction — row upon dark blue row, tilted to the south, marching across a landscape so hot and so barren that little else could ever be done with it. Summer temperatures here climb past fifty degrees Celsius; dust storms roll through with a regularity that once made the place almost worthless. And yet across roughly fifty-six square kilometers of this wasteland, more than ten million solar panels now drink the fierce Rajasthan sun and turn it into something like 2.2 gigawatts of electrical capacity — a gigawatt being a billion watts, the output of a large conventional power station. Bhadla is among the biggest solar farms on Earth, and it is a good place to begin, because it makes visible a shift that has quietly become one of the most important facts about the modern world: the cheapest way to make electricity that humanity has ever discovered is to lay panels flat across empty land, by the million, and wait for the sun to rise.

This is the utility-scale solar farm, and it is a different beast from the rooftop array we met earlier in this book. A rooftop system is a handful of panels, a few kilowatts, wired to serve one home. A utility-scale farm is a power station — tens or hundreds or, at the extreme, thousands of megawatts — built by developers, financed like infrastructure, and plugged straight into the high-voltage transmission grid to feed cities far away. Bhadla is one; so is the sprawling solar base around Golmud on the high, cold Tibetan plateau of China, and so are the great arrays spreading across the American Southwest, the deserts of the Gulf, and the sun-baked plains of Australia. To understand how the world's electricity is changing, you have to understand how one of these farms is built, wired, and paid for — and why it beats almost everything else on cost.

Start with the panel and build outward, because a solar farm is fundamentally the same physics we have followed since Becquerel, just repeated a staggering number of times. Each panel, or module, is a slab of glass and silicon perhaps two meters by one, containing sixty or seventy-two cells wired in series, and producing a few hundred watts in full sun at a modest voltage. That is a trickle. To make a power station you must gather those trickles into a torrent, and the gathering is done in a strict hierarchy. Panels are wired end to end into "strings," raising the voltage the way batteries stacked in a flashlight add up; strings are wired side by side into arrays, raising the current; the arrays feed into combiner boxes and then into inverters, the crucial machines that convert the panels' direct current into the alternating current the grid runs on, all the while performing the constant balancing act called maximum power point tracking — continuously nudging the operating voltage to squeeze the most power out of the array as the sun brightens and fades and the panels heat and cool. From the inverters the power flows to transformers that step the voltage up high, and from there to the substation where the whole farm's output joins the transmission grid and begins its journey to distant homes.

Now, a subtlety that turns out to matter enormously. The

panels could simply be bolted at a fixed angle, facing the midday sun, and many cheaper farms are built exactly that way. But the sun does not stay put; it swings across the sky from east to west through the day, and a panel that faces due south sees the sun's rays strike it obliquely in the morning and evening, losing power to the shallow angle. So the more sophisticated farms mount their panels on single-axis trackers — long horizontal torque-tubes, each carrying a row of panels, that slowly rotate through the day so the panels tilt east at dawn, flatten toward noon, and lean west by dusk, following the sun like a field of mechanical sunflowers. The gain is real and worth the added cost and complexity: a tracker can lift a farm's annual output by something like fifteen to twenty-five percent over fixed panels, simply by keeping the silicon square to the sun for more of the day. Bhadla, and most large farms in strong-sun regions, track. It is one of those unglamorous engineering choices that quietly adds up to a great deal of extra electricity.

Let me make a farm concrete with a worked example, because the abstractions of gigawatts hide how tangible these places are. Imagine a solar farm rated at 100 megawatts — a hundred million watts, a middling size by today's standards, enough to be a serious contributor to a regional grid. How much land, and how many panels, does that take? A modern panel produces very roughly 400 watts at full sun, so to reach 100 megawatts of panel capacity you need on the order of 250,000 panels. But you cannot pack them edge to edge; you must space the rows apart so that a row does not cast its shadow on the row behind, especially in winter when the sun rides low, and you need gaps for maintenance vehicles and the inverter stations. As a rule of thumb, a utility solar farm occupies somewhere around five to seven acres per megawatt of capacity, so a 100-megawatt farm sprawls across roughly five to seven hundred acres — call it a square a couple of kilometers on a side. And what does it deliver? Solar's capacity factor — the fraction of its theoretical maximum it actually produces over a year, given nights, clouds, and low winter sun — runs from around ten percent in cloudy northern latitudes to perhaps twenty-five or even thirty percent in a place like Rajasthan. Take a sunny site at twenty-five percent: 100 megawatts times the 8,760 hours in a year times a quarter gives roughly 219 gigawatt-hours a year. If a household consumes somewhere around 4,000 kilowatt-hours annually — the figure varies widely by country and lifestyle — then that one 100-megawatt farm supplies the electricity of something like fifty thousand homes, from a few hundred acres of ground that may have grown nothing but scrub.

It helps to walk the site in the imagination, because the layout of a solar farm is a study in patient repetition. The land is graded roughly flat and cleared; then pile-drivers march down the rows, hammering steel posts into the soil at fixed intervals, thousands upon thousands of them, each destined to carry a torque-tube or a fixed rack. Onto the posts go the trackers or frames, and onto those the panels, clipped and bolted by crews that install the same module the same way ten thousand times a day. Wiring is strung along the rows and gathered into combiner boxes; the combiner boxes feed inverter stations spaced across the field like little houses, humming as they convert direct to alternating current; and cables run from the inverters underground to the on-site substation, where transformers lift the voltage high enough to travel and the farm's whole output is fed onto the transmission line. There is a stark, almost agricultural rhythm to it — the same unit, the same motion, repeated until a wilderness of scrub has become a power station. That repeatability is not incidental. It is the very thing that makes utility solar cheap.

Why has this become the cheapest power source in history? The answer is scale, and it is worth being precise about why bigness pays. A rooftop installer arrives at your house, sets up scaffolding, works around your chimney and your awkward roof angles, wires a few panels, and packs up — a slow, custom, labor-heavy job for a few kilowatts. A utility developer, by contrast, buys panels by the shipping container at rock-bottom factory prices, clears a flat open site, and installs identical rows by the hundred thousand using pile-drivers and crews that do nothing all day but repeat the same motion. Every cost — the panels, the mounting, the wiring, the labor, the paperwork, the connection — is spread across a vast output and driven down by repetition and bulk buying. This is why, per unit of electricity, utility-scale solar is dramatically cheaper than rooftop solar, and why it has become, as of the mid-2020s, on an unsubsidized basis in sunny regions, the least expensive electricity that has ever been generated by any means — cheaper than coal, cheaper than gas, cheaper than nuclear. The record-setting auctions tell the story starkly: in Saudi Arabia in 2021, a solar tender was won at a bid of just over a cent a kilowatt-hour — roughly one one-hundredth of what solar cost only a decade or so before.

But cheap-when-the-sun-shines is not the same as always-available, and here the modern solar farm has acquired a powerful new companion. Increasingly, the panels are built together with banks of lithium batteries housed in rows of shipping containers at the edge of the site — an arrangement so common it has its own name, "solar-plus-storage." The logic is elegant: the farm generates a flood of power at midday, more than the grid may want just then, and instead of wasting the surplus it pours it into the batteries; then, as the sun sets and demand rises into the evening peak, it draws the stored energy back out and keeps delivering. A solar-plus-storage plant can therefore promise something close to dispatchable power — electricity on demand, well past sunset — from a resource that only shines by day. What was once an exotic pairing is now routine; new solar farms are increasingly built with storage from the start, and the combination is steadily eroding the old complaint that solar is useless after dark.

A word of even-handed caution belongs here, because the very cheapness and abundance of midday solar creates a problem of its own. When a sunny grid fills up with solar farms, they all pour out their power at the same time — noon — and on a mild spring day, when air-conditioners and heaters are both idle, that midday flood can exceed everything the grid can use. The wholesale price of electricity crashes toward zero, and grid operators are sometimes forced to "curtail" the farms, deliberately throttling back panels that could be generating, simply because there is nowhere for the power to go. California grid planners gave the shape of this daily glut a memorable name, the "duck curve," for the way the demand-for-other-power line sags through the solar-soaked middle of the day and then rears up steeply in the evening as the sun sets and the panels fade. This is the honest counterweight to solar's low cost: cheap electricity at noon is not the same as valuable electricity at noon, and at high penetration the challenge shifts from generating the power to storing it and shifting it to the hours that need it. It is precisely this problem that solar-plus-storage, and the batteries at the edge of the farm, are built to solve — banking the noon surplus for the evening ramp when the duck's neck rises.

Which brings us to the most persistent misconception about solar power, and it deserves to be met head-on with an honest number, because the fear behind it is sincere and widely felt. The objection runs: solar is so dilute, so spread out, that powering a whole country would require paving an impossible fraction of the land — you would have to cover the nation in glass. It is worth working through why this is simply false. Sunlight delivers roughly a thousand watts to each square meter at noon; a solar farm, accounting for panel efficiency, spacing, night, and weather, harvests on average a much smaller but still substantial amount per unit of ground over a year. When you actually run the arithmetic for a country's total electricity demand against the yield per square kilometer of solar farm, the land required comes out to a small percentage of the national territory — commonly on the order of one or two percent, and often much less, for a large industrial country to meet a big share of its power from solar. That is not nothing; it is real land and it must be sited with care. But it is a modest fraction, not a continent, and it is frequently land that is marginal, degraded, or otherwise little used — desert, scrub, brownfield, the shoulders of highways, or, as we will see, land that can go on doing double duty as farmland. The honest framing is not "solar needs impossible land." It is "solar needs a modest and manageable slice of land, which must be chosen thoughtfully." Those are very different claims, and the difference matters enormously to how one thinks about the energy transition.

The second misconception is subtler and worth correcting because it muddies a lot of public argument: that utility-scale and rooftop solar are the same thing, two flavors of one product. They are not, and they are not even really competitors — they are complements with opposite strengths, and a sensible energy system uses both. Utility solar wins overwhelmingly on cost, for all the reasons of scale we have seen; a kilowatt-hour from Bhadla is far cheaper than a kilowatt-hour from a suburban roof. But utility solar has two costs that rooftop avoids entirely. It consumes new land, and it needs new transmission lines to carry its power from the remote, sunny, empty places where it is cheapest to the cities where the power is used — long, expensive, slow-to-permit wires strung across the countryside. Rooftop solar uses no new land at all, sitting on surfaces that already exist, and it generates the electricity right where it is consumed, feeding the house or the building directly and sparing the grid the whole journey. So the trade is clean: utility solar is cheaper per unit but needs land and wires; rooftop is dearer per unit but needs neither. A country building its solar future rationally builds both — vast cheap farms in the deserts and plains, and panels on the rooftops of its cities — and the argument over which is "better" usually dissolves once you see that they are answering different questions.

Where do you meet the utility-scale solar farm today? You meet it from an airplane window over the American Southwest, or the Spanish meseta, or the Chinese west, as a great dark rectangle geometrically incongruous against the wild land around it. You meet it in the record-breaking scale of the biggest parks — Bhadla in India, the Golmud and other bases in China, each rated in the gigawatts, larger in output than the coal plants they are helping to retire. You meet it, quietly, in the falling wholesale price of daytime electricity in sunny grids, sometimes pushed so low at noon that power briefly costs almost nothing. And you meet it, increasingly, in the rows of battery containers now standing sentinel at the edge of these farms, holding the noon sun back for the evening.

There is one more question these vast farms raise, and it points directly to where this book goes next. A solar farm covers its ground with panels — but does that ground have to be lost? Must a solar farm be a dead zone of gravel and glass, sterilizing the land beneath it, an either-or choice between growing food and generating power? For a long time that was the unspoken assumption, and it is the last great misconception about solar land use. It turns out to be wrong, and the way it is wrong opens up one of the most surprising and hopeful ideas in the whole story: that the same field might yield two harvests at once — sunlight for the panels above, and crops or grazing for the animals and plants below.

Chapter 50 — Agrivoltaics and the Second Harvest

On a hillside farm in the Japanese countryside, a man named Akira Nagashima did something in the early 2000s that most engineers would have called a mistake. Nagashima was a retired agricultural-machinery engineer who had taken up the study of how plants use light, and he had become convinced of a heretical idea: that a field could grow crops and generate solar power at the same time, on the same ground, if only you spaced the panels correctly. So he built low racks over his plots, mounted narrow solar panels high above the soil with wide gaps between them, and planted crops underneath — and waited to see whether the plants would sicken in the shade. They did not. They thrived. The panels overhead made electricity; the crops below grew nearly as well as they would have in open sun, and some grew better. Nagashima had coined a word for it in Japanese that would spread around the world in translation: agrivoltaics, the marriage of agriculture and photovoltaics, the idea that a farmer's field could bring in a second harvest of sunlight without giving up the first harvest of food.

The chapter that ended a moment ago left us with a worry: that a solar farm must sterilize the land it sits on, forcing an either-or choice between growing food and generating power. Agrivoltaics is the answer to that worry, and to understand why it works — why it is not a cheat or a compromise but a genuine two-for-one — we have to understand something surprising and rather beautiful about how plants use light. Because the whole idea rests on a single biological fact that runs against common sense: many plants do not actually want full sun.

Consider what happens inside a leaf. A plant grows by photosynthesis, using the energy in sunlight to build sugars out of air and water, and it is easy to assume that more light always means more growth. But that is not how the machinery works. A leaf's photosynthetic apparatus can only process photons so fast; feed it light beyond a certain intensity and it simply cannot keep up. This threshold is called the light saturation point, and it varies from plant to plant, but for a great many crops it sits well below the blazing intensity of midday summer sun. Past that point, the extra light does the plant no good at all — the photosynthetic machinery is already running flat out, saturated, unable to use another photon. Worse, all that surplus sunlight brings heat and drought stress: the leaf bakes, the plant closes the tiny pores in its leaves to avoid losing water, and photosynthesis actually slows down. A tomato or a lettuce or a stand of grass in the fierce afternoon sun of a hot climate is often not basking in ideal conditions; it is enduring an excess it cannot use and struggling against the heat and thirst that come with it.

Now the logic of agrivoltaics snaps into focus. If you raise solar panels high above such a field, or space them out in strips so they cast a shifting, dappled shade rather than a solid roof, the panels intercept the surplus sunlight — the light beyond the plant's saturation point, the light the crop could not have used anyway — and turn it into electricity. The crop below still receives all the light it can actually process, and in return it gets shelter from the worst of the heat. The shade lowers the temperature at ground level, reduces the water the plants lose to evaporation and transpiration, and softens the punishing midday peak. For crops in hot, dry, sun-drenched climates, this can be not a penalty but a positive help, and here is the fact that startles people most: some crops actually yield more under partial shade than in full sun. In the right conditions, a shaded tomato plant, or a stand of certain leafy greens, or pasture grass, produces as much or more while using noticeably less water. The panels are not stealing from the crop; they are harvesting the light the crop was going to waste. That is the "second harvest" — two useful yields from one patch of ground, from light that would otherwise have gone begging.

The water saving deserves a moment on its own, because in a warming, drying world it may matter as much as the food or the electricity. A field under partial solar shade loses less moisture to the air; the soil stays damper, the plants transpire less, and a farmer can grow the same crop with meaningfully less irrigation. In arid regions where water, not land, is the binding constraint on agriculture, that co-benefit can be the whole point — the panels make the farming possible, not merely coincidental. It is the same physical gift we met with floating solar, where shading a reservoir cut its evaporation: shade conserves water, and here the shade happens to be made of power-generating glass.

Think of it by analogy with a familiar experience. A gardener knows that many plants — lettuces, spinach, soft fruit, shade-tolerant herbs — actually scorch and bolt in the full glare of high summer, and that they do better in the dappled light beneath a tree or against a north-facing wall. The tree does not starve the lettuce; it filters the excess, and the lettuce is grateful for it. Agrivoltaics is that same intuition rendered in steel and glass: the panels are an engineered canopy, a tree that also happens to generate electricity, throwing a moving shade that spares the crop the worst of the midday sun. A second analogy comes from the kitchen. A pan on a stove can only absorb heat so fast; turn the flame past a certain point and you are not cooking faster, only scorching the food and wasting gas. A leaf is much the same — past its saturation point, extra sunlight is a flame turned up too high, doing no good and some harm. The panels simply carry off the surplus that was going to waste.

There is a second, humbler form of agrivoltaics that is even more widespread, and if you have driven past a solar farm in Britain, or Germany, or the American Midwest and seen sheep grazing placidly among the panels, you have already met it. This is solar grazing, and it has quietly become one of the most common land-management practices in the whole solar industry. A solar farm's ground has to be kept clear — vegetation left to grow wild will eventually shade the lowest panels, and the grass and weeds must be cut back somehow. You can send in crews with mowers and herbicides, burning fuel and money, or you can turn the whole site over to a flock of sheep. The sheep eat the grass, keep it trimmed to the perfect height, fertilize the soil as they go, and shelter from sun and rain beneath the panels; the farmer earns income from the flock and often a fee for the grazing service; and the solar operator gets its vegetation managed for less than machines would cost. Sheep, it turns out, are nearly ideal solar-farm groundskeepers — small enough to walk under the panels, uninterested in chewing cables, content in the partial shade. Cattle are too large and too rough on the equipment, and goats will eat anything including the wiring, but sheep are a natural fit, and solar grazing has grown from a novelty into a standard practice across much of the industry.

The land beneath and between panels can be put to gentler uses too. Rather than covering a solar site in gravel or bare, sprayed earth, developers increasingly sow it with wildflowers and native grasses, creating what are sometimes called pollinator-friendly solar meadows: a carpet of flowering plants beneath the arrays that feeds bees, butterflies, and other pollinators whose populations have been collapsing across the industrialized world. The meadow stabilizes the soil, needs no herbicide, supports the very insects that neighboring farms depend on to set their fruit, and turns what might have been a dead expanse of stone into functioning habitat. A well-designed solar farm, far from being a biological desert, can host more life than the intensively farmed monoculture it replaced.

Which brings us to the misconception this chapter exists to demolish. The charge, made sincerely by people who love the countryside, is that solar farms "sterilize" the land — that they smother good soil under glass and turn living fields into industrial dead zones. It is a real fear and it deserves a fair answer, not a dismissal. The answer is that it need not be true, and increasingly is not: agrivoltaics grows crops beneath the panels, solar grazing runs livestock among them, and pollinator plantings turn the ground into meadow. Land under a thoughtfully built solar array can go on producing food, supporting animals, and sheltering wildlife. The either-or — food or power, life or panels — is a false choice. The honest counterpoint, to keep this even-handed, is that not every solar farm is built this way; plenty are still gravelled and sterile, because that is cheaper and simpler, and agrivoltaics adds cost and complexity, since raising panels higher and spacing them wider to let crops grow means more steel, more land per megawatt, and more careful design. So it is fair to say that solar need not sterilize land, and also fair to say that making it coexist with agriculture takes deliberate effort and money that not every developer will spend. The point is that the sterile solar farm is a choice, not a law of nature.

There is a further, quieter benefit that flows in the other direction, from the plants back to the panels, and it closes a satisfying loop. Recall from the discussion of floating solar that a photovoltaic cell loses efficiency as it heats up — a baking panel makes less power than a cool one. A crop growing beneath and around the panels transpires, breathing water vapor into the air, and the living ground cover keeps the surface cooler than bare gravel or asphalt would. That moisture and shade can lower the temperature around the panels by a useful margin, and a cooler panel is a more productive panel. So the field and the array trade favors: the panels shade the crop and save its water, and the crop, by keeping the ground green and damp, cools the panels and lifts their output. It is the kind of mutual benefit that engineers rarely get for free, and it is one more reason the second harvest can come out ahead rather than merely even.

The second misconception is that agrivoltaics is a fringe curiosity, a few experimental plots for enthusiasts. It was, once — Nagashima's early racks really were a hobbyist's experiment. But it is scaling fast. Japan, where the idea was born, has thousands of agrivoltaic installations threading solar panels above rice paddies and vegetable plots on its scarce farmland. Germany and France have built serious pilot farms growing everything from apples to wine grapes beneath and between panels, testing which crops thrive and how high the racks must stand for a tractor to pass underneath. In the United States, hundreds of sites now combine solar with grazing, pollinator habitat, or crops, amounting to gigawatts of capacity and growing quickly, tracked and studied by a research community that barely existed a decade ago. Across the sun-belts of Asia, Europe, and the Americas, agrivoltaics has moved from oddity to a fast-growing, seriously funded branch of the solar industry — not the mainstream yet, but no longer fringe, and expanding every year.

Let me carry through one worked illustration of the second harvest, because the balance of light is the crux of the whole idea and it repays being made concrete. Picture a field in a hot, sunny region receiving, at midday, that familiar thousand watts of sunlight on every square meter. Suppose the crop growing there — say a leafy green, or a forage grass — reaches its light saturation point at perhaps a third of that intensity: beyond a few hundred watts per square meter, its photosynthesis is maxed out and the extra light is wasted, converted only into heat and water stress. Now raise panels above the field, arranged to intercept, say, half the incoming light on average across the day, leaving the crop with the other half. The crop still receives more than its saturation point for most of the day, so its growth barely suffers — and it now enjoys cooler soil and lower water loss into the bargain. Meanwhile the panels are converting that intercepted half into electricity. You have not divided one harvest in two; you have taken light the plant could never have used and turned it into power, while the plant keeps very nearly all the growth it would have had. Two yields, one field, from the honest accounting of how much light a leaf can actually swallow. That is why the arithmetic of agrivoltaics can come out ahead rather than merely even — the second harvest is drawn largely from waste.

Where do you meet it today? You meet it in the sheep grazing beneath the arrays of a thousand solar farms across Britain and the American countryside, so ordinary now that no one photographs it. You meet it in the wildflower meadows humming with bees under the panels of a well-built modern farm. You meet it in Japanese vegetable plots and French vineyards and German orchards growing their crops in the dappled shade of solar racks, and in the research stations from Arizona to Bavaria measuring, plant by plant, exactly how much shade each crop prefers. And you meet it, most importantly, as a rebuttal — every time someone argues that solar power and food production must compete for the same acres, agrivoltaics stands as the living proof that they need not.

We have travelled a long way in these five chapters, out from the shallow North Sea to the deep-water platforms of the open ocean, across the mirror-fields of the desert and the gigawatt solar parks of Rajasthan and China, and back at last to a Japanese hillside where a retired engineer discovered that panels and plants could share the sun. What ties it all together is a single recurring lesson: the sun and the wind are effectively boundless, and the entire art of harvesting them is an art of finding room — on the deep sea, on the reservoir, in the desert, and now in the very fields that feed us, where a second harvest of light can be gathered above the first harvest of food without either one giving way.

But none of it — not the offshore farms, not the floating platforms, not the desert parks, not the panels over the crops — would exist as it does without a change that happened not in a laboratory or a shipyard but in a parliament. The machines were ready, or nearly so, for decades before the world began building them in earnest. What finally unleashed them was policy: a set of laws, pioneered above all in Germany, that guaranteed a market for clean power and set the whole cascade of falling costs in motion. To understand how the sun and the wind went from curiosities to the cheapest power on Earth, we have to turn from the engineers to the politicians — to feed-in tariffs, and to the remarkable German experiment known as the Energiewende.

Chapter 51 — Feed-in Tariffs and the German Energiewende

On a gray morning in the Bundestag in the winter of 1999, a rumpled Social Democrat named Hermann Scheer was doing something politicians almost never do: he was arguing that the state should promise to overpay for a product almost nobody was buying, for twenty years, no matter what. The product was electricity from the sun and the wind. The critics in the chamber thought he was a dreamer with a calculator that did not work. Utilities lobbied against him. Economists warned of waste. Scheer, who had spent years being told that solar power was a hobby for idealists, kept returning to a single stubborn point. The problem with renewable energy, he said, was not that it was impossible. It was that no bank would finance a technology whose revenue nobody could predict. Take away the uncertainty, guarantee the price, and private money would pour in. Give the sun a contract, and the sun would show up for work.

The law he championed took effect on the first of April, 2000. It was called the Erneuerbare-Energien-Gesetz, the Renewable Energy Sources Act, mercifully shortened to EEG. It was not the first such measure — Germany had passed a smaller Feed-in Law in 1990 — but the EEG was the version that changed the world, and to understand why, you have to understand the elegant, almost embarrassingly simple mechanism at its heart. It is called a feed-in tariff, and it is one of the most consequential policy ideas of the last half-century, even though hardly anyone outside the energy trade has heard the phrase.

Here is how a feed-in tariff works, stripped to first principles. Imagine you own a bakery, but instead of hoping customers wander in, the government passes a law saying: anyone who bakes bread may sell every loaf to the national grocery system at a fixed, generous price, guaranteed for twenty years, and the grocery is legally required to buy it first, before bread from anyone else. You would, of course, start baking. So would your neighbors. That is the whole trick. A feed-in tariff does three things at once. It guarantees access — the grid must accept your renewable power. It guarantees a price — a fixed rate per kilowatt-hour, set above the going market rate. And it guarantees duration — typically twenty years, long enough to pay off the hardware and then some. Take those three guarantees to a bank, and the loan officer stops seeing a gamble and starts seeing a bond.

The genius of the design was that it attacked the real bottleneck, which was never sunlight and never engineering. It was risk. A solar panel in the year 2000 was a known quantity: you could calculate almost to the watt how much power it would make on a German rooftop over its lifetime. What no one could calculate was how much that power would be worth, because electricity prices swing with the price of coal and gas and the moods of the market. The feed-in tariff took the one unknowable variable and nailed it to the floor. Suddenly a homeowner in Bavaria could do the arithmetic on the back of an envelope: this many panels, this much sun, this guaranteed price, this many years, equals a return of perhaps six or eight percent — better than a savings account, and safer than the stock market. It was, in effect, a government-backed annuity that happened to be made of silicon and bolted to your roof.

And here is the part that still astonishes me. The people who signed up were not power companies. They were dentists and farmers and schoolteachers and church congregations. By the peak years of the boom, roughly half of Germany's installed renewable capacity was owned not by utilities but by ordinary citizens and cooperatives — a farmer with a barn roof, a village that pooled its savings to buy a wind turbine, a housing block that carpeted its flat roof in blue panels. This is the first surprising fact worth pausing on: the German energy transition was, to an unusual degree, a citizens' movement with a spreadsheet. People voted for it not only at the ballot box but with their pension savings. A grandmother in Freiburg could become, in a small way, a power producer, feeding electrons into the same grid as a coal baron and getting paid a premium for the privilege.

Who paid the premium? This is where the design gets clever, and also where the trouble starts. The above-market price was not paid out of general taxes. It was spread across every electricity bill in the country as a small surcharge, the EEG-Umlage. Every kilowatt-hour a German household consumed carried a few extra cents that flowed to the owners of solar panels and wind turbines. In effect, all electricity users collectively topped up the guaranteed price. The burden was diffuse — a little on each of eighty million people — while the benefit was concentrated on those who installed the hardware. That diffusion is exactly what let the policy survive politically for years, because no single bill-payer felt robbed. But as installations exploded far beyond anyone's forecast, those few cents grew, and by the mid-2010s the surcharge had climbed to more than six cents on every kilowatt-hour, helping push German household electricity prices to among the highest in Europe. A policy designed to be painless had become, for many families, quite noticeable indeed.

It helps to work a single rooftop through the arithmetic, because the whole national drama is only this sum, repeated a million times. Picture a modest German house in the mid-2000s with room for a solar array of perhaps three kilowatts — a kilowatt being a thousand watts, roughly the draw of a hair dryer running flat out. Under Germany's dishwater skies such an array might yield around three thousand kilowatt-hours a year, not much, given the clouds. But the feed-in tariff in those years guaranteed the homeowner something on the order of fifty cents for every kilowatt-hour fed into the grid — many times the going wholesale price of power — locked in for twenty years. Multiply it out and the roof earned perhaps fifteen hundred dollars a year, every year, for two decades, against an installed cost the homeowner could finance and pay off within the first half of that stretch. After that, it was profit falling from the sky. No stock, no bond, no rental flat offered a return so predictable, and none of them required nothing more of you than that the sun keep rising. The homeowner was not a green idealist making a sacrifice. He was an investor collecting a coupon. That is the alchemy the feed-in tariff performed: it turned an environmental act into a financial no-brainer, and financial no-brainers scale.

Nowhere did that logic take stranger root than in the small Black Forest town of Schonau, where a mini-story unfolded that captures the whole spirit of citizen energy. After the Chernobyl reactor disaster in 1986 dusted parts of Germany with radioactive fallout, a group of Schonau parents, led by a physician's household, grew determined to run their town on power that did not come from atoms. Told by the local utility that this was none of their business, they did something audacious: over years of referendums and fundraising drives — including a national appeal that raised money from strangers across Germany — they bought the town's electricity grid outright and founded their own cooperative power company. The "electricity rebels of Schonau," the press called them. It was a quixotic act by a handful of ordinary people, and it became a template. When a law like the EEG then guaranteed a price for the clean power such cooperatives produced, the rebels multiplied into a movement. By the boom's height, thousands of energy cooperatives dotted Germany, owned by neighbors who had pooled their savings to raise a wind turbine or blanket a field in panels. This is the texture behind the statistic that half of German renewable capacity was citizen-owned: it was not an abstraction but a nation of Schonaus.

Now, the German word for the whole endeavor is Energiewende — literally "energy turn," or energy transition — and it means far more than the feed-in tariff. It is the national project of shifting an industrial economy off fossil fuels and nuclear power toward renewables and efficiency. The word predates the EEG; it goes back to a 1980 book by a German research institute. But the EEG gave the Energiewende its engine. And the results, measured on their own terms, were genuinely remarkable. In 2000, renewables supplied about six percent of German electricity. Two decades later they supplied roughly half, and on sunny, windy afternoons the figure could momentarily exceed the entire country's demand. A cloudy, densely populated, heavily industrialized nation in northern Europe had built one of the largest renewable fleets on Earth.

Which brings us to the most delicious misconception about the whole story, one that catches almost everybody. People assume Germany led the world in solar power because Germany is sunny. It is not. Germany is famously, reliably overcast. Munich receives roughly as much annual sunlight as chilly Seattle; much of the country gets less sun than Alaska's panhandle in the wet season would make you fear. The German solar boom happened not because of the resource but in spite of it. Bavaria became one of the most solar-dense regions on the planet while sitting under skies that any Californian would call miserable. The lesson, and it is a profound one, is that in the early years the decisive variable was policy, not latitude. If you guarantee the price, the panels will come, even to a place where the sun clocks in part-time and leaves early. Germany proved that solar's destiny would be decided in parliaments before it was decided by the weather.

And that German policy, paid for by German bill-payers, is the reason your solar panels — wherever you live — are cheap. This is the second great surprise, and it is the moral center of the chapter. When Germany created enormous, guaranteed demand for solar panels, it created something the industry had never had before: a market big enough to justify building giant factories. Manufacturers, most of them soon in China, raced to supply that demand and discovered they could cut costs dramatically with every doubling of production — the phenomenon we met earlier as Swanson's law, the roughly twenty percent price drop for each doubling of cumulative solar shipments. German subsidies pulled the industry up the early, expensive part of that learning curve, and Chinese manufacturing scale drove the price down the far side. Between them, they turned solar from a luxury into the cheapest source of new electricity in much of the world. A homeowner in India or Nigeria or Texas today buys a panel at a price made possible by the electricity bills of German pensioners twenty years ago. It is one of history's stranger acts of accidental generosity: one nation's ratepayers funded a learning curve the whole planet now rides for free. When critics say the German program was wildly expensive, they are right. When defenders say it was one of the best investments in decarbonization anyone ever made, they are also right — the expense bought a global price collapse. Both things are true, and an honest account holds them together.

The critics, though, deserve their strongest case, because the Energiewende is not a fairy tale. Start with cost. Germans paid, over two decades, on the order of hundreds of billions of euros in feed-in tariff commitments, locked in by those twenty-year contracts at prices that looked reasonable in 2004 and absurdly generous by 2014, once hardware had gotten cheap. Because the tariffs were set in advance and adjusted only slowly, early adopters kept collecting premium rates long after panels had become cheap, and everyone else kept paying for it. Economists point out, fairly, that Germany could have cut its emissions faster and cheaper had it spent the same money differently — and the tariff's slow adjustment let a lot of that money flow as pure profit rather than pure decarbonization.

Then there is the nuclear question, and here even sympathetic observers wince. In the wake of Japan's Fukushima disaster in 2011, Germany accelerated a decision to shut down every one of its nuclear reactors — plants that produced enormous quantities of electricity with almost no carbon emissions. The last ones went dark in 2023. The strongest version of the critics' case is straightforward and hard to dodge: by retiring zero-carbon nuclear plants at the same time it was fighting to reduce carbon, Germany forced itself to lean, for years, on coal — including lignite, the dirtiest fossil fuel of all — to fill the gaps when the wind was calm and the famous clouds rolled in. For a stretch, German electricity remained substantially more carbon-intensive than that of neighboring France, which had quietly decarbonized decades earlier with nuclear power. Critics argue, with real force, that had Germany kept its reactors running and phased out coal first, it would have cut far more carbon for the same money. Defenders answer, also with force, that the nuclear phase-out reflected a democratic choice by a public with deep and long-standing fears of atomic power, that the renewable buildout was real and enormous, and that emissions did fall over the longer arc even so. You can weigh those arguments differently, but you should not pretend the trade-off did not happen.

Which leads to the last and laziest misconception, the one you hear at both extremes: that the Energiewende simply "failed." It did not fail, and it did not triumph unblemished. The record is mixed, and mixed is the honest word. It failed to be cheap. It failed, for a costly interval, to cut coal as fast as its own goals demanded, largely because of the nuclear decision. But it succeeded in building a vast renewable fleet in an unpromising climate; it succeeded in putting ownership of energy into the hands of ordinary citizens; and, most importantly for the rest of us, it succeeded in bankrolling the early, unglamorous, expensive years of an industry that then became cheap enough to spread everywhere. If you want to know whether it was worth it, the answer depends on what you count. Count only German electricity bills, and it looks expensive. Count the global price of solar panels, and it looks like one of the great bargains of the century.

You meet the legacy of that Berlin debate every time you see a rooftop covered in blue panels, whether in Stuttgart or San Diego. You meet it in the small line on some electricity bills labeled a renewable surcharge, the direct descendant of the German EEG-Umlage. You meet it, most of all, in the price tag on a solar module, which has fallen by more than ninety-nine percent across the decades — a plunge that began, in significant part, with a stubborn man convincing a parliament to guarantee a price. Hermann Scheer did not live to see the full result; he died in 2010, just as the industry he had midwifed was about to make solar the cheapest electricity in history. He had insisted that the sun would come to work if only you gave it a contract. He was right, and the contract turned out to be the most productive one Germany ever signed on the world's behalf.

But a policy that guarantees a fixed, generous price forever has a built-in flaw: it keeps paying yesterday's high price long after the technology has gotten cheap. As solar and wind matured, governments needed a way to let the market discover just how low the price had really fallen — and for that, they turned from guaranteed tariffs to open competition, and to the auction block.

Chapter 52 — Auctions and the Race to the Bottom

In April 2021, an envelope was opened in Saudi Arabia that contained a number so small the people who read it assumed there had been a mistake. The Al Shuaibah solar tender — a competition to build a giant photovoltaic plant near the Red Sea coast — had drawn bids from developers around the world, each naming the lowest price at which they would sell the plant's electricity for the life of the contract. The winning bid worked out to about 1.04 cents per kilowatt-hour. As of 2021 it was the cheapest electricity ever contracted anywhere on Earth, from any source, by a wide margin. A cent a kilowatt-hour is roughly the price of running a single dim bulb for an hour and having change left over. For comparison, a new coal or gas plant in most of the world cannot dream of touching that figure; it must pay for fuel every day it operates, forever, and sunlight arrives without an invoice.

That envelope is the emblem of a quiet revolution in how the world buys renewable power. The feed-in tariff of the previous chapter had a fatal charm: the government set the price. Bureaucrats, however well-intentioned, guessed at what solar and wind ought to cost, wrote that guess into law, and then watched the guess go stale as the technology got cheaper faster than anyone could legislate. What the world needed was a way to stop guessing and start asking — to let the developers themselves reveal how low the price had really fallen. The instrument that does this is the auction, sometimes called a tender, and it has become the dominant way large-scale renewables are bought across the planet.

The mechanism is worth understanding from first principles, because it inverts everything about a normal auction. When you picture an auction, you picture bidders shouting ever-higher numbers for a painting, and the highest bid wins. A renewable energy auction runs backward. It is what economists call a reverse auction, or a procurement auction. A government or utility announces that it wants, say, five hundred megawatts of new solar power. Developers then compete not to offer the most money but to accept the least — each submits the lowest price per kilowatt-hour at which it is willing to build the plant and sell its output, usually under a long contract of fifteen or twenty years. The lowest bidder wins the right to build. It is the reverse of the art auction: the floor, not the ceiling, is the prize. And because every developer knows that undercutting a rival by a fraction of a cent can win a project worth hundreds of millions, the pressure drives bids relentlessly downward. Hence the phrase, half admiring and half nervous, that the industry uses for the phenomenon: the race to the bottom.

Think of it another way. A feed-in tariff is like a restaurant with fixed prices printed by the health ministry — every chef gets the same guaranteed sum for a meal, whether they cook well or badly, cheaply or wastefully. An auction is like inviting every chef in the country to bid on a catering contract: whoever can feed the wedding for the least money, at the required quality, gets the job. The first system guarantees that dinner gets cooked. The second discovers, ruthlessly, how cheaply dinner can actually be cooked. Both have their place, and the world moved from the first to the second precisely because renewables had grown up. You use training wheels to teach a child to ride; you take them off once the child can fly down the hill on her own. Auctions were the training wheels coming off.

The numbers that came off the auction block over the last decade read like a series of records at a track meet where the athletes keep smashing their own times. In the mid-2010s, solar bids in the sunniest, most competitive markets — Chile's Atacama Desert, the United Arab Emirates, Mexico — tumbled below three cents a kilowatt-hour, a level that seasoned analysts had confidently predicted would not be reached for another decade or more. Then they fell below two. India's vast solar tenders drew bids near or below two cents. The Gulf states, blessed with searing sun, cheap land, cheap capital, and governments eager to diversify beyond oil, pushed lower still, until that Saudi bid of just over one cent set a mark that, as of 2021, no source of electricity in human history had ever undercut. The first surprising fact of this chapter is simply the existence of that number, and what it means: in the best sites on Earth, sunlight is now, by a wide margin, the cheapest way ever discovered to make electricity.

It is worth walking through how a single such auction actually clears, because the mechanism sounds abstract until you watch the money move. Suppose a government wants five hundred megawatts of solar — a megawatt being a million watts, enough to power a few hundred homes. It opens a window; developers submit sealed bids naming a price per kilowatt-hour and the size of the plant they will build at it. The auctioneer then stacks the bids from cheapest to dearest and walks down the stack, accepting plants until the five hundred megawatts are filled. The last bid accepted — the most expensive one that still fit — often sets the price everyone gets, or else each winner is simply held to the number it named. Either way, the developer who bid a hair too high goes home with nothing, and the developer who bid too low to build goes broke. Between those two cliffs lies a razor's edge, and every bidder must find it blind, guessing at what its rivals will do and at what its own costs will be years hence. The whole apparatus is a machine for squeezing the truth about cost out of people who have every incentive to shade it.

The speed of the plunge caught even insiders flat-footed, and the anecdotes from the mid-2010s have a slightly vertiginous quality. When Mexico and Chile ran some of their first big competitive tenders around 2016 and 2017, the winning solar prices came in so far below what officials had budgeted for that some assumed a keyboard slip or a currency error. In Chile's Atacama, where the sun is as fierce and reliable as anywhere on the planet, bids arrived that undercut not just new coal but the wholesale price of existing power, and a few coal generators found themselves outbid on their own turf. Seasoned energy economists had published careful forecasts predicting these price levels would not arrive until the 2020s or 2030s; the market blew through them in a season. There is a particular humility that comes from watching your professional forecast rendered obsolete by a sealed envelope, and the renewable auctions of that decade dispensed that humility generously.

What auctions really did, beyond crushing prices, was reveal a truth that had been hiding behind the subsidies. For years, the debate over renewables had been tangled in argument about how much support they needed and deserved. Auctions cut through it by forcing developers to put their own money where their forecasts were. When a company bids one and a half cents a kilowatt-hour and signs a twenty-year contract to deliver at that price, it is staking hundreds of millions of dollars on the claim that solar has become genuinely, unsubsidizedly cheap. You cannot bluff a bid like that; if you cannot build the plant for the money, you lose the money. The auction is a lie detector for cost claims. And what it detected, again and again, was that in good locations renewables no longer needed the crutch of a generous fixed tariff. The subsidy could shrink from a fat guaranteed premium to, in the best cases, nothing at all beyond the contract itself.

This marks a genuine philosophical shift in how governments push technology, and it is worth naming plainly. The feed-in tariff was what policy scholars call technology-push: the state reaches into the market, picks up an infant industry, and carries it, paying whatever it takes to keep it alive until it can walk. The auction is market-pull: the state simply says what it wants and lets competing firms fight over who can supply it cheapest, trusting that the industry is now strong enough to compete. The move from tariffs to auctions is, in miniature, the story of solar and wind growing from subsidized experiments into the default choice of hard-nosed procurement officers who care nothing for the environment and everything for the bottom line. When a Gulf petro-state, of all customers, buys solar not out of idealism but because it is the cheapest kilowatt-hour on the menu, you know the technology has crossed a threshold that no amount of advocacy could have talked it across.

But the race to the bottom has a dark verge, and honesty requires driving right up to it. The first danger is the winner's curse, an old idea from auction theory. In any competitive bid, the winner is by definition the party that was most optimistic — most aggressive about future costs, most willing to assume everything goes right. Sometimes that optimism is justified. Sometimes the winner has simply bid too low to survive, and the project quietly dies: the developer cannot line up financing, or a supplier raises prices, or interest rates climb, and the plant that "won" is never built. Underbuilding is a real and recurring problem with auctions. A tender can post a gorgeous record-low price and then deliver, two years later, an empty field, because the number that won the paper contest could not survive contact with reality. Several countries have watched celebrated auction results evaporate this way. A guaranteed feed-in tariff, for all its expense, at least tended to get things built.

The second danger is that relentless price pressure squeezes the whole supply chain until something snaps. Razor-thin margins mean developers shave every cost, sometimes cutting corners on quality, sometimes betting on hardware prices that then rise. When the global cost of money jumped in the years around 2022 and 2023 — interest rates climbing to fight inflation — a wave of renewable projects that had won auctions at rock-bottom prices suddenly no longer penciled out, because renewables, being all upfront capital and no fuel, are exquisitely sensitive to the cost of borrowing. Offshore wind was hit especially hard; developers walked away from contracts they had won, forfeiting deposits, because building at the promised price would have meant losing a fortune. The race to the bottom, in other words, can bottom out below the level at which anyone can actually deliver. A well-designed auction guards against this with qualification requirements, penalties for non-delivery, and indexing to inflation — but designing the auction well turns out to be as much an art as running one.

Not every country buys its renewables by auction, and the world's largest clean-energy economy uses a third mechanism entirely: the tax credit. The United States, for reasons rooted in its politics, has long preferred to subsidize through the tax code rather than through electricity prices. Two instruments have done most of the work. The Investment Tax Credit, or ITC, hands the builder of a solar plant a credit worth a percentage — long set around thirty percent — of the project's upfront cost; it rewards you for building. The Production Tax Credit, or PTC, favored by wind, instead pays a set amount for every kilowatt-hour the plant actually generates over its first years; it rewards you for producing. The distinction matters: the ITC helps a technology that is expensive to build but whose output is predictable, like solar; the PTC helps a technology whose value is in the electrons it reliably churns out, like wind. Both are subsidies, but they pull different levers.

In August 2022, the United States dramatically enlarged this approach with the Inflation Reduction Act, a law whose clean-energy provisions were, by any measure, enormous.

Its headline figure for climate and energy spending ran to hundreds of billions of dollars, and because many of its credits were uncapped — available to anyone who qualified, with no ceiling on the total — independent analysts estimated the true cost could run far higher over a decade, plausibly past a trillion dollars if deployment boomed. The law extended and expanded the ITC and PTC, tied their full value to wage and apprenticeship standards, added bonus credits for using domestic content and for building in communities losing fossil-fuel jobs, and — in a change that unleashed a flood of private capital — let developers sell their tax credits to other companies or collect them as direct cash payments. It also created new credits to lure solar and battery manufacturing back onto American soil. Supporters called it the most significant climate legislation in the country's history and pointed to the factories and projects that promptly broke ground. Critics called it staggeringly expensive, warned that its uncapped credits made the final bill impossible to control, and questioned whether taxpayers should subsidize technologies that auctions elsewhere suggested were already the cheapest option. Both critiques deserve a fair hearing, and I will not pretend to settle them here; I will only note that the law's provisions became politically contested almost immediately, and their long-term durability remains, as of this writing, genuinely uncertain.

That debate points at the two great misconceptions this chapter exists to correct. The first is the belief that renewables still need heavy subsidies to compete. In the best sites — the Atacama, the Gulf, sun-drenched India — the auction results say otherwise, and say it with money on the line: solar and wind now win on price alone, no premium required. The subsidy has become, in those places, less a crutch than a thumb on the scale. But the second misconception cuts the other way, and the balanced observer must hold it too: that subsidies are somehow a peculiar vice of renewable energy. They are not. Fossil fuels have been subsidized, directly and indirectly, for more than a century — through tax breaks for drilling, through the roads and militaries that guard oil supply, and, most enormously, through the uncounted cost of the pollution they emit and the public never charges them for. International agencies routinely tally global fossil-fuel subsidies in the trillions of dollars a year once those unpriced harms are included. To attack renewable subsidies while ignoring fossil ones is not an argument; it is a selective flashlight. An honest ledger notes that every major energy source in modern history has been helped up by the public hand, and asks not whether a source is subsidized but whether the help buys something worth having.

You meet the auction's handiwork every time you read that some country has contracted the cheapest power in its history, or that a new solar farm will sell electricity for less than the coal plant next door spends merely on fuel. You meet the tax credit every time an American solar installer mentions the thirty percent you can knock off your rooftop system, or a new battery factory opens in a state that used to mine coal. These mechanisms — tariffs, auctions, tax credits — are the three great tools by which societies have dragged clean energy from the laboratory to the grid, each suited to a different stage of the journey. The tariff bootstrapped the industry; the auction proved how cheap it had become; the tax credit is one nation's bet on building the factories at home.

And that last phrase — building the factories at home — points to a deeper contest than price. For once solar and wind became the cheapest power on Earth, the question stopped being whether the world would build them and became who would control the machines, the materials, and the leverage that come with them. Energy had always been a source of national power. Now the source was changing, from fuel you must keep buying to technology you build once — and that change was about to rewrite the geopolitics of the entire planet.

Chapter 53 — The Geopolitics of Sun and Wind

On the twenty-fourth of February, 2022, Russian tanks rolled into Ukraine, and within weeks a truth that energy analysts had discussed in seminars became something every European household felt in its bones. Europe bought a large share of its natural gas from Russia, piped westward through steel arteries that had been laid, in part, to bind the two together in mutual dependence. Now that dependence was a noose. As the war ground on and pipelines were throttled and then, in one case, mysteriously blown apart beneath the Baltic Sea, the price of gas in Europe spiked to levels that shuttered factories and terrified finance ministers. Homes that had once paid modest heating bills faced winter with dread. The continent discovered, in the space of a season, that it had outsourced a share of its warmth and its industry to a hostile power, and that the power knew it.

Out of that shock came a phrase that captured the strategic logic of renewable energy better than any think-tank white paper ever had. You cannot embargo the sun. You cannot cut off the wind. No foreign capital can turn a valve and freeze the air moving over the North Sea, or dim the light falling on a Spanish plain. This is the deepest geopolitical fact about solar and wind power, and it upends five hundred years of assumptions about energy and national strength. To see why, you have to grasp a distinction that sits at the root of everything in this chapter: the difference between fuel-based energy and technology-based energy.

For all of the industrial age, energy meant fuel. Coal, oil, gas, uranium — you dug them out of specific holes in specific countries, and whoever owned the holes held power over everyone who needed what came out of them. This is the geology of geopolitics. The Persian Gulf became the pivot of the world not because of its people or its ideas but because a quirk of ancient seabeds had left a disproportionate share of the planet's easy oil beneath its sand. Nations fought wars, propped up dictators, and built globe-spanning navies to keep the fuel flowing. The Organization of the Petroleum Exporting Countries, OPEC, could — as it did in 1973, in the oil shock that opens so much of this book's story — choke the supply and send the whole industrial world into recession. Fuel-based energy is a game of chokepoints: he who controls the wellhead, the strait, the pipeline, controls the customer.

Solar and wind are different in kind, not merely in degree. Their "fuel" is sunlight and moving air, which fall on everyone, cannot be owned, cannot be shipped, cannot be hoarded, and cannot be cut off. The scarce, valuable thing is no longer the fuel; it is the machine that harvests the fuel — the panel, the turbine, the battery. And a machine, unlike an oil field, is something you buy once and then own. This is what I mean by technology-based energy. Once the hardware is bolted down, the sun and the wind arrive for free, forever, immune to embargo, blockade, or the mood of a distant strongman. A solar panel on your roof is a small declaration of energy independence. A nation's worth of them is a strategic revolution.

The energy-security argument for renewables follows directly, and it is the strongest card the enthusiasts hold. A country that runs on imported oil and gas is perpetually exposed: to price spikes it cannot control, to supply cutoffs it cannot prevent, and to the uncomfortable politics of buying its lifeblood from regimes it might otherwise oppose. A country that runs on domestic sun and wind is insulated from all of that. It may still import the hardware, but hardware is a one-time purchase you can stockpile, diversify, and eventually build yourself, whereas fuel is a subscription you can never cancel and your supplier can raise at will. This is why the Ukraine invasion became, paradoxically, one of the great accelerants of the renewable transition. Europe responded to the gas shock not only by scrambling for alternative pipelines and shiploads of liquefied gas but by pouring money into solar, wind, heat pumps, and efficiency at a pace that would have seemed fantastical a year earlier. The European Union rolled out an emergency plan explicitly framed around ending its dependence on Russian fossil fuels, and renewables were its centerpiece. Vladimir Putin, without intending to, may have done more to speed European decarbonization than a decade of climate summits. This is the first great surprise of the chapter: the sharpest acceleration of the clean-energy transition in Europe was triggered not by an environmental awakening but by a war and the naked leverage of a gas supplier.

The scale of Europe's scramble makes the point in hard numbers. Before the war, Russia supplied something like forty percent of the European Union's imported natural gas — a dependence built up over decades and, in hindsight, dangerously unhedged. When that supply was throttled, Europe did not merely shiver; it acted, and the pace astonished even seasoned observers. In the year after the invasion, the continent installed record volumes of solar power, roughly forty gigawatts of it — a gigawatt being a billion watts, about the output of a large conventional power station — in a single year, and it ramped up wind, heat pumps, and efficiency measures alongside. Governments that had debated renewable targets for years suddenly treated them as matters of national survival, because they were. There is a grim historical rhyme here worth naming. The 1973 oil embargo, which shadows the early chapters of this book, was the last time an energy weapon of comparable force was wielded against the industrial democracies, and it too triggered a burst of investment in alternatives. The difference is that in 1973 the alternatives were embryonic; in 2022 they were cheap, proven, and ready to deploy at scale. The weapon was the same shape; this time the target had armor.

Now, if the story ended there — free sun, no embargoes, energy independence for all — it would be a fairy tale, and this book does not deal in fairy tales. The critics and the realists have a powerful rejoinder, and it is the second half of the chapter's argument: renewables do not abolish energy geopolitics. They relocate it. The leverage does not vanish; it moves — from those who own the fuel to those who make the machines and refine the minerals inside them. And here the map of power redraws itself in an unfamiliar shape.

Consider where the machines come from. The overwhelming majority of the world's solar panels are manufactured in China, which as of the mid-2020s produced upward of eighty or ninety percent of the key steps in the photovoltaic supply chain — the polysilicon, the wafers, the cells, the finished modules. China also makes most of the world's batteries and dominates the processing of the critical minerals that go into both. A world that buys its energy hardware overwhelmingly from a single country has not escaped dependence; it has swapped an oil dependence spread across many petro-states for a manufacturing dependence concentrated in one. The Gulf could squeeze oil; Beijing, in principle, could squeeze solar panels, or the lithium inside a battery, or the rare-earth magnets inside a turbine. This is why strategists have taken to calling minerals and manufacturing "the new oil" — not because they burn, but because they are becoming the chokepoints of the coming energy order. It is a striking inversion: the second great surprise of this chapter is that the transition away from fuel may hand as much geopolitical leverage to the world's dominant factory as the age of oil ever handed to the world's dominant wells.

The balanced view holds both truths at once, and refuses to collapse into either the techno-optimist's dream or the skeptic's despair. It is genuinely true that sun and wind cannot be embargoed, and that a nation running on them is far less vulnerable to the classic energy blackmail of the fossil age. It is equally true that the hardware to harvest them comes, for now, from a narrow set of suppliers, and that this creates new vulnerabilities of a different shape. The crucial distinction is between a one-time exposure and a permanent one. If China cut off solar-panel exports tomorrow, it would be a serious shock — but the panels already installed would keep making power for decades, and other countries could, with effort and money, build their own factories, as the United States and Europe and India are now scrambling to do. If a petro-state cuts off oil, by contrast, the pain is immediate and unrelenting, because you need the next barrel tomorrow and the one after that. A dependence on machines is a problem you can eventually engineer your way out of. A dependence on fuel is a problem you must pay for every single day, forever. That asymmetry is the heart of the matter, and reasonable people weigh it differently.

There is a subtler shift, too, in the very shape of the energy trade. Oil and gas move around the planet in a vast circulatory system of tankers, pipelines, and terminals — physical fuel sloshing from producer to consumer, and every chokepoint along the route, every strait and canal, a place where power can be applied or war can be fought. Sunshine and wind do not travel that way; you cannot load a tanker with sunlight. But a curious thing is happening at the edges of the map: nations are beginning to invent ways to move renewable energy across borders anyway. One is the undersea power cable, thousands of miles of high-voltage line proposed to carry solar power from sun-drenched deserts to distant cities, or wind power from windy seas to inland grids. Another is green hydrogen — using cheap renewable electricity to split water into hydrogen fuel that can be shipped like any cargo, a way of bottling sunshine for export. Should these schemes mature, the geography of energy power will shift yet again, and the sun-rich nations of the desert belt may find themselves selling not oil but electrons and hydrogen — the tanker replaced by the cable, the wellhead by the solar field. It is early, and much of it may never pencil out. But it hints that even the seemingly unexportable can be given a passport, and that the map of energy winners is still being drawn.

The transition also sorts the nations of the world into winners and losers in a pattern that scrambles old alignments. The clearest losers, over the long arc, are the petro-states whose entire economic and geopolitical weight rests on the world's thirst for oil and gas. A planet that burns less of both is a planet in which their leverage — and their revenue — slowly drains away. This is why the shrewdest oil producers, the Gulf states foremost among them, are racing to build solar plants and diversify their economies before the tide goes out; the Saudi record-low solar bid we met in the last chapter is, among other things, a hedge against the twilight of the very commodity that made the kingdom rich. The clearest winners are harder to guess. Some are the manufacturing and mineral powers, China above all. But there is another set of winners that gets too little attention, and they are among the poorest countries on Earth.

Here is the third surprise, and to my mind the most hopeful. Much of the developing world — across Africa, South Asia, and Latin America — happens to be drenched in exactly the resource the new energy order runs on: sunlight. A country with abundant sun but no oil, no gas, no coal, and no money to import them was, in the fossil age, condemned to energy poverty, forever buying expensive fuel it could barely afford. In the solar age, that same country holds a first-class energy resource for free, falling out of the sky. Just as many nations skipped the era of copper telephone wires and leapt straight to mobile phones, so many are now positioned to leapfrog the age of centralized fossil grids and build distributed solar power — a panel and a battery for a village that the national grid never reached, a solar mini-grid for a town that never had reliable electricity at all. India has built vast solar capacity at record-low prices and championed an International Solar Alliance to spread the technology across the sun-rich global South. For hundreds of millions of people, the first electricity they ever use may come not from a distant coal plant but from the roof over their heads. The geography of energy fortune is being redrawn, and for once it favors places the old map left in the dark.

There is a human face to this leapfrog that fixes it better than any statistic. Across sub-Saharan Africa, where hundreds of millions of people have never had a reliable connection to any grid at all, the first light in a home has increasingly come not from a wire strung across the savanna but from a small solar panel on the roof, a cheap battery, and a couple of LED bulbs — sometimes sold on a pay-as-you-go plan through a mobile phone, in installments smaller than the kerosene the family used to burn. A child who once did homework by the fumes of a smoky lamp now reads by a clean electric bulb charged by the afternoon sun. That is the leapfrog made flesh: not a distant coal plant and a national grid that may never arrive, but a self-contained scrap of the new energy order, owned outright, fueled by a sky that asks nothing in return. Multiply it across a continent and you have an energy transition that, for the world's poorest, is not a sacrifice imposed by rich countries' guilt but the first electricity they have ever been able to afford.

Which brings us to the misconception this chapter must correct, one popular among the most ardent boosters: the idea that renewables will end energy geopolitics altogether, ushering in a peaceful world where every nation makes its own power and no one need fight over resources again. This is too pretty to be true. Renewables do not end the geopolitics of energy; they transform it. The contest over who controls the fuel becomes a contest over who controls the factories, the mineral mines, and the refineries that turn ore into the guts of a battery or a magnet. The chokepoints move from the Strait of Hormuz to the processing plants of a few dominant nations. There will be trade wars over solar tariffs, export controls on critical minerals, frantic efforts to "friend-shore" supply chains among allies, and diplomatic scrambles over who mines and refines what. The great powers will still maneuver for advantage in energy, because energy is the substrate of everything a modern economy does. The board is being reset and the pieces recut, but the game goes on.

You meet this new geopolitics whenever the news carries a story you might not have filed under "energy": a dispute over solar-panel tariffs, a country restricting exports of a mineral you have never heard of, an alliance of sun-rich nations forming to pool their advantage, a European plan to wean itself off a hostile supplier's gas. Each is a move in the same long game, the reshuffling of power that follows when the world changes what it runs on. The age of oil made deserts strategic because of what lay beneath them; the age of the sun and the wind makes those same deserts strategic again because of what falls upon them — and makes factories and mines strategic in a way wellheads once were.

And that last thread — the mines — is the one we must now follow underground. For if the new leverage lies not in fuel but in materials, then the real contest of the clean-energy age is being fought over a short list of metals with unfamiliar names: lithium and cobalt, neodymium and copper, the ores that must be dug from the earth and, more decisively, refined, before a single clean electron can flow.

Chapter 54 — Critical Minerals and the New Scramble

Two hundred meters underground in the Katanga region of the Democratic Republic of Congo, a man with a headlamp and a hand-held pick is chipping cobalt ore out of a wall, filling a sack he will haul to the surface on his back and sell for a few dollars. He is not employed by any company. He is one of hundreds of thousands of artisanal miners — the industry's polite word for people digging by hand, often in tunnels that collapse, sometimes alongside children — who supply a slice of the cobalt that ends up in the batteries of the world's electric cars and the storage packs that let solar power flow after dark. That sack of ore is the first link in a chain that ends, improbably, in a gleaming clean-energy future that most of its beneficiaries imagine as spotless. The chain does not start clean. It starts here, in the dark, with a pick.

I begin with that image not to indict clean energy but to insist on honesty, because the story of the transition cannot be told truthfully without its underground half. Every solar panel, every wind turbine, every battery is a machine, and machines are made of stuff — stuff that must be dug out of the ground somewhere, by someone. The romance of "clean" energy tends to end at the shiny blue panel and the graceful white blade, as if they materialized from sunlight itself. They did not. They came from mines. And the list of what must be mined is the hidden skeleton of the whole enterprise.

Start with the batteries, the linchpin of a grid that runs on a variable sun and an intermittent wind. A lithium-ion battery earns its name from lithium, the lightest metal, prized because it packs enormous energy into little weight. It also needs, in many chemistries, cobalt and nickel, which stabilize the battery and let it hold more charge. Then the wind turbines: the most efficient of them, especially the direct-drive machines used offshore, spin enormous permanent magnets, and the strongest permanent magnets ever made are forged from an alloy of iron, boron, and a rare-earth metal called neodymium, often seasoned with its cousins praseodymium and dysprosium. A single large direct-drive turbine can contain something on the order of hundreds of kilograms of these magnets, a substantial fraction of it rare-earth metal — a mass of exotic material spinning quietly atop a tower in the sea. The solar panels need their own inputs: high-purity silicon refined from quartz, thin films of silver to carry the current off each cell, and a great deal of aluminum for frames. And threaded through absolutely everything — every panel, every turbine, every battery, every wire that carries their power to a home — is copper, the metal that has conducted human electricity since the beginning, and which the clean transition devours in quantities the mining industry finds frankly alarming.

That copper appetite is the first surprise worth dwelling on, because it is so easy to miss. Renewable systems are far more copper-intensive than the fossil systems they replace. An offshore wind farm, with its undersea cables and generators, can use several times the copper per unit of power as a conventional plant; electric vehicles use several times the copper of a gasoline car; and the sprawling new transmission lines needed to move wind and solar power from where it is made to where it is used are, essentially, thousands of miles of copper and aluminum. Analysts warn that the world may need to mine more copper in the coming few decades than humanity has mined in all of history to date. Copper is not rare and not exotic, but it must be found, dug, and smelted, and opening a new copper mine can take fifteen or twenty years from discovery to production. The humble red metal, not the glamorous lithium, may turn out to be the tightest bottleneck of them all.

Now to correct the misconception buried in the very phrase "rare earths," because it trips up almost everyone. The rare-earth elements — neodymium and its seventeen relatives — are not, geologically speaking, rare. Neodymium is more abundant in the Earth's crust than copper or lead. The word "rare" is a historical accident, coined when these metals were first isolated with great difficulty from unusual minerals. What makes them strategically fraught is not scarcity but two other things. First, they are almost never concentrated; they are smeared thinly through ore, so extracting them means processing mountains of rock. Second, and far more important, refining them from ore into usable metal is a filthy, chemically brutal, and highly specialized business — and one country, China, chose decades ago to dominate it. This is the crucial distinction the whole chapter turns on, and it deserves its own name: the difference between mining and refining.

Here is the pattern that recurs across nearly every critical mineral. The mining is scattered around the world; the refining is concentrated in China. Lithium is dug from hard-rock mines in Australia and pumped as brine from the high salt flats of the "lithium triangle" where Chile, Argentina, and Bolivia meet — but a large majority of it is shipped to China to be processed into the battery-grade chemicals that actually go into cells. Cobalt is mined overwhelmingly in the Congo, but again refined mostly in China. Rare earths are mined in China, yes, but also in the United States and Australia and elsewhere — and then the ore from those Western mines has often been shipped to China anyway, because that is where the refineries are. China's dominance of the clean-energy supply chain, in other words, is not mainly a dominance of holes in the ground. It is a dominance of the chemical plants that turn ore into material. This is the chokepoint that matters, the one we glimpsed in the last chapter, and it is far harder for rivals to break than opening a new mine, because refining is where the hard-won expertise, the environmental burden, and the accumulated investment all sit. You can find a new lithium deposit in a decade. Building the refining industry to rival China's is the work of a generation.

The world got a sharp, early lesson in what that refining chokepoint means in the autumn of 2010, in an episode worth recounting because it was a rehearsal for anxieties now central to the transition. A collision between a Chinese fishing trawler and Japanese coast guard vessels near disputed islands escalated into a diplomatic standoff, and China — which even then dominated rare-earth production — abruptly slowed its exports of the metals to Japan, whose high-tech industry depended on them. Prices of some rare earths spiked severalfold in a matter of months. The message landed like a thunderclap in every capital that had assumed these obscure metals would always be available on the open market: a single supplier could, if it chose, turn a valve, exactly as an oil state might. Japan responded by stockpiling, recycling, and hunting for alternative sources; the United States scrambled to reopen a shuttered rare-earth mine in the California desert at a place called Mountain Pass, which had once led the world before cheaper Chinese ore drove it out of business. The mine reopened — and then discovered the deeper trap, for even after digging the ore from American ground, it long had to ship that ore to China for refining, because the refineries were there and almost nowhere else. Mining, it turned out, was the easy half. The 2010 shock was small and brief, but it was the tremor before the earthquake, and everyone in the business felt it.

The human and environmental costs of all this mining must be stated plainly, without either flinching or exaggerating, because both temptations distort the truth. The Congolese cobalt story is genuinely troubling: artisanal mining that puts children and untrained adults into dangerous pits, communities poisoned by tailings, wealth extracted while poverty remains. Lithium brine operations in the arid Andes draw down scarce water in some of the driest inhabited places on Earth, straining the indigenous communities and fragile ecosystems that share the salt flats. Rare-earth refining generates toxic and mildly radioactive waste that has scarred landscapes where it was done carelessly. Copper and nickel mines gouge open landscapes and can leave acid drainage for generations. None of this is clean. Anyone who tells you the clean-energy transition has no dirty underside is selling something.

But — and this is the balance the chapter must hold, the second temptation resisted — that underside has to be weighed against the thing it replaces, and the comparison is not close. The fossil-fuel system is also an extractive industry, an almost unimaginably larger one. Humanity digs and pumps and burns on the order of fifteen billion tonnes of coal, oil, and gas every single year, and having burned it, must do it all again the next year, forever, because fuel is consumed in the using. The mining for a solar panel or a wind turbine or a battery, by contrast, is largely a one-time affair: you extract the materials once, build the machine, and then it generates power for twenty or thirty years on sunlight and wind that require no further digging. The tonnages are not remotely comparable — the minerals for a clean-energy system are measured in millions of tonnes a year against the fossil system's billions — and much of the mined material can, in principle, be recovered and used again when the machine retires. The honest framing is not "clean energy versus no impact." It is "a large, one-time, and partly recyclable mining footprint versus a colossal, continuous, and irrecoverable one." A wind turbine's neodymium is mined once and could be reclaimed; a gas plant's fuel is burned and gone, and next week it needs more. Weighed that way, the case for the transition survives its dirty underside — but only if we are truthful that the underside exists, and only if we work to clean it up rather than pretend it away.

To feel the weight of that comparison, carry a single worked example through. A large wind turbine's direct-drive generator may hold on the order of a few hundred kilograms of rare-earth magnet material, plus the steel and copper and composite of the machine — mined once, when the turbine is built, and then left to spin for two or three decades. Over that same lifetime, an equivalently rated gas plant will have swallowed a river of fuel: not kilograms but millions of tonnes of natural gas across the years, every cubic meter of it burned and gone up the stack, each day's supply needing tomorrow's replacement. Set the two piles of "stuff" side by side and they are not in the same universe. The turbine's material is a one-time bag of ingredients you could, in principle, dig back out of the retired machine and use again; the gas plant's material is an endless conveyor belt of fuel that vanishes irreversibly into the air. This is the arithmetic that matters, and it is why the balanced verdict comes down where it does. The clean machine has a real and sometimes ugly material footprint, concentrated at the moment of its birth. The fossil machine has a vastly larger one, spread across every day of its life, and it never stops.

The misconception to name here, then, is the belief that renewable energy is "clean" in the sense of having no material footprint at all — that because sunlight and wind are free and pure, the machines that catch them must be too. They are not. They have real footprints, real mines, real human costs, and pretending otherwise does the cause no favors and hands its critics an easy and fair rebuttal. The stronger and more defensible claim is not that clean energy is impact-free but that its impact is smaller, largely one-time rather than perpetual, and increasingly recoverable — and that is a claim honest enough to survive scrutiny.

It is worth pausing on why the refining chokepoint is so

much harder to break than the mining one, because the reflex in every worried capital is to open more mines, and mines are only half the answer. Refining a metal from its ore is chemistry at industrial scale, and it is often nasty chemistry — acids, solvents, high heat, toxic tailings, and a great deal of cheap energy. It is precisely the kind of dirty, low-margin, expertise-heavy business that wealthy countries spent the last half-century happily offshoring, and that China spent the same decades methodically mastering. Rebuilding it elsewhere means not just money but time: permits, pollution controls, trained chemists, and years of trial and error to match yields that a Chinese plant already hits in its sleep. This is why a country can announce a shiny new lithium mine to great fanfare and still find, a few years later, that its ore is being shipped abroad to be turned into anything useful. The bottleneck is not in the ground; it is in the chemistry, and chemistry is stubborn. Any serious attempt to diversify the clean-energy supply chain has to grapple with the unglamorous truth that the hard, decisive step is not digging but refining — and that no one reshoring effort has yet cracked it at scale.

Two answers to the mineral problem are already taking shape, and both point beyond the mine. The first is recycling. Unlike a lump of burned coal, the lithium and cobalt and nickel in a dead battery are all still there, and the copper in a decommissioned turbine, and the silver and silicon in an old panel. A mature clean-energy system becomes, over time, its own mine — an "urban mine" of retired hardware from which materials can be recovered and fed back in, so that each generation of machines is built partly from the bones of the last. Recycling is still immature and, while raw materials remain cheap, often costlier than digging fresh ore, but as the first great wave of panels and batteries and turbines reaches the end of its life, that calculus shifts. The second answer is substitution: engineering the exotic materials out. Battery makers have moved rapidly toward chemistries that use little or no cobalt, sidestepping the Congo problem; turbine designers offer magnet-lean and geared designs that use far less rare earth; researchers hunt for sodium-based batteries that trade lithium for one of the most common elements on Earth, dissolved in every ocean. The scarce mineral of today has a way of becoming the substituted-away mineral of tomorrow, because human ingenuity, faced with a bottleneck and a price signal, is very good at finding another road.

You meet these minerals, invisibly, every day. The lithium from an Andean salt flat rides in the phone in your pocket and the car in your neighbor's driveway. The neodymium from a refinery you will never see spins in the turbines on the horizon and hums in the tiny motors that vibrate your phone and drive its speakers. The copper is everywhere current flows, which is everywhere. A modern life is wired together with a short list of metals dug from a scattered handful of places and refined, mostly, in one — and the scramble among nations to secure, diversify, and reshore that supply is one of the defining industrial contests of our century, the mineral echo of the oil rushes of the last one.

The mine, then, is the hidden foundation of the clean machine, with costs that are real but bounded, and answers that are emerging. Yet minerals and machines are not the only foundation. Behind every panel installed and every turbine raised stands a person who built it, and beyond every mine that opens or coal town that closes stands a community whose fortunes rise or fall with the transition. Having gone underground for the metals, we must come back to the surface for the people — to the jobs the transition creates, the communities it unsettles, and the promise, so much easier to make than to keep, of a transition that leaves no worker behind.

Chapter 55 — Jobs, Communities, and the Just Transition

In the hills of eastern Kentucky, in a county where the word "economy" had for a century meant one thing — coal — a former miner named Jimmy sat in a folding chair in a converted warehouse, learning to write code. The mine where his father and grandfather had worked was closed, its seams played out or priced out, the company gone, the paychecks with it. The town's grocery store had shuttered; the school's tax base had thinned; the young people had mostly left. Jimmy was in his forties, with hands that knew a continuous miner and a roof bolter but not a keyboard, and he was being asked, in effect, to become a different person in a different economy that had not asked his permission to arrive. He is one face of the phrase that hovers over this entire chapter, a phrase easy to say and brutally hard to deliver: the just transition.

Let me start, though, with the numbers that make the transition look like an unambiguous triumph, because they are real and they are large. Renewable energy has become one of the great job-creating industries of the age. Around the world, the sector employs on the order of tens of millions of people — building panels and turbines, installing them, wiring them, maintaining them, and running the factories that make their parts. In the United States, two of the single fastest-growing occupations, year after year in the government's own projections, are solar photovoltaic installer and wind turbine service technician. Not fastest-growing "green" jobs — fastest-growing jobs, period, ahead of nurses and software developers in percentage terms. This is the first surprising fact, and it is worth sitting with: at the very moment critics warn that the energy transition will destroy livelihoods, the specific trades of climbing turbines and bolting panels to roofs are expanding faster than almost any other work a person can do.

Why so labor-intensive? Here we can reason from first principles. A coal or gas plant concentrates its work: a relatively small crew runs a large, centralized machine, and the fuel does much of the "work" of making power by simply burning. Solar and wind flip this. Their "fuel" is free and does no digging or hauling, so the labor shifts from feeding the machine to building and spreading the machines themselves — and there are a great many machines, dispersed across a great many rooftops and ridgelines and fields, each needing hands to install and hands to maintain. A single gas plant might power a city from one site with a hundred workers. The equivalent in solar might be a hundred thousand rooftops, each requiring installers, or a wind farm sprawling across a county, each turbine climbed and serviced by technicians. The energy transition is, in labor terms, a shift from a few big jobs at the fuel end to many smaller jobs at the building-and-maintaining end. That is why the employment figures look the way they do.

But an aggregate is a cruel abstraction when you are the one it abstracts away. And this is the hard truth the boosters too often skate past: the jobs created and the jobs destroyed are not the same jobs, are not in the same places, and are not held by the same people. A solar-installer boom in sunny Arizona and a battery factory in Georgia do precisely nothing for a laid-off coal miner in West Virginia or a roughneck in a depleting oilfield. The transition can be a net job creator nationally and still be a catastrophe locally, and it frequently is. This is the crux of the just-transition problem, and reasoning about it clearly requires a principle borrowed from economics: the costs of a transition and the benefits of a transition tend to fall on different people. The benefits — cheaper power, cleaner air, new industries — are broad and diffuse, spread thin across a whole society. The costs are narrow and concentrated, landing hard on specific workers in specific towns whose entire identity and tax base rested on the fuel now being left in the ground.

Consider what actually happens to a fossil-fuel community when its mine or plant closes. First to go are the direct jobs, which were often the best-paid work for a hundred miles — unionized, skilled, generational. Then the indirect ones: the suppliers, the diners, the hardware stores that lived off the miners' wages. Then, quietly and devastatingly, the tax base. A coal plant or a mine is frequently the largest taxpayer in its county by an enormous margin, funding the schools, the roads, the ambulance service, the library. When it closes, the town does not merely lose jobs; it loses the money that made it a functioning town at all. Teachers are laid off, the fire department goes volunteer, the roads crack and stay cracked. The young leave. The old stay and watch. This is the phenomenon economists call "stranded workers" and "stranded communities" — people and places whose skills and prospects were, through no fault of their own, tied to an asset the world decided to strand. To wave this away, to tell a fifty-five-year-old miner that the national job numbers are excellent, is not analysis. It is contempt dressed as optimism.

The idea of a "just transition" arose precisely as an answer to this, and it is worth understanding on its own terms rather than as a slogan. The phrase originated decades ago in the North American labor movement, among trade unionists who argued that workers should not be asked to bear, alone, the cost of environmental progress that benefits everyone. A just transition, in its fullest sense, means that the society reaping the gains of cleaner energy owes something concrete to the workers and towns paying the price for it: retraining programs that actually lead to comparable-paying jobs rather than to a certificate and a shrug; investment to bring new industry to old fossil regions so people need not abandon their homes; support for the pensions and healthcare that closing companies often leave in doubt; and help for local governments to bridge the loss of their tax base. It is, at bottom, a promise that no one will be treated as acceptable collateral damage. Whether that promise is kept is another matter entirely — retraining programs have a mixed and often disappointing record, and a coding class in a warehouse is cold comfort if there are no coding jobs within three hundred miles when it ends. Skeptics of these programs are not being callous; they are being empirical. The gap between the promise of a just transition and its delivery is one of the genuine open wounds of the whole enterprise, and honesty demands admitting the record is uneven at best.

Yet there is a quieter side of the transition's local economics that gets far too little attention, and it points in a hopeful direction, especially for rural America and rural Europe. When a wind farm or a solar installation comes to farm country, it does not just employ technicians — it pays rent. A farmer who leases a corner of a field to host a wind turbine can earn thousands of dollars per turbine per year in lease payments, for twenty or thirty years, while continuing to farm the ground right up to the base of the tower; the turbine's footprint is tiny, and cattle graze and corn grows around it undisturbed. For a family farm running on thin and weather-beaten margins, that lease income can be the difference between keeping the land and losing it — a drought-proof, price-proof cash crop that yields whether or not it rains. This is the second surprising fact of the chapter: for many rural landowners, the most reliable "crop" their land now produces is electricity, harvested by a machine they did not buy, paying rent that arrives on schedule no matter what the weather does to everything else. Solar leases work similarly, and across the American Midwest and Great Plains, turbine and panel payments have become a significant and stabilizing source of rural income, along with new property-tax revenue that can revive the very county budgets that fossil closures elsewhere have gutted.

Work the farmer's arithmetic and the appeal becomes obvious. A single large turbine might pay its host landowner somewhere in the range of several thousand to well over ten thousand dollars a year in lease income, depending on the turbine's size and the wind, and a good-sized field might carry several turbines while losing only a sliver of its plantable acreage to the towers and access roads. For a family farming a few hundred acres of corn or wheat on margins that a bad season can erase, three or four turbines can throw off more reliable income than the crop itself — and unlike the crop, the checks do not fail in a drought, do not rot in a wet autumn, and do not collapse when commodity prices swing. The farmer keeps farming right up to the base of each tower. The land does double duty, growing food below and harvesting wind above, and the second harvest is the one that pays the mortgage in the lean years. There is a quiet dignity in this that the political debate often misses: the transition, at its best, does not take the farm from the farmer. It gives the farmer a second, weatherproof crop.

There is also a longer historical example that ought to temper both the optimists and the despairing, and it comes from Germany's Ruhr Valley. For more than a century the Ruhr was Europe's coal-and-steel heartland, its skyline a forest of pit-head towers and blast furnaces, its identity welded to the mines. Over the second half of the twentieth century, as cheaper coal and steel came from elsewhere, that economy died — slowly, painfully, across decades. What is instructive is not that the pain was avoided; it was not. Whole towns lost their reason for being, and the wound is still visible in the region's older industrial cities. What is instructive is that, with sustained public investment — in universities, in new industries, in retraining, in cleaning up the poisoned land — the Ruhr eventually became something else, a region of technology and services and culture rather than coal. The transition took two generations and a great deal of money and was never painless. But it demonstrated that a fossil heartland need not simply be abandoned to rot, and that a wealthy society, if it chooses to spend and to plan, can help a region become something new. It is the closest thing we have to a proof that a just transition is possible — and also a sobering reminder of how long and costly it is to do even passably well.

This local dimension is also where the transition meets its fiercest resistance, and the resistance deserves a fair hearing rather than a sneer. Not everyone wants a turbine on the ridge they look at from their kitchen window, or a sea of panels where a meadow used to be. Concerns about the view, about noise, about the character of a rural place, about whether the profits flow to distant developers while the disruption stays local — these are real, and dismissing them as mere obstruction ("not in my backyard") has repeatedly backfired, hardening opposition into organized political movements that can block projects for years. The communities that have made peace with renewables are usually the ones that were given a genuine stake in them: local ownership, community benefit funds, a share of the revenue, honest consultation before the bulldozers rather than after. The lesson, learned the hard way across many countries, is that a transition imposed on a community tends to be fought, while a transition that pays the community and asks its leave tends to be embraced. People do not object to progress so much as to being its object rather than its partner.

Layered atop all of this is the newest and most politically charged dimension: the fight over where the machines themselves are made. For years, the West was content to let the manufacturing of panels and batteries migrate to Asia, keeping mainly the installation and service jobs at home. The strategic anxieties of the last chapter — dependence on a narrow set of foreign suppliers — have now fused with the economic anxieties of this one to produce a powerful push to "reshore" clean-energy manufacturing: to build the solar factories, the battery plants, the turbine works on home soil. The argument for it braids together national security, good manufacturing jobs, and the political appeal of building things again in regions that lost their industry. The argument against it is that home-built hardware costs more than the cheapest imports, raising the price of the very clean energy the world is trying to make abundant, and that protecting domestic factories can slow the transition it means to serve. Both cases are strong; both are made in good faith; and the balance between cheap clean energy from abroad and secure clean-energy jobs at home is one of the defining economic-policy arguments of the decade, unresolved and probably unresolvable in the abstract.

The misconception this chapter exists to correct sits, as usual, at both extremes. From one side you hear that the energy transition is a job-killer, a machine for throwing miners and drillers onto the scrapheap. From the other you hear that it is a pure job-bonanza, millions of green jobs and nothing but winners. Neither is true, and the truth is more demanding than either. The transition is, in aggregate and by the best measures, a net creator of jobs — the numbers on installers and technicians are not propaganda, they are real and large. And yet it is also, unavoidably, a source of concentrated and often brutal local disruption for the workers and towns tied to the fuels being left behind. To be fair to the displaced coal miner and the anxious oil town is not to oppose the transition; it is to insist that "net positive" is an accountant's verdict, not a moral one, and that a society which reaps the broad gains owes a real and specific debt to those who bear the narrow costs. A transition can be both genuinely good for the many and genuinely painful for the few, and any account that denies either half is not to be trusted.

You meet all of this closer to home than you might think. It is in the wind technician, harnessed and helmeted, ascending a tower on a cold morning to service a gearbox. It is in the solar crew on your neighbor's roof, a trade that barely existed a generation ago. It is in the lease check that keeps a family farm in the family, and in the shuttered store on the main street of a coal town that the checks never reached. It is in the political argument, playing out in capitals and county halls, over whether the machines of the new energy should be built here or bought cheaper there. The transition is not an abstraction happening to someone else. It is remaking the map of work and the fortunes of communities in real time, and it is doing so unevenly, generously in some places and cruelly in others, exactly as large economic transformations always have.

So we have followed the sun and the wind through the physics that governs them, the machines that harvest them, the money that built them, the politics that fights over them, and the people whose livelihoods rise and fall with them. There remains the hardest question of all, the one every skeptic asks and every honest advocate must answer without flinching. It is the question a child asks and a grid operator loses sleep over, and it is where we turn next: when the sun goes down and the wind goes still, does any of this actually work?

Chapter 56 — Does It Work When the Sun Goes Down?

Just before nine o'clock on a still October evening, a control-room engineer at the Australian Energy Market Operator watched a number on her screen do something that, twenty years earlier, would have been dismissed as a fantasy. Across the state of South Australia, wind and rooftop solar were meeting nearly all the demand of more than a million people, and the gas plants that once formed the backbone of the grid sat idle, their turbines cold. Then, as the last light drained from the sky and the wind slackened, the same engineer watched the picture reverse. Solar fell to nothing within the hour, as it always does. The wind wandered. And the grid did not so much as flicker, because the operators had spent the day preparing for exactly this moment — charging batteries, holding gas in reserve, reading the weather like sailors. The lights stayed on. They always had.

That ordinary evening contains the whole of the argument this chapter must settle, because the single most common objection to solar and wind is also the most intuitive. The sun sets. The wind drops. A society cannot run on a power source that comes and goes with the weather. It is a serious objection, and it deserves to be stated at its strongest before it is answered, because the honest answer is neither the skeptic's "it can't be done" nor the enthusiast's "it's already done and it was easy." The truth lives in the harder, more interesting country between them.

Let me put the skeptic's case as forcefully as I can, because a myth is best corrected by first granting it everything it is owed. A modern grid is a machine of astonishing fussiness. It must match supply to demand not once a day but continuously, second by second, keeping the whole synchronized network humming at a fixed frequency — fifty cycles a second in most of the world, sixty in the Americas. Let supply and demand drift apart by even a little and that frequency sags or spikes; let it drift far and protective relays trip, generators disconnect to save themselves, and the failure cascades into darkness across whole regions. For a century we ran this machine with sources we could command: coal, gas, hydro, nuclear — dispatchable power, meaning power you turn up or down on request, like a tap. Wind and solar are not taps. They are non-dispatchable. Their output follows the sky, not the operator's instruction. The sun delivers nothing at midnight and little on an overcast afternoon; the wind can fall calm across an entire country for days when a stubborn high-pressure system parks itself overhead, a condition the Germans have given the marvelous name Dunkelflaute — "dark doldrums," the windless, sunless stretch of a winter anticyclone. So, the skeptic concludes, every megawatt of wind and solar you build must be shadowed by a megawatt of something firm, sitting in reserve, waiting for the weather to betray you. You are not replacing the old grid. You are building a second one and calling the first a spare.

It is a clean argument, and parts of it are simply true. Wind and solar are variable; a grid cannot be run on hope; firmness has value. What the argument gets wrong is subtler and more important: it imagines the variable source in isolation, one lonely wind farm feeding one city, and reasons as if the only remedy for a gap were an identical fossil twin. The real toolkit for running a grid on weather is far richer than that, and it is worth laying out in full, because each tool chips away at the problem from a different angle, and their combined effect is far greater than any one alone.

Begin with geography. A single wind turbine is wildly erratic; its output can swing from nothing to full power and back within an hour. But spread turbines across a thousand kilometers and the erratic individual becomes a steady crowd. When it is calm in Kansas it is often blowing in the Dakotas; the lulls of one region are filled by the gusts of another, and the summed output of a continent-spanning fleet is dramatically smoother than any single farm. This is nothing more mysterious than the law of large numbers, the same statistical smoothing that lets an insurance company predict, with uncanny accuracy, how many houses will burn this year while knowing nothing about which ones. It is why transmission lines — the unglamorous copper and steel that carry power over distance — are quietly one of the most powerful tools in the whole kit. A wire that lets Denmark sell surplus wind to Norway, and buy hydropower back at night, is worth more to a renewable grid than almost any single power plant.

There is a companion to geographic spread that the skeptic almost never mentions, and it is quietly one of the most powerful tools of all: knowing what the weather will do before it does it. The wind and the sun are variable, but they are not random. A gust that will reach a wind farm in the North Sea at four o'clock this afternoon is, at ten in the morning, already a measurable pattern of pressure and temperature that the forecasters can see coming. Over the past two decades weather prediction for the specific needs of grid operators has improved beyond recognition, folding in satellite data, on-site sensors, and increasingly machine-learning models trained on years of a particular region's quirks. A grid operator who knows, with fair confidence, that the wind will fall by nightfall can begin warming a gas plant, topping up a battery, and nudging flexible demand hours in advance. Variability you can see coming is a fundamentally different animal from variability that ambushes you, and forecasting turns much of the first kind into the routine, manageable second. It is the difference between sailing with a barometer and sailing blind.

Then there is a happy accident of physics that the two technologies share almost as a gift. The sun shines by day and, in most temperate climates, the wind blows harder by night and in winter — the very seasons and hours when solar is weakest. They are, to a useful degree, complementary, each covering the other's absences the way two shift workers cover a single job. It is not a perfect fit; nature was not so kind as to make them exact opposites. But a grid built on both is far steadier than a grid built on either, and the combination is one of the quiet reasons high-renewable systems work better in practice than the pessimists predicted on paper.

Next comes storage, the tool everyone thinks of first and understands least. A grid battery does not need to power a city for a week; it needs, most of the time, to move energy a few hours across the day — to catch the flood of cheap solar at noon and pour it back into the evening peak, when families come home and switch on everything at once. That is the shape of the daily problem, and lithium-ion batteries are extraordinarily good at exactly that shape. Picture the standard unit as a shipping container packed with battery cells; a modern grid installation is a field of these containers in neat rows, and the fleet has grown at a pace that has startled even its champions. As of 2024 the world was adding grid storage at record speed — on the order of two hundred gigawatt-hours in a single year, roughly a twelvefold increase in installed power over just four years from 2020. In California and Texas, batteries charged by midday solar now routinely carry the grid through the steepest part of the evening ramp, doing the exact job the skeptics said was impossible.

It helps to make the battery concrete, because "storage" is an abstraction that hides its own scale. Take that shipping container full of cells and ask what it holds. A large grid battery container might store on the order of a few megawatt-hours — enough, very roughly, to power a few hundred to a thousand average homes for an hour. So carrying a city of a million homes through, say, four hours of evening peak is not a matter of one heroic super-battery; it is a matter of thousands of these containers, laid out in fields the size of a few soccer pitches, beside the substations that already exist. That is a large undertaking, but it is a boringly buildable one — no new physics, just steel boxes and concrete pads multiplied — and it is being built, at the pace of that record two hundred gigawatt-hours a year. The image to hold is not a single miraculous device but an orchard of ordinary containers, each doing its small shift, summing to a reservoir that fills at noon and empties at dusk. The daily rhythm of a solar grid is, in the end, a plumbing problem, and plumbing is something civilization is very good at.

There is a subtler service the old power plants provided that deserves a word, because the sophisticated skeptic raises it and the honest advocate should not dodge it. A big coal or nuclear generator has an enormous spinning mass — tonnes of steel turning in lockstep with the grid's frequency — and that spinning mass acts as a flywheel, an automatic shock-absorber that resists sudden changes and buys the operator precious seconds when something trips. This is called inertia, and wind and solar, whose power flows through electronic inverters rather than massive spinning shafts, do not naturally provide it. For years this was cited as a hard physical wall: a grid without spinning giants, the argument ran, would be too twitchy to hold together. It turned out to be another cost rather than a wall. Modern inverters can be programmed to mimic inertia — "grid-forming" inverters that supply the same stabilizing push electronically and, in some respects, faster than a real flywheel can. South Australia installed synchronous condensers, essentially spinning flywheels with no fuel, to top up its inertia, and its inverter-heavy grid now rides through disturbances that engineers once swore it could not. The wall, on inspection, was a solvable engineering problem wearing a physicist's disguise.

The fourth tool inverts a century-old assumption. For a hundred years we treated demand as a fixed thing the grid must chase, supply forever running to catch a customer who could ask for anything at any instant. But demand can be made flexible — nudged, with the right prices and a little automation, toward the hours when clean power is abundant. An electric car does not care whether it charges at midnight or three in the morning, so long as it is full by dawn. A water heater, a factory's freezer, a building's air conditioning can all shift their appetite by an hour or two without anyone noticing. This is demand response, and in aggregate it is like giving the grid operator a soft, invisible reservoir that fills and empties on command — not by generating power, but by moving the moment we use it.

The fifth tool sounds wasteful until you look at the arithmetic. Because wind and solar have become so astonishingly cheap — cheaper, by 2024, than any fossil source on a levelized basis — it can pay to simply build too much of them and throw the surplus away. Overbuild the fleet to twice or three times the average need, and even a weak, cloudy, half-calm day still produces enough; on good days you curtail the excess, spilling power you did not have to burn anything to make. Curtailment sounds like a failure and is often reported as one, but when the fuel is free, spilling it is no more scandalous than a river running past a mill that has already ground its grain. The waste is real; it is also cheap, and cheap waste is a perfectly good engineering answer.

And finally, firmness. Here the honest advocate must simply concede the skeptic's core instinct: a grid running on very high shares of weather does still want some firm, dispatchable capacity in reserve for the worst stretches — the multi-day Dunkelflaute that geography and storage cannot fully cover. But firm does not have to mean a fleet of coal plants running around the clock. It can mean hydropower, which is stored sunshine in a lake and dispatchable on a moment's notice. It can mean a modest amount of gas that runs only a few days a year, as insurance rather than backbone. It can mean geothermal heat drawn from the deep earth, or nuclear plants humming steadily beneath the variable layer, or someday cheap hydrogen made from surplus renewables and burned when the wind fails. The point is not that firmness is unnecessary. It is that the amount of it you need, and how often it runs, both shrink dramatically once the other six tools are doing their work.

So does the grid run on weather? Look at what high-penetration systems already achieve and the answer is plainly yes, up to a point that keeps rising. Denmark generates more than half its electricity from wind across a full year and exports the surplus. South Australia, which not long ago was warned it was courting blackouts, now runs for stretches on wind and solar alone and has become a laboratory the world studies. Across 2024, wind supplied about 8 percent of global electricity and solar nearly 7 percent — together approaching 15 percent of the entire planet's power, and rising by the largest annual increments in history. Several grids now handle instantaneous shares of wind and solar that engineers a generation ago would have called reckless, and they do it without the sky falling.

Where, then, is the real ceiling? Here is the fact that both sides tend to get wrong, and it is the crux of the honest answer. Getting a grid from zero to something like 70 or 80 percent wind and solar turns out to be surprisingly manageable and, increasingly, cheap — the tools above do the work and costs mostly fall. It is the last stretch, the final 10 or 20 percent to a fully weather-powered grid, that grows expensive and hard, because covering the rare, prolonged, continent-wide lull — the deep winter Dunkelflaute — requires either enormous overbuild, or vast quantities of storage that sits idle most of the year, or firm backup you rarely run but must still pay to keep. The cost curve is not a straight line; it bends sharply upward near the top. This is why serious analysts speak of the difficulty of the "last mile" to 100 percent, and why the cheapest, most robust deep-decarbonization plans almost always keep some firm low-carbon source in the mix rather than chasing the final purist percentages with brute force.

And there is one genuinely unsolved piece, which I will not paper over. The daily storage problem is essentially cracked; the seasonal one is not. We do not yet have a cheap way to store the energy of a sunny, windy summer and release it across a dark, calm winter month. Lithium batteries are far too costly to hold energy for that long; the leading candidates — green hydrogen, vast pumped-hydro reservoirs, novel long-duration chemistries — are promising but not yet cheap at the scale a continent would need. Cheap seasonal storage is the frontier's great open question, and anyone who tells you it is already solved is selling something.

Which lets us dispatch two opposite misconceptions in a single stroke, for the truth cuts down the myth from both sides. The first, beloved of critics, is that every unit of renewable capacity must be backed one-for-one by an idling fossil plant, so nothing is really displaced. This is simply false; geographic spread, complementarity, storage, demand flexibility, and transmission mean the firm reserve needed is a fraction of the renewable fleet, and it runs a fraction of the time. But the opposite over-claim, beloved of the most fervent enthusiasts, is just as wrong: that we can flip to 100 percent renewables trivially, tomorrow, at no extra cost. We cannot. The last mile is genuinely hard, the seasonal problem genuinely open, and the costs genuinely rise near the top. A grid can run on weather far better than the skeptics feared and rather less easily than the boosters promise — and holding both of those truths at once is the beginning of thinking clearly about it.

Strip it back to first principles and the shape becomes simple. Electricity must balance every second; wind and solar are cheap but variable; and variability is not an unbreakable wall but a cost — a cost you buy down with wires, weather-hedging, batteries, flexible demand, cheap overbuild, and a slim reserve of firmness. Each tool is imperfect; together they carry a grid remarkably far, and the honest debate is not whether the sun setting ends the story but how much the final percentages are worth paying for.

That South Australian engineer, watching the sun vanish and the grid hold, was not witnessing a miracle. She was watching a machine do exactly what a hundred small, unglamorous decisions had prepared it to do. The lights stayed on because someone had built the wires, charged the batteries, and kept a little gas in reserve — not because the weather cooperated, but because the system no longer needed it to. Intermittency, it turns out, is a solvable problem right up until the last stubborn stretch, and even there it is a question of price, not possibility. But solving it on paper is one thing; solving it on the land is another, and the machines that keep our lights on must stand somewhere real — on ridgelines and in deserts, among birds and bats and the people who live in their shadow — which is where the next objections wait.

Chapter 57 — The Land, the Birds, and the Blades

At dawn in the Altamont Pass, east of San Francisco Bay, the golden hills are combed with old steel lattice towers, and on the wind that funnels through the gap turn thousands of small, fast rotors — some of the earliest utility wind machines in the world, survivors of California's 1980s wind rush. A biologist named Shawn Smallwood spent years walking those hills at first light, counting bodies. Beneath the turbines he found the feathers of golden eagles, red-tailed hawks, burrowing owls, and American kestrels — raptors that hunted the ground squirrels the towers seemed almost designed to lure them toward. His counts, published through the 2000s, put the annual toll at thousands of birds, a substantial share of them protected raptors. Here, at last, was the wind industry's original sin, documented in feathers on a hillside. Any honest reckoning with the environmental cost of renewable energy has to start here, on the ground where the damage is real, before it can put that damage into any kind of proportion.

Because proportion is the whole of this chapter. There is a temperament, common and understandable, that wants an energy source to be either clean or dirty, either the answer or the problem. But nothing that powers a civilization is free of cost. Coal blackens lungs and warms the planet; nuclear leaves waste that outlives empires; hydropower drowns valleys and blocks the runs of migrating fish. The right question is never whether wind and solar harm anything — everything at scale harms something — but how their harms compare, in kind and in size, with the harms of the alternatives and with the harm we are trying to prevent. First principles here means comparative risk, not zero risk, and the reader who insists on zero has simply opted out of the real choice.

Start with land, because it is the objection you can see from an airplane window. Wind and solar are, unavoidably, diffuse. The energy in sunlight and moving air is spread thinly across the sky, so gathering a large amount of it means occupying a large area — far more land per unit of energy than a coal plant or a nuclear station, whose fuel is concentrated into a black rock or a uranium pellet dug from a small hole. This is real physics and there is no wishing it away. A solar farm producing as much energy as a compact nuclear plant may spread across many square kilometers of panels; a wind farm's turbines march across ridgelines and prairies for miles. The critic who says renewables are "land-hungry" is telling a genuine truth about their nature.

But the figures that reach the public are almost always exaggerated, and in two specific ways. The first is that the alarming numbers usually count the entire footprint of a wind farm — every acre inside the fence — when the turbines themselves, their pads and access roads, occupy only a small fraction of it. The land between the towers is not consumed; it is farmed, grazed, and lived on. A wheat field with turbines is still a wheat field, and the farmer collects a lease check besides. This is dual use, and for wind it is the rule, not the exception: cattle graze in the shade of the rotors, combines harvest around the bases, and the "land use" of a wind farm, properly measured as land actually taken out of other service, is startlingly small. Solar is more totalizing — a field of panels does shade out most conventional crops — but here too the picture is softening, as agrivoltaics matches shade-tolerant crops and pollinator habitat and grazing sheep with the panel rows; by late 2024 the United States alone had several hundred such dual-use sites. And the second correction is one of sheer scale: powering an entire country from solar requires a genuinely small percentage of its land, often a single-digit fraction, much of it marginal desert or rooftop that competes with nothing. The rooftops are the quiet answer to the whole objection — a solar array on a warehouse roof or a suburban home occupies land already occupied, stealing nothing from field or forest, and the rooftops of the world's buildings could carry a staggering amount of the world's power.

It is worth doing the arithmetic once, because a worked number defuses the "they'll pave the whole country" fear more thoroughly than any assurance. Recall the physics from earlier chapters: at solar noon under a clear sky, roughly a kilowatt of sunlight falls on every square meter of ground, and a modern panel turns something like a fifth of that into electricity. Run those figures across a nation's yearly appetite for power and you find that a country could meet a very large share of its total electricity from solar occupying only a low single-digit percentage of its land — and a good deal of that could sit on roofs, over parking lots, along highway margins, and on desert that grows nothing. The mental image of panels stretching horizon to horizon is a trick of perspective; a solar farm is vast to the person standing in it and vanishingly small against the map of a nation. The land genuinely used, netted against rooftops and marginal ground, is a rounding error beside the land already given over to growing animal feed, or to the sprawl of roads, or — the pointed comparison — to the mining and drilling that feed the fossil plants solar would replace.

Now to the birds and the bats, where the argument is fiercest and the emotion strongest, and where the honest verdict cuts squarely down the middle. It is true, and must be said plainly, that wind turbines kill birds and bats. Smallwood's feathers on the Altamont hills were not a fabrication of the fossil lobby; they were real animals, and raptors especially, drawn to hunt the open grassland the turbines command. Bats die too, and in even stranger numbers — not always struck by the blades, but killed by a phenomenon called barotrauma, in which the sudden drop in air pressure in the low-pressure zone behind a spinning blade tip ruptures the delicate blood vessels of their lungs. A bat can die near a turbine without ever being touched by it, its lungs failing in the invisible wake. This is real, it is documented, and the industry spent years reluctant to admit its full extent.

And yet. Set the toll in the company of the other things that kill birds, and the picture reorders itself completely. In the United States, wind turbines kill on the order of hundreds of thousands to perhaps a million birds a year, and every one is a loss. But domestic and feral cats kill birds in the United States by the billions — estimates run to over two billion a year, a number so large it dwarfs every industrial cause combined. Collisions with buildings and glass windows kill on the order of a billion more. Cars and trucks kill hundreds of millions. Set beside these, the entire wind fleet's toll is a rounding error, a fraction of a percent of the birds lost to housecats alone. This is not a rhetorical trick; it is the single most important fact in the whole debate, and it is why the sober conservation organizations, the ones that count carefully, support well-sited wind even as they push hard to reduce its toll. The turbine is a real killer of birds in exactly the way a picture window is, and no one proposes we ban windows.

There is a deeper comparison still, and it is the one that ought to settle the matter for anyone who cares about birds as birds. The gravest threat to bird populations is not any machine but the changing climate — the shifting ranges, failing food supplies, drowned coastal habitats, and intensifying storms that a warming world is already inflicting on wildlife, and that fossil combustion drives. Fossil-fuel infrastructure kills birds directly too, in oil-pit drownings and power-line collisions and coal-tailing ponds. To oppose wind on behalf of birds while burning the fuel that threatens birds most is to strain at a gnat and swallow a camel. The comparison is not close.

I want to be careful, though, not to let the comforting aggregate erase a real and specific harm, because the two live together and both are true. The reassuring fact — that turbines kill a tiny fraction of what cats and windows do — is a statement about totals, about the vast anonymous mass of common songbirds. But conservation does not only count totals; it counts particular populations, and here the wind industry's harm is sharper than the aggregate suggests. Cats kill sparrows and robins by the billion, but they do not kill golden eagles; turbines, sited badly on a raptor's hunting ground, can. A machine that kills a handful of eagles a year may matter more to a slow-breeding, long-lived, already-thin eagle population than a housecat that kills a thousand sparrows matters to the endlessly abundant sparrow. This is why the reassurance must never curdle into complacency, and why the sober conservationists who defend wind on the numbers also fight hardest over exactly where each turbine goes. The aggregate exonerates the technology; it does not excuse a careless siting decision on a ridgeline full of hawks. Both halves of that sentence are the truth.

Solar has its own version of the land-and-wildlife question, quieter but real. A large desert solar installation grades and clears habitat that looks empty to a passing eye but is not the desert tortoise, the kit fox, the rare plants of a specific washland all live in ground a bulldozer treats as blank. And there was for years a vivid, alarming story about the Ivanpah solar tower in the Mojave, where the intense concentrated sunlight of the central-receiver design was said to singe birds in flight, "streamers" of smoke marking their passage. The phenomenon was real but its scale was heavily sensationalized, and it belongs mostly to that particular, now largely superseded tower technology, not to the flat photovoltaic panels that make up virtually all solar being built today. Siting a solar farm on already-disturbed land, on rooftops, over reservoirs, or paired with agriculture answers most of the habitat objection before it arises.

The newest front is offshore, and it deserves the most careful handling of all, because it is where honest concern and organized misinformation are hardest to tell apart. As offshore wind expanded along the eastern coast of the United States in the early 2020s, a wave of dead whales washing ashore prompted loud claims that the surveys and construction were killing them. The claim spread fast and felt plausible. But the scientific agency charged with protecting marine mammals, the National Oceanic and Atmospheric Administration, found — as of 2024 — no evidence linking the whale deaths to offshore wind activity; the elevated strandings tracked far better with shifting prey, warming waters drawing whales into busy shipping lanes, and vessel strikes that long predated any turbine. To say there is no established link is not to say there is nothing to worry about, and here the balance matters: the noise of pile-driving foundations into the seabed genuinely can harm marine life, the geophysical surveys emit sound, and the long-term effects of vast turbine arrays on fisheries and currents are real subjects of study, not settled questions. The honest position is neither "offshore wind is slaughtering whales," which the evidence does not support, nor "it touches nothing," which is complacent. It is that the documented harms are addressable and the dramatic ones are, so far, unproven.

And addressable they largely are, which is the hopeful half of the story, because the mitigations are clever and cheap. Bats fly most on warm, low-wind nights, so operators can simply raise the turbine's cut-in speed on those nights — letting the blades idle when a light breeze would earn little power but kill many bats — and studies find this "curfewing" cuts bat deaths dramatically for a trivial loss of generation. Radar and camera systems can spot an incoming flock or an approaching eagle and briefly stop the rotor. Careful siting keeps turbines off the great migratory flyways and away from raptor-rich passes like the one at Altamont, whose worst old machines have been steadily replaced by fewer, larger, slower-turning modern turbines that sweep the same air with far less lethal frequency. Offshore, builders now use bubble curtains — a ring of rising air bubbles around a pile-driver that muffles the underwater sound — and restrict noisy work to seasons when whales are elsewhere.

There is a further point in wind and solar's favor that rarely makes the wildlife debate but belongs in any honest ledger, and it concerns water. A coal or nuclear or gas thermal plant is, at heart, a machine for boiling water, and it drinks prodigiously — pulling vast volumes from rivers and lakes to cool its condensers, warming and returning it in ways that stress the fish and ecosystems downstream. A wind turbine and a photovoltaic panel consume essentially no water to run; the sun and the wind cost the rivers nothing. In a warming, drying world where water is becoming the scarcer resource, this is no small environmental credit, and it partly offsets the land the panels take. The one exception proves the rule: concentrating solar plants that boil water with mirrored heat do need cooling water, which is awkward precisely because the best places for them are deserts. But the flat panels that make up the overwhelming bulk of solar today sit in the sun and sip nothing, an environmental virtue quiet enough that almost no one thinks to count it.

My favorite of all these fixes is also the simplest, and it is worth telling as its own small story. For years, engineers assumed a spinning rotor was invisible to a bird because the fast-moving blades blur into a transparent disc — a phenomenon called motion smear, the same reason you cannot see the individual blades of a running fan. A Norwegian study on the island of Smola tested a startlingly low-tech idea: paint a single one of the three blades matte black. The black blade breaks the smear, giving the bird's eye something solid to register and dodge. Over years of before-and-after counts, turbines with one black blade killed dramatically fewer birds — on the order of seventy percent fewer at that site. A can of black paint, it turned out, may do more for raptors than any amount of litigation. The result is now being tested more widely, and it stands as a small emblem of the whole chapter's lesson: these harms are not the immovable nature of the technology but engineering problems, and engineering problems yield to attention.

So let me name the two misconceptions this chapter must correct, because the truth again lies between them and swallows both. The first, the doom version, is that wind turbines are a bird apocalypse, mincing the skies clean of raptors. They are not; their toll is real but tiny beside cats, glass, cars, and above all the climate change that fossil fuels drive. The second, the boosters' version, is that renewables harm nothing at all, that a solar farm or a wind ridge is a pure green good with no cost to the living world. That is false too; the birds and bats and tortoises and grated desert are real losses, owed real mitigation and honest accounting. To correct only the first myth is to become a salesman; to correct only the second is to become a concern troll. A grown-up weighs both.

Reduced to first principles, the whole matter is a comparison, not an absolution. Wind and solar do use more land and do kill some wildlife, and pretending otherwise forfeits your credibility. But measured against the land degraded by mining and drilling, against the birds killed by fossil infrastructure and by a warming world, and against the diffuse enormity of climate harm, their footprint is smaller by orders of magnitude — and shrinking, as siting improves and a can of black paint teaches us how much a small idea can save. The choice is never between an energy source that harms and one that does not. It is between larger harms and smaller ones, and clear sight means keeping the scale in view.

The bodies beneath the Altamont towers were real, and so is the wheat that grows between the turbines and the black blade that now spares the hawk. Every machine that gathers the sun and the wind eventually wears out, though — the panel dims, the blade cracks, the rotor stills for the last time — and what becomes of it then has become an objection all its own, complete with a haunting image of blades buried in the ground.

Chapter 58 — What Happens to a Dead Panel?

In a windswept lot outside Casper, Wyoming, a photographer in 2019 captured an image that would circle the world. Row upon row of wind-turbine blades, each longer than a passenger jet's wing, lay stacked in the bare earth of a landfill, sawn into segments and waiting to be buried. The picture was arresting precisely because the blades looked so wrong lying down — these elegant white airfoils, engineered to slice the sky, dumped like the bones of some grounded flock. "The wind turbine blade graveyard," the headlines called it, and the image became the single most powerful piece of ammunition in the case that renewable energy is not clean at all, merely deferring its filth to a landfill down the road. It is a fair place to begin, because the end of a machine's life is a real cost, and the blades in that Wyoming field are a genuine, unsolved problem — just not the one the picture implies.

Everything wears out. A solar panel bolted to a roof will generate for a very long time — modern modules are warrantied for twenty-five to thirty years and typically keep working, at slowly declining output, well beyond that, so that a panel installed today may still be feeding a grid in the

2060s. A wind turbine runs for roughly twenty to twenty-five years before the economics favor replacing it. But eventually the panel dims below usefulness and the blade develops the microscopic cracks that make it unsafe, and then a society that has built millions of them must answer a question it has barely begun to face at scale: what do you do with the corpse? The honest answer has three parts — the blades, which are genuinely hard; the panels, which are genuinely recyclable; and the two great "gotcha" myths about panels, which are genuinely false — and they must be taken in turn.

The blades are the hard part, and the Wyoming graveyard tells no lie about that. To understand why, you have to understand what a blade is. It is not metal. It is a composite — long fibers of glass or carbon laid in a matrix of hardened resin, the same family of materials as a boat hull or a Formula One chassis. Composites are miraculous for a blade because they are stiff, light, and strong in exactly the directions an engineer specifies. But the very chemistry that makes them tough makes them nearly impossible to un-make. The resins are usually thermosets, meaning that once they cure they are cured for good; unlike a thermoplastic such as a milk jug, which you can melt down and remold, a thermoset does not melt. It is like the difference between butter, which you can soften and reshape at will, and a boiled egg, which having once set will never flow again no matter how you heat it. You cannot melt a blade back into fiber and resin any more than you can un-boil an egg. So for years the cheapest thing to do with a dead blade was to cut it up and bury it, and that is what those Wyoming pictures show.

But even here the situation is better than the graveyard image suggests, and improving fast, and the story of the fixes is worth telling because it shows an industry problem-solving in real time. One approach is co-processing in cement kilns: a shredded blade is fed into the furnace that makes cement, where the glass fiber becomes raw material for the cement itself and the resin burns as fuel, so that the blade is not buried but consumed and transformed, a technique already running at commercial scale in Europe. Another approach attacks the chemistry directly, with new "recyclable resins" designed from the start to be dissolved back into their ingredients at end of life — turning the boiled egg, in effect, back toward butter by chemical trickery the original thermosets never allowed. Manufacturers have built prototype blades from these resins and begun installing them. Neither solution is yet universal, and blades remain the genuinely immature corner of the renewable recycling story, the one place where the critic's complaint retains real force. But "hard and being solved" is a very different verdict from "impossible," and the buried blades of the 2010s are already looking like a transitional embarrassment rather than a permanent feature.

It is worth pausing on the sheer scale of a blade to feel why the problem is real and why it is also bounded. A single modern blade can run to sixty, eighty, even more than a hundred meters — longer than the wing of the largest airliner — and a retiring turbine sheds three of them. Multiply by the tens of thousands of early turbines now reaching old age and you have a genuinely large stream of tough, bulky composite to deal with. And yet, set against the waste streams a society already manages, it is modest. The composite from all the world's retiring blades in a given year is dwarfed by the plastic packaging thrown away in an afternoon, or by the fly ash a coal fleet produces without anyone photographing it for a headline. The blade graveyard is photogenic in a way that a coal ash pond is not — a white airfoil in a field reads as tragedy, a grey slurry behind a fence reads as nothing — and part of the objection's power is simply that composite blades make a more haunting picture than the far larger, far more toxic wastes of the system they replace. That is not an argument for burying blades; it is an argument for keeping the camera's drama in proportion.

Now to the panels, where the tide of coming waste is real and the good news is much larger. There is a wave building. The great expansion of solar began in earnest in the 2010s, which means the great retirement will crest in the 2040s and beyond, and the projections run to tens of millions of tonnes of dead panels worldwide by 2050 — a genuinely large stream of material that will have to go somewhere. Left to a landfill, that is a waste and, at the margins, a hazard. But a solar panel, unlike a blade, is a fundamentally recyclable object, and here the physics is on the recycler's side. Take the panel apart and it is mostly glass — by weight the great bulk of a module is ordinary glass, which is straightforwardly recoverable. Around it sits an aluminum frame, one of the most valuable and endlessly recyclable metals we have. Inside are the working materials: silicon, the purified sand of the whole industry; silver, a genuinely precious metal laid down as the fine conductive lines on each cell; and small amounts of copper and other metals. Nearly everything of value in a panel can, in principle, be recovered and sold, and the economics that make recycling worthwhile improve every year as the waste stream grows large enough to build real factories around.

This is why a recycling industry is rising to meet the wave rather than being caught flat by it. Dedicated panel-recycling plants have opened in Europe, the United States, and Asia, some able to recover the overwhelming majority of a panel's mass and the high-value silver and silicon along with it. The infrastructure is still immature — I will not pretend a global system exists today, and much of the recovery is still crude glass-and-aluminum stripping that leaves the silver on the table — but the trajectory is unmistakable, and the resource logic is compelling. A retired panel is not trash; it is an ore, and a richer ore of silver than most rocks you could mine. As the volumes climb toward mid-century, the incentive to mine the panels rather than bury them climbs with them.

There is a lovely inversion buried in that idea, and it is worth drawing out because it reframes the whole "waste wave" from a threat into a resource. When we mine silver from the earth, we dig through tonnes of rock to win a few grams of metal, because natural silver ore is dilute. A solar panel, by contrast, has already had its silver refined and laid down in neat, findable lines by the manufacturing process — it is, in effect, pre-concentrated ore, delivered above ground and sorted by type, sitting in a field with a known address. The mid-century flood of dead panels is therefore not only a disposal problem; it is a silver mine, and a copper mine, and an aluminum mine, and a source of high-purity silicon, all pre-assembled and waiting. The reason it is not fully exploited yet is not that it cannot be but that the flood has not yet crested — you do not build a smelter for a trickle. The economics of recycling scale with the size of the stream, and the stream is about to become a river. The picture to hold is not of panels piling up as garbage but of a future recycling industry treating the retiring fleet the way a scrapyard treats old cars: as a rich, convenient, above-ground resource that it would be foolish to bury.

Which brings us to the two great myths, the "gotchas" that circulate wherever renewables are argued about, and both dissolve under a single number each. The first myth is that a solar panel never repays the energy it took to build it — that making the panel, with all its furnace-hot polysilicon refining, burns more energy than the panel will ever produce, so the whole enterprise is an elaborate way of losing energy while feeling virtuous. This is one of the most durable falsehoods in the entire debate, and it is simply, quantitatively wrong. The relevant idea is energy payback time: how long a panel must operate before it has generated as much energy as went into manufacturing it. For a modern silicon panel that figure is now on the order of one to two years, depending on the climate it sits in — sunnier places pay it back faster. After that year or two, the panel goes on generating clean electricity for another twenty-five or thirty years, all of it pure energy profit. The panel does not fail to repay its energy; it repays it many times over. Over a full lifetime a panel produces something like ten to thirty times the energy it cost to make, a ratio engineers call the energy return on investment, or EROI. Think of energy payback time in the plainest terms: it is like a shop that costs you a year's rent to open and then trades profitably for three decades. No one would call that shop a loss because of the first year's rent, and no one who has looked at the numbers calls solar an energy loss.

There is a second way to picture it that I find even clearer, borrowed from farming. A grain of seed corn holds a little stored energy, and a farmer plants it not to get that grain back but because one planted seed yields a whole ear of many grains — the return on investment is the whole point of agriculture, and a seed that yielded only itself would be no seed at all but a stone. A solar panel is a manufactured seed. You spend a year or two of energy planting it, and then it yields, across its long life, ten to thirty ears for the one you sowed. The myth that a panel never repays its energy is the exact equivalent of insisting that a farmer loses money by planting corn — it mistakes the cost of sowing for the whole of the harvest, and it survives only because most people have never seen the harvest figures. Once you have, the claim simply evaporates. And note where the myth came from: in the earliest days of solar, decades ago, when cells were crude and the manufacturing wasteful, the payback really was slow, and honest analyses said so. The falsehood is a true fact that expired — a snapshot from the 1970s waved around as if the intervening half-century of relentless efficiency gains had never happened.

The second myth is that panels are "toxic waste" — that they are laced with poisons, that a solar farm is a hazmat site in waiting, that we are trading smokestacks for a slow leach of heavy metals into the soil. This one has a grain of truth wrapped in a great deal of exaggeration, and precision matters. The dominant technology, crystalline silicon, is made of silicon, glass, aluminum, copper, and silver — none of them exotic poisons — with only trace amounts of anything worrying, and studies testing whether standard silicon panels qualify as hazardous waste generally find they do not.

There is a distinct and much smaller family of thin-film panels using cadmium compounds, and cadmium is genuinely toxic, so those specific panels warrant careful handling and dedicated recycling; but they are a minority of the market and the cadmium is bound in a stable compound within the module, not sprinkled loose. To call the typical rooftop panel "toxic waste" is to take the real, narrow case of one minority chemistry and smear it across an industry that is overwhelmingly made of sand and metal. The panel on a suburban roof is, by the standards of the electronics in the house beneath it, remarkably benign.

Set both myths against the alternative and the comparison is not close, which is the deeper point beneath the arithmetic. The lifecycle emissions of solar electricity — counting every gram of carbon dioxide from mining the quartz to refining the silicon to trucking the finished panel — are a small fraction of the emissions from burning coal or gas to make the same kilowatt-hour. Manufacturing a panel does emit carbon, chiefly because purifying polysilicon is energy-hungry and much of it is still done on coal-heavy grids in China; that is a real cost and a real reason to clean up the manufacturing. But the panel pays back that carbon debt on roughly the same timescale as its energy debt, a year or two, and then generates clean power for decades. A solar panel is not a net emitter any more than it is a net energy sink; both charges rest on ignoring the twenty-eight good years after the first two.

Reduced to first principles, then, the end-of-life question turns on two plain ideas that anyone can carry. The first is energy payback time and its cousin EROI — the simple accounting of energy out over energy in, which for solar comes out strongly, lopsidedly positive. The second is the distinction between hard-to-recycle and impossible-to-recycle: blades are the former, a genuine and immature problem yielding slowly to cement kilns and new resins; panels are eminently recyclable, an ore of glass and silver awaiting the factories that the coming waste wave will justify building.

To hold both honestly is to concede that the recycling infrastructure is not yet built at the scale the mid-century wave will demand — that concession is owed and I make it plainly — while refusing the false leap from "not yet built" to "cannot be done" or "poisons the earth."

So what happens to a dead panel? For too many of them today, honestly, not enough — they are landfilled because the recycling plant is not yet down the road and the silver inside is not yet worth the trouble to a system still learning. But that is a description of an industry in its infancy, not a verdict on its nature. The panel that dims on a roof in the 2050s will most likely be mined for its glass and aluminum and silver and silicon, its atoms cycled back toward the next generation of panels, having already repaid its energy thirty times over. The blade in the Wyoming field will more and more often be routed to a kiln or built from a resin that can be un-made. The graveyard image was true, and it was also a snapshot of a problem being outgrown.

None of which, of course, addresses the objection that has nothing to do with birds or blades or silver at all — the one that begins not with the machine but with the neighbor, who looks up at the ridgeline or out at the field and simply says: not here, not in my backyard.

Chapter 59 — Not in My Backyard

On a raw evening in a village hall in the English countryside, a developer's representative stood before a hundred folded arms and a slideshow of turbine renderings, and lost. He had the numbers. The wind resource on the ridge above the village was excellent; the turbines would power tens of thousands of homes; the grid connection was already there. What he did not have was a single friend in the room. One after another the villagers rose — a retired teacher who had walked that ridge for forty years and could not bear to see it changed, a couple who feared for the value of the cottage that was their whole retirement, a farmer worried about the flicker of shadow across his fields — and by the end of the night the project was, for all practical purposes, dead. No law of physics had stopped it. No cost curve had turned against it. It had run into the oldest and most underrated obstacle in the whole of the energy transition, the one engineers are worst at anticipating because it appears nowhere in their equations: the people who live there did not want it.

This is the objection that the previous chapters' answers cannot touch, and it may be the most important of them all. We have spent this part of the book weighing intermittency, land, wildlife, and waste — the technical and environmental objections, each of which yields, more or less, to engineering and evidence. But the binding constraint on how fast the world can build wind and solar is, increasingly, none of these. It is siting: the sheer difficulty of getting permission to put a specific machine on a specific piece of ground near specific human beings. In many wealthy countries a wind or solar project is now likelier to die in a planning hearing than to fail on cost or physics. The technology is ready. The neighbors are not. And understanding why — fairly, without either sneering at the objectors or romanticizing them — is essential, because this bottleneck is social, and social problems do not fall to better silicon.

Begin with aesthetics, the objection most often dismissed and least deserving of dismissal. People do not want a two-hundred-meter machine on the horizon they love. Critics of wind call it "visual pollution," and it is easy for an advocate to wave this away as mere sentiment — but it is not nothing to change a skyline that a community has looked at for generations, and pretending the objection is illegitimate only hardens it. Beauty is real to the people who feel it, and a ridgeline or a coastal view is a genuine common good that a turbine genuinely alters. Whether that alteration is a scar or a signal of progress depends entirely on the eye, and the eye depends, as we will see, enormously on who owns the thing.

Then there is noise, where legitimate concern and dubious claim have become tangled, and honesty requires separating them carefully. A modern turbine is not silent; the blades make a rhythmic swish as they pass the tower, and for a household a few hundred meters away, on a quiet night with the wind in the wrong direction, that sound can be intrusive and can, for some people, genuinely disturb sleep. This is real, it is measurable, and it is worth taking seriously in siting rules. But layered on top of the real complaint is a much larger and much shakier claim: that turbines emit infrasound — sound below the threshold of hearing — that causes a syndrome of headaches, nausea, dizziness, palpitations, and anxiety, dubbed "wind turbine syndrome." Here the evidence must be stated plainly, because it is one of the defining controversies of the field. Repeated reviews have failed to find robust evidence that turbine infrasound, which is weaker than the infrasound our own bodies and ordinary environments produce, directly causes these illnesses. What the evidence does suggest is subtler and, in its way, more interesting: that much of the reported harm is a nocebo effect — the dark twin of the placebo, in which the expectation of harm produces real symptoms. Studies have shown that people told to expect ill effects from infrasound report them even when the infrasound is switched off, while people given no such warning, exposed to the same conditions, feel fine. In one striking experiment, researchers played volunteers a stretch of genuine turbine infrasound and a stretch of pure silence, having first shown one group alarming material about the supposed dangers; the primed group reported symptoms during both the real infrasound and the sham, tracking their expectations rather than the sound. The body, it turns out, is exquisitely suggestible, and a vivid warning about invisible waves can manufacture the very headache it predicts. This does not mean the sufferers are lying or that their symptoms are unreal — a nocebo headache hurts exactly like any other. It means the causal chain runs through fear and expectation rather than through the air, which is why the alarming publicity itself may do more harm than the turbines. Annoyance and sleep disturbance for some nearby residents: real, and worth mitigating. A distinct physiological disease caused by infrasound: not supported by the weight of evidence.

Property values raise a similar tangle of the real and the feared. Homeowners near a proposed wind or solar project worry, understandably, that it will make their houses harder to sell, and the fear is powerful because a house is most families' largest asset. The research on whether turbines actually depress nearby property values is genuinely mixed — some studies find a small effect close to the largest projects, many find no measurable effect once you account for everything else that moves house prices — but the fear operates whether or not the effect is real, and a fear that moves votes is a political fact even when it is not an economic one. The advocate who tells a frightened homeowner that the studies are reassuring has, in that moment, usually already lost, because he has answered a feeling with a footnote.

Underneath all of these specifics lies the true engine of local opposition, and naming it clearly is the key that unlocks the whole chapter. It is a mismatch between where the benefits go and where the costs land — a textbook problem of collective action and the distribution of benefits. The benefits of a wind farm are diffuse and global: cleaner air somewhere downwind, a slightly cooler planet a century hence, cheaper power spread across a whole grid, profits banked in a distant developer's headquarters. The costs are concentrated and local: this view, this shadow, this sound, this stretch of road torn up by construction trucks, borne by the few hundred people who happen to live nearest. When a person is asked to absorb a sharp local cost so that strangers far away may enjoy a diffuse global benefit, "not in my backyard" is not irrational selfishness. It is the predictable response of a rational person to a raw deal. The problem is not that the objectors are bad; it is that the arithmetic of benefit has been laid out to make them losers.

Which points directly at the solution, and it is the most hopeful fact in this entire part of the book, demonstrated most vividly in Denmark. The Danes built one of the world's densest wind fleets with remarkably little of the bitter opposition that stalls projects elsewhere, and the reason is structural, not cultural. For much of their build-out, wind turbines were owned not by distant corporations but by the people who lived beside them — through cooperatives, through laws that gave locals the right and the encouragement to buy shares in the turbine on their own horizon. When you own the machine, the whole equation inverts. The swish of the blades becomes the sound of your dividend; the tower on the ridge becomes a thing you point out to visitors with pride rather than resentment; the local cost is now paired with a local benefit, and the raw deal becomes a fair one. Study after study finds that community-owned projects enjoy dramatically higher local acceptance than identical machines owned from afar. The turbine did not change. The ownership did, and with it the answer to "what's in it for me?" This is the deepest lesson of the whole siting problem: acceptance is not won by better arguments about climate or better data on infrasound; it is won by making the neighbors partners rather than victims.

The contrast with how much of the rest of the world built its wind is instructive, and a little rueful. Where a distant company arrives, wins a permit, erects its turbines, and wires the profits back to a headquarters the villagers will never see, it has, without meaning to, staged a small drama of extraction — the outsider taking the resource and the view and leaving behind only the shadow flicker. It should surprise no one that such a project draws resistance, or that the same machine, offered to the same village as a share to be owned, draws none. The psychology is not subtle. People do not resent a benefit they hold a piece of; they resent a cost imposed for someone else's gain. When Germany opened its wind build-out to citizen cooperatives and local investors, whole regions came to see the turbines as theirs, and a remarkable fraction of the country's early renewable capacity ended up in the hands of ordinary people and farmers rather than utilities. The turbines multiplied because the profits stayed home. This is not charity or public relations; it is the alignment of who pays with who benefits, and it is the single most reliable predictor of whether a community will welcome or fight the future arriving on its ridge.

The same logic explains an obstacle that will only grow: opposition to transmission lines. We saw in an earlier chapter that long-distance wires are among the most powerful tools for running a grid on weather — the copper that lets one region's wind cover another's calm. But those wires must cross somebody's land, and a transmission tower confers even less visible local benefit than a turbine; it does not even spin. Landowners along a proposed corridor see all of the cost — the pylons, the cleared easement, the eminent-domain letter — and none of the reward, which flows to consumers hundreds of miles away. Major transmission projects now routinely take a decade or more and die by the dozen in exactly this way, and the grid we need to integrate all that renewable energy is being throttled less by the difficulty of stringing wire than by the difficulty of persuading the people beneath it. The remedy is the same as for turbines: share the benefit locally, compensate the burdened honestly, and route with respect rather than arrogance.

The transmission bottleneck deserves a hard, unsentimental look, because it may prove the tightest constraint of all, and it exposes an irony at the heart of the transition. The very interconnection that makes a weather-powered grid reliable — the long wires that let a windless region borrow another's gale, that turn a continent's scattered turbines into one steady crowd — is precisely the piece hardest to build, because it delivers its benefit everywhere and its burden somewhere. A wind farm at least gives its host village a landmark, a lease, an argument for local pride; a transmission line offers the farmer whose fields it crosses nothing but a pylon and a letter from a lawyer. Whole clean-energy plans that pencil out beautifully on the engineer's spreadsheet founder on this, stranding gigawatts of cheap wind behind a permitting queue a decade long. It is a sobering thing to realize that the limiting reagent in decarbonizing a grid may not be the turbine or the panel or even the battery, all of which we can now build faster than we can connect, but the humble, contested, decade-delayed wire — and that the fix for it is not a technology at all but a fairer bargain with the people it crosses.

There is one more strand that a fair account cannot omit, and it must be handled with care in both directions. Some local opposition is genuine grassroots feeling — neighbors, sincerely worried, organizing to protect the place they love. But some of it is not what it appears. Investigations have repeatedly found that fossil-fuel interests and allied groups have quietly funded, coached, and amplified anti-renewable campaigns, seeding local groups with talking points, spreading the wind-turbine-syndrome literature, and manufacturing the appearance of a spontaneous uprising where the reality is astroturf — fake grass. This is real and it is documented, and the advocate is right to point it out. But the point must be made carefully, because the lazy version of it — dismissing all opposition as paid fossil stooges — is both false and self-defeating. The retired teacher who loves her ridge is not on anyone's payroll, and treating her as if she were is the surest way to make an enemy of her. The honest picture is that some opposition is legitimate concern deserving a real answer, and some is manufactured misinformation deserving exposure, and the hard, unglamorous work is telling them apart in each specific fight rather than assuming all of one kind.

That even-handedness is the whole discipline this chapter demands, and it cuts down two opposite errors. The first error, common among frustrated developers and their champions, is that all opposition is irrational — NIMBYism pure and simple, selfish people blocking the common good over a view. That is false: the collective-action problem is real, the sleep disturbance is real for some, the property fear is understandable, and the mismatch of costs and benefits is a legitimate grievance, not a character flaw. The second error, common among opponents, is that all opposition is therefore legitimate, that every concern deserves a veto. That is false too: the infrasound-disease claim is not supported by the evidence, the property-value fear often outruns the data, and some of the loudest campaigns are funded misinformation. The truth requires holding a scalpel, not a hammer — conceding what is real in the objection while refusing what is not.

Strip it to first principles and the siting problem is not, at bottom, a problem of turbines or panels at all. It is a problem in the distribution of benefits: a technology whose gains are shared by everyone and whose costs are borne by a few will always struggle for the consent of those few, and no amount of engineering can substitute for solving that. The Danish answer — make the neighbor an owner — is not a trick of public relations but a genuine realignment of who wins, and it works because it is fair. The bottleneck on the energy transition, more and more, is not the cost of the machine or the physics of the sky. It is the far older human question of whether the people asked to host the future can be given a real stake in it.

The developer in that English village hall never grasped this. He came armed with the wattage and the climate math and the reassuring studies, and he was routed by a room full of people who had been offered a cost and no share of the gain. Had he come instead offering the village a stake in the turbine on its own ridge, the folded arms might have loosened. The lesson of this whole part of the book — of intermittency and land and birds and blades and backyards — is that the objections to solar and wind are neither the fatal flaws the skeptics claim nor the trivial nuisances the boosters wave away, but real problems with real, mostly available answers, if we are honest enough to name them and humble enough to share the benefits. Having weighed the objections, we can at last turn from what holds the technology back to what may carry it forward — and the frontier begins with a crystal that promises to catch the light silicon cannot.

Chapter 60 — Perovskites and the Tandem Cell

In a clean room outside Xi'an, in the last weeks of 2024, a technician slid a small square of coated glass into a solar simulator — a lamp engineered to mimic the noon Sun — and watched a number climb on a screen. The number was an efficiency, the fraction of the light landing on the cell that came out the other side as electricity, and it settled at 34.85 percent. To anyone outside the field it was an unremarkable figure, a decimal in a trade journal. To the people in that room, and to the National Renewable Energy Laboratory in Colorado that would certify it, it was a wall coming down. The company was LONGi, one of the giants of Chinese solar manufacturing, and the little coated square was not an ordinary silicon cell at all. It was two solar cells stacked one atop the other, a sandwich of old material and new, and together they had just done something a single silicon cell can never do. They had climbed past a ceiling that physics itself had seemed to bolt shut.

To understand why that number mattered, we have to go back to a limit we met many chapters ago, when we first followed a photon from the Sun into a wafer of silicon. Recall the trouble with any single solar cell. Sunlight is not one color but a whole crowd of them, a spectrum running from the high-energy blue and ultraviolet, through the yellows and greens our eyes love, down into the sluggish, low-energy reds and the invisible infrared beyond. A solar cell is built from a semiconductor with a particular threshold, its band gap — the minimum kick of energy a photon must deliver to knock an electron loose and set it flowing as current. Silicon's band gap is about 1.1 electron-volts, which happens to sit down in the near-infrared. And here is the cruelty of the arrangement. Any photon carrying less energy than the band gap sails straight through the silicon as if it were glass; the cell never feels it, and that light is simply wasted. Any photon carrying more energy than the band gap does knock an electron loose — but the surplus energy, everything above the threshold, is thrown away instantly as heat, the electron shaking off the excess like a diver who jumped from far too high a board only to end up at the same pool.

Add those two losses together — the reds too feeble to be caught, the blues too energetic to be used efficiently — and you arrive at a hard number that two physicists, William Shockley and Hans Queisser, worked out on paper back in 1961. For a single-junction cell under ordinary sunlight, the theoretical maximum efficiency is roughly 33 percent. Not a manufacturing limit, not a matter of cleaner factories or better wiring, but a ceiling written into the physics of using one material to catch a spectrum of many energies. The best silicon cells rolling off production lines today reach the low-to-mid twenties in percent; the very finest laboratory silicon has crept toward 27 percent, close enough to that Shockley–Queisser ceiling that engineers speak of silicon as approaching the end of its own road. You cannot polish your way through a wall. And this is the first misconception worth naming and correcting right away, because it is repeated confidently and often: that silicon solar is "maxed out," a mature technology with nowhere left to go. It is true that single-junction silicon is near its limit. It is not true that silicon is finished. The way past the ceiling is not to make silicon better. It is to give silicon a partner.

That is the whole idea of a tandem cell, and once you see it you cannot unsee its elegance. If one material must compromise — set its band gap low to catch the reds and waste the blues, or set it high to use the blues cleanly and miss the reds — then use two materials. Stack them. Put on top a cell with a wide band gap, tuned to feast on the high-energy blue and green light, converting it efficiently while letting the low-energy reds pass through untouched. Then place beneath it a second cell, silicon, with its lower band gap, waiting to catch precisely those reds and infrared photons that slipped through the top layer. Each cell handles the slice of the spectrum it is best suited to, and neither is asked to do the impossible job of handling all of it. The blues are converted without so much heat wasted; the reds are no longer thrown away. Two junctions, working in series, add their voltages together, and the combined stack can in principle reach efficiencies in the mid-forties — the theoretical tandem ceiling sits somewhere around 45 to 47 percent, comfortably above the single-junction wall.

I find the cleanest way to picture it is a rainwater analogy. Imagine trying to collect a downpour of mixed drops — some fat and heavy, some fine as mist — using a single sieve. Make the mesh coarse and the fine mist blows through and is lost; make it fine and the fat drops splash and overflow. But stack two sieves, a fine one beneath a coarse one, and each catches the drops it is built for. Or think of a fishing crew that cannot decide between a wide net and a narrow net, and finally does the obvious thing: they run the wide net first for the big fish, then the fine net behind it for the small ones. Nothing new is invented in either case. You simply stop asking one tool to do two incompatible jobs. That, at bottom, is why a tandem beats a single junction, and why the LONGi number was not a lucky fluctuation but the predictable payoff of sound physics.

It is worth pausing to say that the tandem is not a brand-new invention dreamed up for the perovskite era; it is an old and proven idea that has simply been too expensive for the everyday market until now. For decades, the most efficient solar cells in the world have been multi-junction cells — not two layers but sometimes three or four, each of an exotic semiconductor from the gallium-and-indium family, stacked with exquisite precision. You will find them in an unlikely place: bolted to satellites and spacecraft. The rovers that crawl across the surface of Mars, the telescopes that hang in orbit, the communications satellites relaying this planet's chatter — many run on multi-junction cells reaching efficiencies well above 30 percent, and under concentrated sunlight in the laboratory some have topped 45 percent. Out there, where every kilogram launched costs a fortune and there is no repairing a dead panel, efficiency is worth almost any price, so the engineers spared none. The proof that stacking works has therefore been orbiting over our heads for a generation. What perovskites promise is to bring that proven trick down from the exotic economics of the space program to the price point of a suburban roof — to make the tandem cheap enough to be ordinary.

The reason tandems have gone from a laboratory curiosity to an industrial obsession in just the last few years comes down to the top layer — to a family of materials called perovskites, and here the story turns genuinely surprising. The name "perovskite" originally belonged to a specific mineral, a calcium titanium oxide named in the nineteenth century after the Russian mineralogist Lev Perovski. What matters for solar is not that particular mineral but its crystal structure — a certain repeating cage-like arrangement of atoms that many different compounds can adopt. The perovskites now electrifying solar research are mostly built around lead and halogens like iodine and bromine, laced with small organic molecules, and they turn out to be astonishingly good at absorbing light. A film of perovskite only a few hundred nanometers thick — a fraction of the width of a human hair, hundreds of times thinner than the silicon wafer beneath it — can soak up as much sunlight as a far thicker slab of silicon. They are strong absorbers, they carry their freed electrons well, and, crucially, their band gap is tunable: by adjusting the recipe, changing the ratio of iodine to bromine, chemists can dial the perovskite's threshold up or down to whatever value best complements the silicon underneath. It is as if you could order silicon to a custom band gap by changing the seasoning. That tunability is exactly what a tandem's top layer wants, which is why perovskite-on-silicon has become the leading candidate to carry solar past the Shockley–Queisser wall.

And then there is the fact that stops most people who hear it for the first time: you can print them. Silicon demands furnaces near 1,500 degrees Celsius, ultra-pure crystals grown with painstaking slowness, clean rooms and vast capital. Perovskites can be laid down from a liquid ink, spun onto a surface, slot-die coated, in some experiments literally printed roll-to-roll like a newspaper, at temperatures low enough that flexible plastic can survive the process. The prospect this raises — solar cells unspooling off a press, cheap and bendable, coated onto rooftiles or windows or the curved skin of a vehicle — is the kind of thing that sounds like marketing until you watch it happen in a lab. It is a real and radical departure from the way we have made solar cells for seventy years, ever since the Bell Labs three coaxed six percent out of a silicon wafer in 1954. Cheap, printable, tunable, and a superb catcher of blue light: on paper the perovskite is the ideal partner silicon has been waiting for.

On paper. And here we come to the catch, because a book that told you only the first half of this would be a sales brochure, not an honest account, and the honest account is that perovskites have a serious flaw. They are delicate. The very softness of chemistry that lets you print them at low temperature also lets the world take them apart. Moisture creeps in and the crystal decomposes. Heat — the ordinary heat of a cell baking on a rooftop through a summer afternoon, which can reach sixty or seventy degrees — degrades them over time. Ultraviolet light, the same energetic blue-and-beyond radiation the top cell is built to harvest, slowly attacks the material that harvests it. A silicon panel is warranted to hold most of its output for twenty-five or thirty years, a workhorse that shrugs off decades of weather. Early perovskites could lose their performance in weeks or months of real exposure. A solar cell that dies in two years is not a product; it is a science experiment. Closing that durability gap — encapsulating the perovskite against moisture, engineering the crystal and its interfaces to endure heat and light — is the single largest reason perovskite tandems are not yet on your neighbor's roof, and it is where much of the field's effort now goes.

There is a second hurdle, quieter but real, and it deserves a straight word: most high-performing perovskites contain a little lead. Not a lot — the amount in a panel's worth of cells is small, and it is bound up in the crystal — but lead is a toxin we have spent a century learning to keep out of paint, pipes, and gasoline, and the idea of coating rooftops with it makes people reasonably uneasy. The industry's answers are threefold: seal the lead so it cannot escape in use, build recycling and take-back into the product's life so it does not end up in a landfill, and pursue lead-free perovskite chemistries, which so far perform worse but are improving. It is not a solved problem. It is an acknowledged one, and acknowledging it plainly is part of the difference between engineering and hype.

Which brings us to the second misconception, and it is the more important of the two, because it colors how the public reads every announcement in this field. It is the belief that a laboratory record is a preview of next year's market — that 34.85 percent certified in a Chinese clean room means 34.85 percent on sale shortly after. It does not. A record cell is often a tiny thing, a square centimeter or a few, made once, under ideal conditions, and measured in the hour of its glory before anyone asks it to survive a decade on a roof in Arizona. Between that champion cell and a shippable product lies a long valley: scaling from a thumbnail to a full-sized module without the efficiency crumbling at the larger area, holding the performance through years of weather, and doing it all at a cost that competes with silicon that keeps getting cheaper underneath you. Every one of those steps has killed promising solar technologies before. The lesson is not cynicism — perovskite tandems are advancing with real momentum, and pilot production lines are being built as I write — but calibration. A record is a signpost pointing down a road, not the destination itself. When you next read a breathless headline about a solar efficiency record, hold both thoughts at once: the physics is genuinely thrilling, and the durability gap is genuinely unsolved.

Let me make the payoff concrete with the kind of worked example this book has leaned on throughout, because "34.85 percent" only means something once you translate it into a roof. Take a good silicon panel today, converting about 22 percent of the sunlight that strikes it. On a decent-sized residential roof — say a stretch that can hold twenty such panels, each about two square meters — you might install something like eight kilowatts of capacity, enough to cover a typical household's yearly electricity in a sunny climate with a bit to spare. Now suppose that same roof, that same twenty panels of the same physical size, were fitted instead with perovskite-silicon tandems converting, once they mature into commercial products, perhaps 30 percent of the light rather than 22. That is roughly a third more electricity from an identical patch of roof — the same racking, the same wiring, the same installer spending the same day on your shingles, but around eleven kilowatts where there were eight. On a large solar farm the logic compounds: a third more energy from every acre of land, every mounting post, every kilometer of access road and cabling. Because so much of the cost of solar is no longer the cell itself but everything around it — the land, the steel, the labor, the grid connection, what the industry calls the balance of system — squeezing more watts through the same fixed footprint is worth far more than the efficiency number alone suggests. That is the quiet economic engine behind the whole tandem race.

Where, then, do you meet this frontier today? Mostly you don't, not yet, not on the open market — and that itself is the honest answer. What exists in 2024 and 2025 is a wave of laboratory records tumbling one after another (LONGi's 34.85 percent was one of a rapid series that pushed past the old 33.9 percent it had set only a year before), pilot lines standing up in China and Europe, and startups and giants alike racing to be first with a durable, full-sized perovskite-silicon panel. The mainstream module on your neighbor's roof is still silicon, quietly excellent, near its ceiling. But the tandem is coming out of the lab in stages, and when it arrives it will most likely not announce itself as something exotic. It will arrive as an ordinary-looking panel that simply produces a third more power from the same patch of roof — the same footprint, the same racking, more electricity — because the extra work is happening invisibly in a film thinner than a wavelength of light, stacked atop the silicon you already trust.

So here is the recap, in plain terms. A single solar cell is trapped by the Shockley–Queisser limit near 33 percent, because one material cannot both catch the feeble reds and use the energetic blues without waste. Silicon has very nearly reached that ceiling and cannot be polished through it. The way past is the tandem: stack a wide-band-gap top cell over silicon so each layer harvests the slice of sunlight it handles best, and the theoretical ceiling jumps into the forties. Perovskites — cheap, printable, and tunable — are the leading candidate for that top layer, but they degrade in moisture, heat, and ultraviolet, and they usually carry a little lead, so durability and toxicity are the hurdles now being worked. And a laboratory record, however genuine, is a promise still to be kept, not a product on a shelf. The frontier of the solar cell, in other words, is no longer about wringing more from silicon. It is about giving silicon a partner and teaching that partner to last.

But squeezing more electricity from each panel is only one half of the frontier. The other half asks a harder question:

what do we do with electricity in the places electricity cannot easily go — the furnaces of a steel mill, the hold of a cargo ship, the long dark stretch of a winter with no sun and no wind? For that, the frontier turns from electrons to molecules, and to the oldest trick for storing the wind's work: splitting water into hydrogen.

Chapter 61 — Green Hydrogen and the Molecules of the Future

On a windswept hill in the small Danish town of Askov, sometime in the 1890s, a schoolteacher-turned-scientist named Poul la Cour stood beside a windmill he had wired to do something no windmill had done before. We met la Cour earlier in this book as the father of Danish wind electricity, the man who ran the world's first wind-power laboratory and worked out that a rotor with a few slender, fast-moving blades beats one crowded with many slow ones. But la Cour had a problem that would sound very familiar to any grid operator today: the wind does not blow on schedule. It gusts and dies, races through the night when no one needs it, and falls calm on the still afternoon when everyone does. So la Cour took the electricity his windmill made when the wind was generous and used it to run a current through water, splitting the water into its two elements — hydrogen and oxygen — and he stored the hydrogen. For a time he lit the Askov schoolhouse with lamps burning that wind-made gas. There was even, on at least one occasion, an explosion from a leak, which tells you something about how the story of hydrogen has always braided together promise and hazard. What matters is the idea: more than 130 years ago, a Danish schoolteacher was already turning surplus wind into hydrogen so that the wind's work could be kept for a windless hour. Almost everything in this chapter is a modern, industrial-scale reprise of what la Cour did on that hill.

To see why the world is returning to his idea now, we have to be honest about a limit of the whole solar-and-wind project we have been tracing. Cheap clean electricity is a magnificent thing, and if every use of energy were simply a matter of electricity, the transition would be far simpler than it is. But it is not. A great deal of what industrial civilization actually does with energy is stubbornly resistant to being run off a wire. Consider steelmaking, which traditionally uses coal not merely for heat but chemically, to strip oxygen from iron ore — you cannot do that with a power cord. Consider cement, fertilizer, the ammonia that grows much of the world's food, the fuel that drives cargo ships across oceans and lifts aircraft into the sky, and the problem of storing energy not for four hours but for four months, from a sunny summer into a dark winter. Batteries are superb for the short haul, but a lithium battery that could hold a nation's electricity through a windless January would be unimaginably vast and expensive. These are the hard-to-electrify corners of the energy system, the places where electrons alone cannot reach. And the leading candidate to reach into those corners is a molecule: green hydrogen.

The physics is beautifully simple, which is part of the appeal. Water is two atoms of hydrogen bound to one of oxygen. Pass an electric current through it — a process called electrolysis, which any schoolchild can demonstrate with a battery and two pencils in a glass — and the current tears the molecules apart, hydrogen bubbling up at one electrode, oxygen at the other. A device built to do this at scale is an electrolyzer, and it is, in a sense, a solar panel run backwards: where the panel turns light into electric current, and the current could turn a motor, the electrolyzer takes electric current and stores it in the chemical bond of a hydrogen molecule. Later, when you want the energy back, you can run the process in reverse in a fuel cell — combine hydrogen with oxygen from the air, and instead of an explosion you get a controlled reaction that yields electricity, heat, and pure water as the only exhaust. Or you can simply burn the hydrogen for its heat, or feed it into a chemical process, using it not as fuel but as an ingredient. Split water with electricity; store the hydrogen; recombine it for power, or use it to make steel or ammonia. That is the entire arc, and it is exactly la Cour's arc, only with better machines.

The word doing the heavy lifting in "green hydrogen" is green, and to understand it you need the color code the industry uses, which sounds whimsical but carries a serious point. Most of the hydrogen the world makes today — and the world already makes an enormous amount, tens of millions of tonnes a year, mostly for oil refining and fertilizer — is grey hydrogen. Grey hydrogen is made not from water but from natural gas, by cooking methane with steam in a way that strips out the hydrogen and releases carbon dioxide straight into the atmosphere. It is cheap, and it is dirty; producing it is one of the larger industrial sources of emissions few people have ever heard of. Then there is blue hydrogen: the same gas-based process, but with equipment bolted on to capture much of that carbon dioxide and bury it underground rather than venting it. Blue is cleaner than grey if the capture works well and the gas does not leak too much along the way, though both of those are real ifs. And then there is green hydrogen: made by electrolysis, splitting water using electricity from solar or wind, so that the only inputs are water and sunlight-or-wind, and the only local output besides the hydrogen is oxygen. No fossil fuel, no carbon dioxide. This is the crucial distinction, and it leads straight to a misconception that must be corrected firmly: hydrogen is not "clean" by nature. Hydrogen is only as clean as the electricity or the gas used to make it. Grey hydrogen is a climate problem wearing a futuristic name. The whole environmental case rests on the word green — on the assumption that the power splitting the water is itself clean and, ideally, surplus.

That word "surplus" is where solar and wind re-enter the story, and where the economics finally start to make sense. Remember the strange thing that happens on a grid heavy with renewables: at midday, when the Sun is high and the panels are pouring out power faster than anyone can use it, prices can crash toward zero, and operators sometimes curtail — deliberately throttle — perfectly good solar because there is nowhere for it to go. The same happens on a blustery night with the wind farms. That surplus, that embarrassment of cheap or nearly free electricity at the wrong moment, is precisely what an electrolyzer loves. It can run when power is abundant and cheap, sit idle when power is scarce and dear, and convert those wasted midday and midnight gluts into hydrogen that can be stored and used any time. In that light, green hydrogen is not really a competitor to batteries so much as a companion to them — batteries for the daily rhythm, hydrogen for the seasonal one and for the industrial jobs electricity cannot do.

So much for the promise. Now the honest problems, and they are substantial enough that I want to be blunt about them, because hydrogen has a long history of inspiring more enthusiasm than it has yet earned. The first and largest problem is inefficiency. Every conversion of energy from one form to another loses some of it, and the hydrogen path involves several conversions in a row. Start with a hundred units of clean electricity. Run it through the electrolyzer to make hydrogen, and you lose perhaps a quarter or a third of it as waste heat. Compress or cool the hydrogen for storage and transport, and lose more. Then, if your goal was electricity again, run it back through a fuel cell, and lose a third or more once more. By the time you have your electricity back, you may be holding thirty to forty units out of the original hundred; the rest has escaped as heat along the way. Compare that with a battery, which returns eighty-five or ninety of every hundred units you put in, and you see the difficulty at once. This "round-trip" penalty is not a flaw that better engineering will entirely erase — some of it is thermodynamics, the tax nature charges on every energy conversion. It is the single strongest argument against using hydrogen for jobs that electricity or batteries can do directly. Why throw away sixty percent of your clean power to run a car on hydrogen when a battery would waste only ten to run it on electricity?

Which is exactly the misconception most worth correcting in this chapter, because it has been sold hard for decades: the idea that hydrogen will soon, cheaply, power your car and heat your home. For most cars, it almost certainly will not, and the reason is the arithmetic just described — the battery electric vehicle is simply far more efficient with scarce clean electrons, and it has won that particular race in the marketplace. For home heating, the same logic favors the electric heat pump over piping hydrogen into your boiler. The mistake is to imagine hydrogen as a universal replacement for fossil fuels, a drop-in that lets everything continue as before with a cleaner gas in the tank. It is not that. Its round-trip losses make it a poor choice wherever a wire or a battery will serve. Its value lies precisely in the places where wires and batteries will not serve — the steel and the ammonia and the ships and the aviation and the seasonal storage — the hard-to-electrify sectors. Green hydrogen is best understood not as a fuel for everyone but as a specialist tool for the jobs nothing else can do. Deployed there, it is close to indispensable. Deployed everywhere, it wastes the very clean power it depends on.

The other problems are more prosaic but no less real. Green hydrogen today is expensive — costing several times what grey hydrogen costs, somewhere in the range of a few dollars to six dollars a kilogram in 2024 depending on where and how, against targets of one to two dollars a kilogram that the industry hopes to reach around 2030 as electrolyzers get cheaper and clean electricity gets cheaper still. Whether it hits those targets is genuinely uncertain. And then there is the maddening physical nature of the molecule itself. Hydrogen is the lightest element in the universe, its atoms so small they wriggle through seals and materials that would contain any other gas, and prone to making some metals brittle over time. It is bulky to store — even compressed to enormous pressures or chilled to a liquid colder than almost anything on Earth, a given volume of it holds far less energy than the same volume of gasoline. It leaks, and because it is so light it is hard to keep and hard to move. Building the pipelines, tanks, and ships to handle it at scale is a formidable undertaking, and leaked hydrogen is not entirely harmless in the atmosphere either. None of these is a reason to abandon it. All of them are reasons to be sober about how fast and how far it goes.

A word is owed here to hydrogen's fearsome reputation, because the mention of a leaky, flammable gas summons one image above all others: the airship Hindenburg, wreathed in flame over New Jersey in 1937, the newsreel that seared into a century of minds the idea that hydrogen is a bomb waiting for a spark. It is worth correcting the misconception gently, because fear is a poor guide to engineering. Hydrogen is indeed flammable — so is the gasoline in every car's tank, so is the natural gas piped into millions of kitchens, and we have learned to handle both with valves and sensors and sensible design. Hydrogen has some hazards the others lack: it burns with a nearly invisible flame, and being so light it leaks readily. But it also has a virtue the others lack: being the lightest thing there is, it does not pool and linger the way heavier fuel vapors do; it rushes upward and disperses into the sky within seconds of escaping. The engineers building today's electrolyzers and tanks treat it as what it is — an industrial gas with a known and manageable set of dangers, handled safely by the millions of tonnes in refineries for a century — not as the villain of a newsreel. The real obstacles to hydrogen are not that it might explode. They are the duller villains of efficiency, cost, and the sheer awkwardness of storing a whisper-light molecule.

Let me make that awkwardness concrete, because it is the fact that most surprises people. Hydrogen holds a tremendous amount of energy for its weight — kilogram for kilogram it packs nearly three times the energy of gasoline, which is why rockets burn it. But energy for its volume is another matter entirely, and volume is what a tank must hold. At ordinary pressure, the hydrogen carrying as much energy as a single liter of gasoline would fill something like three thousand liters — a small room's worth of gas for a cupful of petrol's worth of energy. To make it manageable you must either crush it to hundreds of times atmospheric pressure, which costs energy and demands heavy, expensive tanks, or chill it to a liquid at around minus 253 degrees Celsius, a cold barely twenty degrees above the absolute zero of the universe, which costs even more energy and constant vigilance to keep it from boiling away. Picture the difference between storing a day's water in a jug versus storing the same water as steam: the substance is the same, but the room it demands is not. That volumetric problem, more than any fear of fire, is why hydrogen is so hard to pump through pipes, load onto ships, or fit into the fuel tank of an ordinary car.

There is a clever partial answer to the storage-and-transport headache, and it is another idea with deep roots: turn the hydrogen into ammonia. Ammonia is a compound of nitrogen and hydrogen, a liquid that is far easier to store and ship than hydrogen gas, and one the world already moves around by the millions of tonnes for fertilizer, with all the infrastructure that implies. You can make green ammonia by combining green hydrogen with nitrogen pulled from the air, then carry it across oceans as a dense, liquid carrier of hydrogen — cracking it back into hydrogen at the far end, or better yet burning the ammonia directly as a carbon-free fuel in a ship's engine. Much of the serious talk about hydrogen for global shipping is really talk about ammonia, the more portable cousin. Here is a surprising fact to sit with:

the fertilizer that feeds a large share of humanity is already made from hydrogen, today, almost all of it grey; simply cleaning up how that existing hydrogen is made — turning the world's ammonia plants green — would be one of the largest single wins available, before hydrogen ever powers a single new thing.

Where do you meet green hydrogen today? Mostly in pilot plants and pledges rather than in daily life — this is a frontier still being built, and you should be wary of anyone who tells you otherwise, because hydrogen has been perpetually five years from transforming everything since at least the 1970s oil shocks, and the graveyard of over-hyped hydrogen visions is well populated. But real projects are rising: demonstration steel mills in Sweden and elsewhere making iron with hydrogen instead of coal, green ammonia plants breaking ground, electrolyzers being manufactured at growing scale, with China now producing a large share of the world's electrolyzer capacity much as it dominates solar panels. The honest picture is of a technology emerging in its proper niches, neither the miracle its boosters promise nor the fantasy its critics dismiss, but a specialist tool being sharpened for specialist work.

To recap in plain language: some things — steel, cement, fertilizer, shipping, aviation, and storing energy across seasons — are very hard to run on electricity directly. Green hydrogen, made by splitting water with cheap surplus solar and wind, exactly as Poul la Cour did with his windmill in the 1890s, offers a way to reach those hard corners with a clean molecule instead of an electron. But it is only clean if it is green; it wastes a large fraction of the energy put into it, which makes it a poor choice wherever a battery or a wire will do; it is costly and physically awkward to store and move; and it has a long record of being over-promised. Used narrowly, for the jobs nothing else can do, it may be essential. Used broadly, it squanders the very clean power that gives it any point at all.

The panel and the turbine, the battery and the electrolyzer

— we now have the whole toolkit in hand. The question that remains is one of sheer scale: how big has this grown, how big must it grow, and can it possibly grow fast enough? To answer that we have to step back and take the measure of the entire enterprise, now counted not in megawatts or gigawatts but in terawatts — the trillion-watt age.

Chapter 62 — The Terawatt Age

Let me try to give you a number you can actually feel, because the numbers in this chapter are so large that they slide off the mind. A watt is a small thing — a bright LED bulb sips a handful of them. A kilowatt is a thousand watts, roughly what a hairdryer draws. A megawatt is a million watts, enough for a few hundred homes. A gigawatt is a billion watts, the output of a large power station, the kind of plant that dominates a region's skyline and shows up on maps. And a terawatt is a thousand gigawatts — a trillion watts — a thousand great power stations' worth of capacity. For most of the history of electricity, the terawatt was a unit for describing the entire world's power system, not any one kind of generation within it. As I write, at the close of 2023 and running into 2024, the world has installed roughly 1.6 terawatts of solar panels and roughly 1.13 terawatts of wind turbines. Two technologies that a single lifetime ago were laboratory curiosities and desert follies — Fritts's selenium wafer, Shuman's mirrors on the Nile, Brush's cedar-bladed dynamo in a Cleveland backyard — now stand, together, at nearly three terawatts of generating capacity spread across the planet. The schoolteacher's windmill has become a civilization-scale machine.

The pace of the thing is even harder to grasp than its size.

In the single year of 2024, the world added something on the order of 450 to 550 gigawatts of new solar — the estimates vary by how you count, with IRENA tallying around 452 gigawatts of additions and the IEA reckoning closer to 550, the difference coming down to whether you count panels connected to the grid or panels shipped and installed — and roughly 117 gigawatts of new wind. Pause on the solar figure. Half a terawatt of new solar in twelve months. That is more generating capacity of one kind added in a single year than the total electrical capacity of most large countries. Solar was about three-quarters of all the new renewable capacity the world built that year. And it is not slowing; each recent year has set a fresh record, the additions climbing rather than leveling off. It is fair to say, and specialists do say it, that solar photovoltaics is now the fastest-growing source of energy in human history — faster in its rise than coal in the age of steam, faster than oil in the age of the automobile. Nothing in the long story of how humans have powered themselves has scaled like this.

Set against total demand, the share is still modest but no longer marginal, and this is where honesty about the numbers matters most. As of 2024, wind supplied about 8 percent of the world's electricity and solar about 7 percent — together right around 15 percent, roughly one electron in seven coming from the Sun and the wind. Add in the older clean sources, the hydroelectric dams and the nuclear plants, and all clean power together crossed a threshold in 2024 that is worth marking: more than 40 percent of the world's electricity now comes from sources that burn no fossil fuel, according to the energy analysts at Ember. Two facts held together give the true shape of the moment. Solar and wind are growing explosively and are still, between them, only about a seventh of global electricity — and electricity, remember, is itself only a portion of all the energy humanity uses, alongside the fuel we burn for transport, heat, and industry. The transition is simultaneously the fastest energy build-out in history and nowhere near finished. Both things are true at once, and any account that emphasizes only one of them is misleading you.

How did we arrive here, and why now rather than twenty years ago? The answer is the single most important idea in this entire book, and it is not an idea about physics or heroism but about cost. Recall Swanson's law, named for the SunPower founder Richard Swanson — the observation that every time the world's cumulative production of solar panels doubles, their price falls by about 20 percent. That is not a law of nature like the cube law of the wind or the Betz limit; it is a law of manufacturing, the same learning-by-doing that made every mass-produced thing cheaper over time, from the Model T to the transistor. But it has held, doubling after doubling, for four decades, and its cumulative effect is almost beyond belief: the price of a solar module has fallen by something like 99.9 percent since the 1970s, a thousand-fold collapse. The consequence is a crossover that quietly changed everything. Around the mid-2010s, in more and more places, newly built solar and wind became not merely competitive with new coal and gas plants but cheaper than them. By 2024, the global average cost of electricity from a new utility-scale solar farm had fallen to roughly four cents a kilowatt-hour, and from a new onshore wind farm to roughly three and a half cents — cheaper than the cheapest new fossil-fueled power on Earth. In the sunniest places with the fiercest competition, solar has been bid at extraordinary lows: Saudi Arabia's Al Shuaibah tender in 2021 came in near one cent a kilowatt-hour, a price that would have seemed like a typographical error a decade earlier.

It is worth dwelling on how total, and how recent, that collapse has been, because numbers this steep are hard to believe without a comparison you can hold. In 1975, a watt of solar cell cost something on the order of a hundred dollars. Today a watt of solar module can be had for a dime or less — a fall of roughly a thousand-fold in half a century. Try to think of anything else that has cheapened like that. If a family car that cost twenty thousand dollars in 1975 had followed solar's curve, it would now sell for about twenty dollars. If a loaf of bread had done the same, you would buy it for a fraction of a cent. Almost nothing in the entire economy behaves this way; the closest cousin is the computer chip, whose relentless cheapening built the digital age, and it is no accident that solar, another child of the semiconductor, followed a similar path. This is the deep reason the terawatt age arrived when it did and not before. It was not a sudden discovery or a single breakthrough. It was a slow, compounding, decades-long slide down the learning curve, each doubling of production shaving another fifth off the price, until one day the cheapest way to make a new kilowatt-hour anywhere on Earth was to point a piece of silicon at the sky.

This cost crossover is the hinge of the whole story, and it corrects the most stubborn misconception of all — the belief that solar and wind are expensive indulgences propped up by subsidies, that they survive only on the taxpayer's generosity. That was once true. It is no longer. In most markets today, an investor building new power generation chooses solar or wind not out of virtue or mandate but because it is the cheapest electricity available, full stop. The transition has changed its own engine. For its first decades it was pushed uphill by policy — by Carter's tax credits, by Germany's feed-in tariff and Hermann Scheer's Energiewende, by California's wind rush, by every subsidy and mandate that kept a young and costly technology alive long enough to grow up. Those policies mattered enormously; without them the learning curve would never have been climbed. But the machine now rolls under its own weight. It is driven increasingly by economics rather than idealism, which is precisely why it has become so hard to stop: you do not need to persuade a hard-nosed utility to love the planet to get it to buy the cheapest kilowatt on the menu.

And yet — and this is where the sober voice in this book has to speak clearly — cheap electricity from the Sun and the wind is not the same as a solved problem, and it would be a serious error to close this chapter in a glow of inevitability. Several genuinely hard problems remain, and pretending otherwise does the reader no favors. The first is storage, which we have circled throughout these pages. Lithium batteries have grown astonishingly fast — grid storage added a record 200 gigawatt-hours in 2024 and now handles the daily rhythm of shifting midday solar into the evening peak. But the daily rhythm is the easy rhythm. The seasonal one — carrying energy from a bright, breezy summer across a dark, still winter — is not solved cheaply by any technology we have, which is why green hydrogen and long-duration storage remain frontiers rather than settled facts. The second hard problem is the grid itself: the wires. The best sunshine and the strongest winds are often far from the cities that need the power, and building the long-distance transmission lines to carry it — across jurisdictions, past objections, through permitting that can take longer than building the power plants themselves — has become the bottleneck of the transition in country after country. A cheap panel in a field connects to nothing without a wire, and wires have proven harder to build than panels.

The third problem is that electricity is only part of the job. To truly ride on the Sun and the wind, we have to electrify things that today run on fuel — cars, which means the shift to electric vehicles; home heating, which means heat pumps; and as much of industry as can be made to run on electrons. Every furnace converted, every gas boiler swapped for a heat pump, every petrol car replaced by a battery one, moves a slice of energy demand out of the fossil column and into the electric column where clean generation can reach it — but each of those transitions is its own slow, expensive, contested project, and the hardest slices, the steel and the cement and the aviation we met in the last chapter, resist electrification altogether. The fourth problem is materials and manufacturing at a scale never before attempted: the polysilicon and the copper and the rare-earth magnets, the mines and the refineries, and the uncomfortable fact that the supply chain for all of this is concentrated to a remarkable degree in a single country, China, which as of the mid-2020s makes the overwhelming majority of the world's solar panels and processes most of its critical minerals. That concentration is a strategic vulnerability and a geopolitical fault line, even as it is also the very thing that made solar so cheap so fast. There is no tidy resolution to that tension, only a set of hard trade-offs the world is now negotiating.

There is a fifth objection one hears constantly, and it deserves a worked answer because it sounds so intuitively damning: that solar and wind, to reach these scales, must pave the world — that a civilization run on sunlight would carpet every field and hillside in glass and steel. The arithmetic says otherwise, and it is worth doing. Recall the foundational fact from the very first pages of this book: enough sunlight strikes the Earth in a single hour to meet all of humanity's energy needs for a year. Sunlight is not scarce; it is staggeringly abundant, arriving at roughly a kilowatt on every square meter at noon. When you run the numbers for a whole country, they are surprisingly modest. To generate all the electricity a large nation uses, you would need solar panels covering something like a few tenths of one percent of its land — a fraction so small it can be met many times over with rooftops, parking lots, roadsides, reservoirs, and the marginal, unfarmed land that every country has in abundance. Wind is even gentler on the land, because a wind farm is mostly empty air: the turbines stand far apart, and the ground between them keeps growing wheat or grazing cattle, the farmer collecting rent from the sky while still working the soil below. The picture of a paved-over planet is a failure of arithmetic. The truth is closer to the opposite — that the Sun's income is so lavish we need only a modest corner of the world to live on it.

I sometimes think of the whole challenge as the problem of the last ten or twenty percent. Getting a grid from zero to something like half solar and wind turns out to be the comparatively straightforward part; the machines are cheap, the sunshine and wind are free, and the existing gas and hydro plants can flex to fill the gaps. It is the final stretch — the last fifth of the way to a fully clean grid, the windless winter fortnight, the run of dark days when neither the panels nor the turbines deliver — that demands the expensive storage, the long transmission, the backup capacity that sits idle most of the year waiting for its rare moment. That last stretch is genuinely unsolved, and it is where the real cost and difficulty of a fully decarbonized system now live. Anyone who tells you the transition is finished has not looked at the last ten percent. Anyone who tells you it is hopeless has not looked at the first ninety.

So let me correct both errors squarely, because the public conversation swings uselessly between them. On one side stands the counsel of despair: that renewables are too little and too late, that they can never scale, that the whole effort is a feel-good gesture against a problem too vast. That view is simply refuted by the numbers — by three terawatts installed, by half a terawatt of solar added in a year, by the cost crossover that has made clean power the default choice of hard-nosed capital. This is not a gesture; it is the largest and fastest energy build-out humanity has ever undertaken, and it is accelerating. On the other side stands the counsel of complacency: that the transition is solved, inevitable, already won, that we can relax because the market will finish the job on its own. That view is refuted just as squarely — by the unsolved seasonal storage, the lagging grids, the unelectrified furnaces, the material bottlenecks, and above all by the clock. Emissions must fall on a timeline set by physics and the atmosphere, not by the comfortable pace of markets, and there is no guarantee the build-out goes fast enough. The honest position is the narrow ridge between those two errors: it is really happening, at astonishing speed, driven now by cost as much as by conviction — and it is a race against time whose outcome is not assured.

Let me gather the whole arc of this book into a single line, because we have come a very long way from where we began. It starts with the Sun, that ordinary star pouring out more energy onto the Earth in a single hour than all of humanity uses in a year. Some of that light we catch directly, with the photovoltaic cell — the trick Becquerel glimpsed in 1839 and the Bell Labs three made practical in 1954, a photon knocking an electron loose across the band gap of a doped silicon wafer. Some of that same sunlight the Earth catches for us and hands on in another form: the uneven heating of the globe stirs the air into wind, so that wind is really sunlight one step removed, and we catch it with the turbine's blade, an airfoil generating lift, bound by Betz's limit, its power rising with the cube of the wind's speed. Both cell and turbine were once absurdly expensive, and both were driven down the learning curve — Swanson's law and its windier cousin — by decades of policy and manufacturing until they became the cheapest power humanity has ever known. Cheap generation then meets its hardest test at the grid, which must be rebuilt with storage and transmission and flexibility to run on power that comes and goes with the weather. And beyond the grid lies the future we have been sketching: the tandem cell wringing more from every photon, the green hydrogen reaching the corners electrons cannot, the terawatts multiplying toward the tens of terawatts that a fully solar-and-wind civilization would require by the middle of this century. Sun to wind, cell to turbine, cheap to grid, grid to future — that is the shape of the thing, and now you can see all of it at once.

To recap the terawatt age in plain words: solar and wind have reached a scale — nearly three terawatts together — that would have seemed like science fiction a generation ago, with solar now the fastest-growing energy source in the history of our species and clean power as a whole crossing 40 percent of the world's electricity. The reason is a cost collapse that made them the cheapest new power on Earth, shifting the transition's engine from policy to economics. But storage across seasons, the wires to move the power, the electrification of everything, the industrial scale-up, and the stubborn last tenth of the way all remain genuinely hard, and the whole effort is a race against a clock that does not wait. It is neither too late nor already won. It is happening, unevenly and unfinished, faster than the pessimists believed possible and slower than the crisis demands.

What would it mean, then, if we ran that race and largely won it? What would the world actually look and feel like on the far side of this transition — the air, the map of power, the texture of an ordinary day — for the generation that inherits it? To imagine that, we have to lift our eyes from the machines and the megawatts one last time, and picture a world remade.

Epilogue — A World Remade

Imagine a morning somewhere near the middle of this century. A child wakes in a city — it could be Lagos or Lisbon, Chengdu or Chicago — and the first thing to notice is what she does not notice. The air outside her window is clean. Not clean in the abstract, statistical way that a report might call clean, but clean in the way you feel in your chest, the way the air felt on a mountainside a hundred years ago, before the cities filled with the haze of burning things. There is no coal smoke, no diesel exhaust hanging blue over the intersections, no acrid tang at the back of the throat on a still afternoon. The great public-health scourge of the fossil age — the millions of lives cut short each year by the simple act of breathing near combustion — has receded, not because of a single dramatic cure but because, quietly, over decades, the world stopped setting fire to things in order to move and heat itself. That, more than any statistic about carbon, might be the change a person actually feels: the return of clean air as an ordinary condition of life.

I want to be careful here, because it would be easy to slide into a utopia, and this book has tried very hard not to do that. So let me ground the vision. This mid-century world is not a paradise. It still has poverty and inequality and politics; it still has winters when the wind dies and the grid strains, still has arguments over where to run a transmission line and who profits from a mine. The transition, if it succeeds, will not have abolished hardship. What it will have changed is the substrate — the physical basis on which the whole human enterprise runs. And that change is deep enough to reshape the texture of daily life in ways both large and small. The roof of the child's building generates its own power, quietly, with no moving parts; a battery in the basement carries the household through the evening; the car at the curb, if there is one, charges overnight on cheap midnight electricity and never visits a fuel station because there are few left. Heat comes from a pump that moves warmth rather than a flame that makes it. None of this is exotic in her world. It is simply how things are, the way a light switch is simply how things are for us, its century-old novelty long forgotten.

There is a place in this imagined world where the change runs deeper than clean air, and it is not in the wealthy cities at all. It is in the villages that the old energy system never reached. For the whole fossil age, electricity meant vast central plants and thousands of kilometers of wire, an enormous capital undertaking that simply never arrived for hundreds of millions of people living beyond the grid's edge — families who lit their evenings with kerosene lamps and charged nothing because there was nothing to charge. The economics of a central coal plant could not justify running a line to a remote village of a few hundred souls. But a solar panel does not require the wire. It makes its power where it stands, on the roof of a single house or the shared frame of a village micro-grid, with a battery to carry the light past sunset. And so, in the mid-century world, the leapfrog that mobile phones performed a generation earlier — skipping the copper telephone lines entirely and going straight to the wireless — happens again with power. A child who studies now by a bright lamp in a village her grandparents lit with a smoky wick is not living in a utopia. She is living in a world where the falling cost of catching sunlight finally made electricity cheap enough to reach the last and poorest, not through the grand infrastructure of the past but through a panel small enough to carry on a motorbike. If the transition has a single face that is neither the smokestack nor the subsidy, it might be hers.

Ground the vision, too, in the texture of work, because a change this large does not come without cost to somebody. The fossil age employed millions — in the coal seams and the oil fields, on the rigs and in the refineries, in whole regions whose towns were built around a mine or a plant. A world that stops burning things does not need those hands in those places, and it would be a lie to pretend the transition is painless for the people whose livelihoods it unmakes. There is new work, and more of it — installing panels, building and servicing turbines, laying cable, manufacturing the machines — but it is not always in the same towns, and a laid-off miner cannot always become a solar installer without dislocation and grief. The honest mid-century vision includes that dislocation: the closed mine, the retrained worker, the region that either found a new footing or did not. Whether the transition is remembered as just or as merely cheap will turn on how well that human cost was met, and that is a question of politics and decency, not of physics, and it is not yet answered.

Look up from the household to the map, and the geopolitics have shifted on their foundations. For a century and a half, the great leverage of nations flowed from beneath the ground — from who sat atop the oil and the gas, from the tankers and pipelines that carried ancient carbon across the world, from the wars and alliances and fortunes built on the fact that energy came from a few blessed and cursed places. A sun-and-wind world redraws that map, because sunlight and wind fall, unevenly but universally, on everyone. No nation can embargo the Sun. A country that once imported every drop of its fuel can, with enough panels and turbines and wire, make much of its own power from the sky above its own territory — a kind of energy independence that was impossible when energy meant a substance you had to dig up or buy. But the vision must stay honest: the leverage does not vanish, it migrates. It moves from those who hold the oil to those who make the machines and refine the minerals — to the manufacturing and mineral powers, above all, as we have seen, to China. The new world has its own chokepoints, its own strategic vulnerabilities in the supply of lithium and copper and rare earths, its own contests over who builds the panels the rest of the world depends on. It trades the old dependencies for new ones. It is freer in some ways and entangled in others. That is not a failure of the vision; it is simply what it means to be honest about power, which never disappears but only changes its address.

And here, at the end, I want to return to the single thread that has run beneath every chapter of this book, because it is the thread that gives the whole story its meaning. Wind, we have said, is sunlight one step removed — the Sun heats the Earth unevenly, the uneven heat stirs the air, and the moving air is what the turbine catches. But the deeper truth reaches further than that. Nearly every source of energy humanity has ever used is sunlight in some disguise. The coal we burned was sunlight that fell on forests three hundred million years ago, pressed into black rock. The oil was sunlight that fell on ancient seas, on the tiny drifting lives that sank and were buried and cooked slowly into crude. Even the food that powers our own bodies is sunlight, caught by leaves last summer. For the few centuries of the fossil-fuel age — and against the long span of human history it was only a few centuries, a brief and extraordinary interlude — we did something unprecedented: we stopped living on the sunlight of the present and began drawing down the sunlight of the deep past, spending in a handful of generations a savings account that took the Earth hundreds of millions of years to fill. It was an astonishing inheritance, and it built the modern world, and we should not be ashamed of having spent it; it lifted billions out of darkness and toil. But it was, unmistakably, a drawing-down of savings — finite, and unrepeatable, and with a bill attached in the form of the altered air.

Sit with that image of the savings account a moment longer, because it reframes the whole of what we have done. A barrel of oil represents the concentrated sunlight of an unimaginable stretch of ancient time — one calculation suggests that the fossil fuel humanity burns in a single year took something like a million years of that ancient sunlight to lay down. We have been drawing on the account, in other words, roughly a million times faster than it was ever filled. No household could run its finances that way for long, and no civilization can either; the only questions were ever how long the savings would last and what the withdrawals would do to the air in the meantime. What makes our moment extraordinary is not that we noticed the account was finite — thoughtful people have said so for over a century — but that we finally built machines cheap enough to live on the daily deposit instead. For the first time since the fires were lit, the choice to stop drawing down the savings does not require accepting a poorer, colder, darker life. That is genuinely new under the Sun, and it is the quiet miracle at the center of this whole account: that living on the light of the present day has become, of all things, the cheapest option on the table.

What the story in these pages describes, then, is not really the invention of something new. It is a homecoming. For all but that brief fossil interlude, humanity lived on the Sun's daily income — on the wind that turned the miller's sails and drained the Dutch polders, on the water wheel and the wood fire and the muscle of animals fed on grass, all of it sunlight arriving fresh each day. The great project of this century is to learn to live that way again — on the sun's daily income rather than its ancient savings — but this time armed with everything the fossil interlude taught us. This time we return to the daily sunlight not with windmills of cedar and canvas but with silicon that catches a photon and hands over an electron, with rotors longer than a football field spinning in the steady offshore wind, with batteries and hydrogen and grids of continental reach, with a mastery of the physics of light and air that Becquerel and la

Cour could only have dreamed of. It is the oldest energy source rediscovered through the newest technology — the daily sunlight, caught this time by high technology. That is the whole of it, and it is why the book you are holding is not, in the end, a story about machines. It is a story about a civilization deciding whether to keep spending its inheritance or to learn, again, to live within its means.

Let me honor, one last time, the long cast who carried this idea across the generations, because none of them lived to see what they set in motion, and most died thinking their work a failure or a curiosity. Edmond Becquerel, nineteen years old in his father's Paris laboratory in 1839, noticing that light striking his electrodes made a current stir, with not the faintest idea of what he had begun. Charles Fritts in the 1880s, coating selenium with gold so thin it was translucent, convinced his one-percent cell would one day rival the coal-fired stations of Edison, and mocked for it. Poul la Cour on his Danish hill, splitting water with the wind, lighting a schoolhouse with hydrogen, running the first laboratory ever devoted to the electricity of the air. Frank Shuman on the banks of the Nile in 1913, his mirrors pumping water to irrigate the desert, certain that solar power would soon rival coal — before the Great War swept his plant and his dream away. The three at Bell Labs in 1954 — Daryl Chapin, Calvin Fuller, Gerald Pearson — who made the first silicon cell that could actually power something, and whose invention found its first real home not on Earth but circling it, aboard a satellite. Richard Swanson, whose stubborn cost curve nobody believed until it had bankrupted the doubters. Hermann Scheer, the German parliamentarian who fought to make his country pay a premium for sunlight so that the rest of the world would not have to, and who did not live to see solar become the cheapest power on Earth. They worked, most of them, in obscurity or ridicule, each adding a stone to a wall whose top they could not see. The terawatts are their monument, though not one of their names is carved on it.

And so I come to the honest tension that this book has refused to resolve, because to resolve it falsely would be to betray everything the story is about. Consider how far it has come. In a single human lifetime — the span of one long life — solar and wind have gone from selenium wafers and cedar-bladed dynamos, from desert curiosities and government follies, to nearly three terawatts of the cheapest electricity our species has ever generated, growing faster than any energy source in history. That is not a small thing. It is, arguably, the most hopeful industrial fact of our age, a genuine and hard-won achievement of human ingenuity, and anyone who has followed the tale this far has earned the right to feel some wonder at it. And yet. And yet the seasonal storage is unsolved, the grids lag, the last hard tenth of the way remains genuinely difficult, the furnaces of heavy industry still burn, and the atmosphere keeps its own indifferent clock that answers to physics and not to our progress. How far and how fast it has come is real. How much remains genuinely unsolved is equally real. Both truths sit together, and neither cancels the other, and honesty means holding them both without flinching.

Perhaps that is the right note to end on — not triumph and not despair, but the particular seriousness of a task that is neither hopeless nor finished. The Sun has been pouring its light on this planet for four and a half billion years, and it will pour it for billions more, wholly uninterested in whether we learn to catch it well. The wind has stirred the air since there was air to stir. These are the oldest facts in the world, older than life, older than the oceans. What is new — what is astonishingly, precariously new — is a species that has finally built the tools to live on that daily light again, and that now must decide, in the narrow window of a few decades, whether it will finish what it has so improbably begun. The machines are ready. The physics is settled. The cost has fallen. What remains uncertain is the only thing that was ever truly in our hands: whether we will choose, in time, to live on the sun's income rather than its savings — to inherit the light, as every generation before the fossil age did, and as every generation after us now can.

Timeline

A chronological tour of the milestones that carried sunlight and moving air from folk technology to the cheapest electricity humanity has ever made. Dates for fast-moving recent figures are given "as of" their year; where sources count differently, the text says so.

Antiquity to 1800: Wind as the First Engine

- **c. 7th–9th century CE** — Vertical-axis windmills appear in Sistan, on the Persian–Afghan borderlands, grinding grain and pumping water. They are the first machines to put the wind to systematic work.
- **c. 1100s–1200s** — Horizontal-axis post windmills spread across Europe, from England and northern France through the Low Countries, milling grain and driving early industry.
- **1200s onward** — Dutch engineers turn windmills into a national technology, using them to drain polders and reclaim farmland from the sea.

The 1800s: Foundations of a Science

- **1839** — Nineteen-year-old Edmond Becquerel, experimenting with electrodes in an electrolyte, discovers the photovoltaic effect: light shining on the right material produces electric current.
- **1860s–1878** — Augustin Mouchot builds a series of solar-powered steam engines in France, culminating in a prize-winning machine exhibited at the 1878 Paris Universal Exhibition.
- **1876** — William Grylls Adams and Richard Day show that selenium generates a current when exposed to light, the first demonstration of the solid-state photovoltaic effect.
- **1882** — Abel Pifre, Mouchot's assistant, runs a printing press in a Paris garden powered entirely by a solar concentrator.
- **1883** — Charles Fritts builds the first solid-state solar cell, coating selenium with a film of gold. It converts roughly one percent of the light that strikes it.
- **1887–1888** — Charles F. Brush erects the first automatically operating wind turbine for electricity in Cleveland, Ohio, a 17-metre rotor with 144 cedar blades charging batteries in his mansion's cellar.
- **1891** — Poul la Cour begins his wind-electricity experiments at Askov in Denmark, founding the first laboratory dedicated to wind power.

1900–1939: The Quantum and the Megawatt Dream

- **1905** — Albert Einstein explains the photoelectric effect, supplying the quantum theory beneath all photovoltaics; the work wins him the 1921 Nobel Prize in Physics.
- **1912–1913** — Frank Shuman builds the world's first solar-thermal power station at Maadi, Egypt, a field of

parabolic troughs driving a roughly 55-to-60-kilowatt engine that pumps Nile water for irrigation.

- **1918–1930s** — A wind-charger industry supplies rural American farms with small battery-charging windmills, bringing electric light to homes beyond the reach of the grid.
- **1919** — Albert Betz publishes the limit that bears his name: no turbine can capture more than about 59.3 percent of the wind's kinetic energy.

1940s–1950s: The Modern Cell and the First Megawatt Turbine

- **1941** — Russell Ohl at Bell Labs patents the silicon p–n junction after noticing that a cracked silicon sample produced current in light, the electrical heart of the modern solar cell.
- **October 1941** — The 1.25-megawatt Smith–Putnam turbine, engineered by Palmer Putnam, begins feeding power to the grid from Grandpa's Knob in Vermont, the first megawatt-scale wind machine.
- **March 1943** — A blade fails on the Smith–Putnam turbine; wartime material shortages prevent repair, and the project is later abandoned.
- **25 April 1954** — Bell Labs' Daryl Chapin, Calvin Fuller, and Gerald Pearson unveil the first practical silicon solar cell, about six percent efficient, hailed as the "solar battery."
- **1957** — Denmark's Gedser turbine, built by Johannes Juul, pioneers the three-blade, upwind design that remains the template for modern turbines.
- **March 1958** — Vanguard 1 becomes the first solar-powered satellite, and space becomes photovoltaics' first paying market.

1960s–1970s: Oil Shocks Spark Policy

- **1968** — Elliott Berman founds Solar Power Corporation with Exxon backing and drives down the cost of photovoltaics, opening the first terrestrial markets.
- **1973** — The first oil shock jolts governments into launching renewable-energy research programs.
- **1977** — The United States enacts federal solar tax credits and establishes the Solar Energy Research Institute, forerunner of the National Renewable Energy Laboratory.
- **1979** — Amid the second oil shock, President Jimmy Carter installs solar thermal panels on the White House roof.

1980s: The First Commercial Scale-Up

- **1980s** — California's "wind rush" fills the Altamont, Tehachapi, and San Gorgonio passes with thousands of turbines, spurred by tax credits and federal purchase rules.
- **1982–1991** — The SEGS parabolic-trough plants rise in California's Mojave Desert, reaching about 354 megawatts of concentrating solar power.
- **1986** — The Reagan administration removes Carter's White House panels, and federal renewable incentives lapse.

1990s: Feed-in Tariffs and Danish–German Leadership

- **1990** — Germany passes the Electricity Feed-in Law, the world's first national feed-in tariff.
- **1991** — Vindeby, off the coast of Denmark, becomes the world's first offshore wind farm with eleven turbines.

- **1990s** — Denmark and Germany build the modern turbine industry, giving rise to Vestas, Enercon, and Bonus (later Siemens Wind).
- **1994** — Richard Swanson founds SunPower; the cost trend later named "Swanson's law" begins to take clear shape.

2000s: The Take-Off

- **1 April 2000** — Germany's Renewable Energy Sources Act, championed by Hermann Scheer and Hans-Josef Fell, takes effect and ignites the Energiewende.
- **2004** — Spain adopts strong feed-in tariffs, touching off a solar and wind boom.
- **2005** — The US Energy Policy Act establishes and expands the solar Investment Tax Credit.
- **2006–2010** — China's Suntech, LDK, and their peers scale up manufacturing, and global module prices begin their steep descent.
- **2008** — David MacKay publishes *Sustainable Energy — Without the Hot Air*, framing the arithmetic of the transition for a generation of readers.

2010s: Cost Collapse and Grid Parity

- **2010** — Global utility-scale solar costs about 41.7 US cents per kilowatt-hour, the baseline for the roughly 90 percent decline to come.
- **2011–2013** — A module oversupply crashes prices, bankrupts many Western manufacturers, and consolidates Chinese dominance.
- **2012** — *The Economist* coins the term "Swanson's law."
- **2015** — The Paris Agreement is adopted, accelerating national renewable targets.
- **2015–2017** — Record-low solar auction prices tumble below three US cents per kilowatt-hour in Chile, the United Arab Emirates, Mexico, and Saudi Arabia.

- **2016** — Cumulative global solar passes roughly 300 gigawatts, and onshore wind becomes competitive with new fossil plants.
- **2017** — Hywind Scotland, the first commercial floating offshore wind farm, begins generating.
- **2019** — The Nobel Prize in Chemistry goes to John Goodenough, Stanley Whittingham, and Akira Yoshino for the lithium-ion battery.

2020s: The Terawatt Era

- **2020** — Renewables set annual records despite the pandemic, and the IEA calls solar "the cheapest electricity in history."
- **April 2021** — Saudi Arabia's Al Shuaibah tender sets a world-record solar bid of about 1.04 US cents per kilowatt-hour.
- **2022** — Cumulative global solar passes roughly 1.2 terawatts.
- **16 August 2022** — The US Inflation Reduction Act is signed, extending the Investment and Production Tax Credits and adding domestic-manufacturing credits.
- **2023** — Cumulative global solar reaches about 1.6 terawatts; in November, LONGi sets a 33.9 percent perovskite–silicon tandem record.
- **2024** — A record year: solar adds roughly 452 to 553 gigawatts and wind about 117 gigawatts, bringing wind's total to roughly 1,136 gigawatts. Solar supplies 6.9 percent and wind 8.1 percent of global electricity, and all clean sources pass 40.9 percent (as of 2024; sources vary by counting method).
- **2024** — Global average solar cost falls to about 4.3 US cents per kilowatt-hour and onshore wind to about 3.4 cents, the cheapest of any source (as of 2024, estimates vary).
- **October 2024** — China's Dongfang unveils the world's largest offshore turbine, 26 megawatts with a rotor

roughly 310 metres across.

- **Late 2024** — LONGi pushes the NREL-certified perovskite–silicon tandem record to 34.85 percent, beating the single-junction ceiling.
- **2024** — Grid batteries add a record roughly 200 gigawatt-hours, for a world total near 375 gigawatt-hours, with China leading deployment.
- **2025** — China holds more than 80 percent of the solar manufacturing supply chain, and battery storage grows on the order of 40 percent year on year (as of 2025; refresh from the latest IEA, IRENA, GWEC, and Ember releases).

Who's Who

The people whose curiosity, stubbornness, and occasional luck built the sun-and-wind age. They are grouped by the field where their work landed hardest, though many crossed the lines.

Solar / Photovoltaics

Edmond Becquerel (1820–1891). A French physicist working in his father's laboratory, Becquerel was only nineteen when, in 1839, he noticed that light falling on electrodes in an electrolytic cell produced a small electric current. That observation, the photovoltaic effect, is the seed from which every solar panel on Earth eventually grew. He spent the rest of his career on the physics of light and phosphorescence, never seeing what his youthful discovery would become.

Charles Fritts (1850–1903). An American inventor who, in 1883, built the first solid-state solar cell by coating the semiconductor selenium with a whisper-thin layer of gold. It converted barely one percent of the sunlight that hit it and could never compete with coal, but it proved that a slab of material with no moving parts could turn light directly into electricity. Fritts predicted, correctly and a century early, that such devices might one day rival fossil-fuelled power stations.

Albert Einstein (1879–1955). Einstein belongs to solar's story for his 1905 explanation of the photoelectric effect, which showed that light arrives in discrete packets of energy, later called photons, each able to knock an electron loose from a metal. This quantum picture, not his relativity, won him the 1921 Nobel Prize, and it is the theoretical foundation on which the working of every photovoltaic cell rests.

Augustin Mouchot (1825–1912). A French mathematics teacher who feared that coal would one day run out, Mouchot devoted himself to solar-concentrating steam engines. His prize-winning machine at the 1878 Paris Universal Exhibition ran a printing press and made ice from the desert sun. Cheap coal ended his funding, but he is the true ancestor of concentrating solar power.

Abel Pifre (1852–1928). Mouchot's assistant and successor, Pifre gave solar power one of its great early spectacles in 1882 when he ran a printing press in a Paris garden entirely on sunlight gathered by a parabolic mirror, printing copies of a newspaper he aptly called the *Journal Soleil*.

John Ericsson (1803–1889). The Swedish-American engineer, already famous for the ironclad warship *Monitor*, built a series of solar-concentrating hot-air and steam engines in the 1870s and 1880s, convinced that the sun would eventually have to power industry when the coalfields were exhausted.

Frank Shuman (1862–1918). An American inventor who built the world's first solar-thermal power station at Maadi, Egypt, in 1912–13, using long parabolic troughs to raise steam for a pump that lifted Nile water onto the fields. It worked, and Shuman dreamed of vast solar plants across the Sahara, but the First World War destroyed his backing and he died before he could resume.

Russell Ohl (1898–1987). A Bell Labs engineer who, in 1941, patented the silicon p–n junction after noticing that a particular cracked silicon rod generated a voltage when exposed to light. The accidental discovery that impurities on either side of the crack created a built-in electric field is the physical principle at the core of the modern silicon solar cell.

Daryl Chapin (1906–1995), Calvin Fuller (1902–1994), and Gerald Pearson (1905–1987). The Bell Labs trio who, on 25 April 1954, demonstrated the first practical silicon solar cell, roughly six percent efficient. Chapin, a physicist, was hunting for a power source for remote telephone equipment; Fuller, a chemist, mastered the doping of silicon; Pearson, a physicist, built and tested the cells. Their "solar battery" is the direct ancestor of today's panels.

Elliott Berman (born 1930s). A chemist who founded Solar Power Corporation in 1968 with backing from Exxon and set out to make photovoltaics cheap enough for use on Earth rather than only in space. By simplifying the cell and its packaging he cut the cost per watt roughly fivefold, opening the first terrestrial markets: oil rigs, navigation buoys, and remote installations.

Richard Swanson (born 1943). A Stanford engineer who founded SunPower and pushed high-efficiency silicon cells toward the commercial mainstream. His name is attached to "Swanson's law," the observation that solar module prices fall about twenty percent for every doubling of cumulative production, a solar-specific case of the broader learning curve.

Martin Green (born 1948). The Australian physicist often called the father of modern photovoltaics. At the University of New South Wales he and his students set a long string of silicon efficiency records and, crucially, developed the PERC cell design that underpins the majority of panels made in the world today. Many of the engineers who built China's solar industry trained in his laboratory.

Zhengrong Shi (born 1963). A student of Martin Green who returned to China and founded Suntech in 2001, becoming for a time the world's largest solar manufacturer and, briefly, one of China's richest people. His rise, and Suntech's later bankruptcy in the 2011–13 shakeout, encapsulates China's turbulent surge to solar-manufacturing dominance.

Jimmy Carter (1924–2024). The US president who, during the energy crises of the late 1970s, installed solar thermal panels on the White House in 1979 and expanded federal support for renewable research. The panels' removal under his successor turned them into a lasting political symbol of how national commitment to the technology could wax and wane.

Hermann Scheer (1944–2010). A German member of parliament and tireless evangelist known as the "solar king," Scheer was the driving political force behind Germany's 2000 feed-in-tariff law. More than any other single politician he engineered the guaranteed-price mechanism that made rooftop and field solar bankable and triggered the global demand that later crashed prices.

Hans-Josef Fell (born 1952). A Green politician who co-drafted Germany's 2000 Renewable Energy Sources Act with Scheer. The law's elegant design, a fixed, long-term price for every kilowatt-hour fed into the grid, became the template that dozens of countries copied.

Wind

Charles F. Brush (1849–1929). An American inventor and industrialist, wealthy from arc lighting, who in 1888 built the first automatically operating wind turbine to generate electricity, a giant rotor of 144 cedar blades that charged batteries in the cellar of his Cleveland mansion. *Scientific American* called it the only successful wind-powered electric-lighting plant of its day.

Poul la Cour (1846–1908). A Danish scientist and teacher regarded as the father of wind electricity. In the 1890s he electrified windmills at Askov, founded the first wind-power laboratory, and showed through careful experiment that a rotor with fewer, faster-moving blades is more efficient, a finding that still shapes turbine design.

Palmer Cosslett Putnam (1900–1984). An MIT-trained engineer whose Smith–Putnam turbine on Grandpa's Knob in Vermont was, in 1941, the first megawatt-scale wind machine, rated at 1.25 megawatts. It fed the Vermont grid until a blade failed in 1943 and wartime shortages ended the experiment, but it proved that giant turbines were mechanically possible.

Ulrich Hütter (1910–1990). A German engineer who in the 1950s and 1960s pioneered lightweight fibreglass composite blades and much of the aerodynamic theory behind slender, efficient rotors. His ideas about blade flexibility and materials fed directly into the machines that later dominated the industry.

Johannes Juul (1887–1969). A Danish engineer and former pupil of Poul la Cour whose 200-kilowatt Gedser turbine of 1957 established the three-blade, upwind, stall-regulated layout, the "Danish concept," that remains the ancestor of virtually every large turbine built since.

Henrik Stiesdal (born 1957). A Danish inventor who as a young man helped define the modern three-blade turbine and went on to lead engineering at Bonus and Siemens, driving turbines out to sea. In later life he has worked on floating offshore foundations and other clean-energy technologies, a living thread connecting the industry's origins to its frontier.

James Dehlsen (born 1937). An American entrepreneur who founded Zond during the California wind rush and later Clipper Windpower, an important figure in building a domestic US wind industry and in the continual push toward larger, more reliable machines.

Robert Righter. A historian of the American West whose

scholarly histories of wind power, from Brush through the California boom, preserved a story that the industry itself had half forgotten, and became a foundation for later writers.

Science & Policy

Albert Betz (1885–1968). A German physicist and aerodynamicist who in 1919 proved that no wind turbine can extract more than 16/27, about 59.3 percent, of the kinetic energy in the air passing through its rotor. The Betz limit is to wind engineering what the speed of light is to relativity: an unbreakable ceiling that every real machine is measured against.

John Goodenough, Stanley Whittingham, and Akira Yoshino. The three scientists awarded the 2019 Nobel Prize in Chemistry for developing the lithium-ion battery. Their work, spanning the 1970s and 1980s, produced the rechargeable cell that now stores solar and wind energy on the grid and in electric vehicles, making variable renewables far easier to live with.

David J. C. MacKay (1967–2016). A Cambridge physicist and later chief scientific adviser to Britain's energy department, whose 2008 book *Sustainable Energy — Without the Hot Air* insisted that the transition be discussed in honest numbers, land, watts, and area, rather than slogans. He died before the great cost collapse vindicated much of the optimism he had cautiously entertained.

Vaclav Smil (born 1943). A prolific energy scholar whose data-dense, deliberately skeptical histories of energy transitions serve as the essential counterweight in any honest account. Smil stresses how slow and material-hungry past transitions have been; engaging his cautions keeps the story of solar and wind grounded rather than triumphalist.

Glossary

Plain-language definitions of the field's essential terms, alphabetical. Cross-references point you to the fuller entry.

Actuator disk — An idealized model that treats a wind rotor as a thin disk extracting energy from the air flowing through it; the starting point for deriving the Betz limit.

AC (alternating current) — Electricity whose direction of flow reverses many times per second, fifty or sixty times depending on the country. It is the form the grid and most appliances use.

Agrivoltaics — Growing crops or grazing livestock on the same land as solar panels, so a single plot both farms and generates power.

Air mass (AM1.5) — A standard describing how much atmosphere sunlight has passed through before reaching a surface. AM1.5 is the reference spectrum used to rate solar panels.

Albedo — The fraction of sunlight a surface reflects, relevant for bifacial panels that harvest light bouncing up from the ground.

Anemometer — An instrument that measures wind speed, used both on turbines and in surveying a site's wind resource.

Array — A group of solar modules wired together; the full field of panels in an installation.

Auction (renewable) — A competitive process in which developers bid to sell power at the lowest price. Auctions have driven record-low solar and wind prices.

Balance of system (BOS) — Everything in a solar installation other than the modules themselves: racking, wiring, inverters, combiner boxes, and labour.

Band gap — The minimum energy a photon needs to knock an electron loose in a semiconductor. Silicon's band gap is about 1.1 electronvolts.

Baseload — The traditional term for always-on power such as coal or nuclear; increasingly seen as an outdated way to describe a flexible modern grid.

Becquerel effect — Another name for the photovoltaic effect, after Edmond Becquerel, who discovered it in 1839.

Betz limit (Betz's law) — The theoretical maximum share of the wind's kinetic energy a turbine can capture: about 59.3 percent, or 16/27.

Bifacial gain — The extra energy a bifacial panel produces from light striking its rear side, expressed as a percentage.

Bifacial module — A solar panel that generates power from light hitting both its front and its back.

BIPV (building-integrated photovoltaics) — Solar cells built into roofing, façades, or windows, so the building material itself generates power.

Blade pitch — Rotating a turbine blade about its long axis to control how much power it captures, or to feather it in dangerous winds.

Capacity (nameplate) — The maximum power a plant can produce under ideal conditions, measured in watts, kilowatts, megawatts, or gigawatts.

Capacity factor — Actual energy produced over a period divided by the maximum possible if the plant ran flat out; roughly 10 to 30 percent for solar and 25 to 50 percent or more for wind.

Capacity value — How much reliable, firm capacity a variable source contributes toward meeting peak demand.

Cell (solar / photovoltaic) — The basic light-to-electricity unit, usually a thin wafer of doped silicon producing about half a volt.

CdTe (cadmium telluride) — A thin-film solar technology, most associated with First Solar, that is cheaper per watt than silicon in some markets.

Concentrating solar power (CSP) — Systems that use mirrors to focus sunlight into heat, driving a steam turbine; the heat can be stored for use after dark.

Cp (power coefficient) — The fraction of the available wind energy a turbine actually captures, capped by the Betz limit.

Curtailment — Deliberately reducing available wind or solar output when the grid cannot use or carry it.

Cut-in speed — The minimum wind speed, around three to four metres per second, at which a turbine starts generating.

Cut-out speed — The wind speed, around twenty-five metres per second, above which a turbine shuts down to avoid damage.

DC (direct current) — Electricity that flows in one direction only, the raw output of solar cells and batteries.

Direct-drive turbine — A turbine with no gearbox; the rotor turns a large multi-pole generator directly, often using rare-earth magnets.

Direct normal irradiance (DNI) — The sunlight arriving straight from the sun's disk, the resource concentrating solar power depends on.

Dispatchable — Power that can be turned up or down on command, such as gas, hydro, or batteries. Solar and wind are not inherently dispatchable.

Doping — Adding tiny amounts of impurities such as phosphorus or boron to silicon to give it the electrical properties a solar cell needs.

Doubly-fed induction generator (**DFIG**) — A common geared turbine generator that handles varying rotor speeds efficiently.

Duck curve — The distinctive daily demand shape, first charted by California's grid operator, caused by a midday solar surplus and a steep evening ramp.

EEG — Germany's Renewable Energy Sources Act (Erneuerbare-Energien-Gesetz) of 2000, the landmark feed-in-tariff law behind the Energiewende.

Efficiency (**module / cell**) — The percentage of sunlight energy converted to electricity, about 20 to 23 percent for commercial modules today.

Electrolyzer — A device that splits water into hydrogen and oxygen using electricity, producing "green" hydrogen when powered by renewables.

Energiewende — Germany's national "energy transition" toward renewables and away from nuclear and fossil fuels.

Energy payback time (**EPBT**) — How long a panel or turbine must operate to generate the energy used to make it, now roughly one to two years for solar.

Energy storage — Any system that saves energy for later use, including batteries, pumped hydro, thermal stores, and hydrogen.

Feathering — Turning turbine blades edge-on to the wind so they stop capturing energy in dangerous conditions.

Feed-in tariff (**FiT**) — A policy guaranteeing renewable generators a fixed, above-market price per kilowatt-hour over a long contract.

Firm power — Electricity reliably available on demand; solar paired with storage is increasingly marketed as firm.

Floating offshore wind — Turbines mounted on floating platforms anchored in deep water, where fixed foundations are impractical.

Floating solar (floatovoltaics) — Panels mounted on floats on reservoirs, lakes, or the sea, saving land and cooling the panels.

Gearbox — A gear system in many turbines that steps up the slow rotor speed to the higher speed the generator needs.

Generator — The machine that converts the rotor's mechanical rotation into electricity.

Gigawatt (GW) — One billion watts, a large-power-plant-scale unit. A gigawatt-hour (GWh) is that power sustained for one hour.

Green hydrogen — Hydrogen made by splitting water with renewable electricity, with no carbon emissions.

Grid parity — The point at which renewable power costs the same as, or less than, grid electricity from conventional sources.

Heliostat — A mirror that tracks the sun and reflects its light onto a central receiver in a concentrating-solar power tower.

Heterojunction (HJT) — A high-efficiency silicon cell design that layers crystalline and amorphous silicon.

Horizontal-axis wind turbine (HAWT) — The standard propeller-style turbine, with a horizontal spin axis facing the wind.

Hub height — The height of a turbine's rotor centre above the ground; taller hubs reach stronger, steadier wind.

Insolation — The amount of solar energy received on a surface over time, often given in kilowatt-hours per square metre per day.

Intermittency — The variable, weather-dependent nature of solar and wind output.

Inverter — Electronics that convert a solar array's or battery's direct current into grid-compatible alternating current.

Investment Tax Credit (ITC) — A US tax credit, 30 percent under the Inflation Reduction Act, for the capital cost of solar and other clean-energy projects.

Irradiance — The power of sunlight hitting a surface, in watts per square metre; "one sun" is 1,000 watts per square metre.

Islanding — When a distributed generator keeps energizing a section of grid during an outage. Inverters carry anti-islanding protection for safety.

Kilowatt-hour (kWh) — The standard unit of energy on an electricity bill: one kilowatt used for one hour.

LCOE (levelized cost of electricity) — The average lifetime cost per unit of electricity from a plant, used to compare technologies.

Learning curve (rate) — The consistent cost drop each time cumulative production doubles; solar's is about twenty percent (see Swanson's law).

LFP (lithium iron phosphate) — A durable, cobalt-free lithium battery chemistry now dominant in grid storage.

Lift — The aerodynamic force on a blade acting as an airfoil that makes a modern turbine rotor spin.

Lithium-ion battery — The dominant rechargeable battery for short-duration grid storage and electric vehicles.

Megawatt (MW) — One million watts, the scale of a single large wind turbine or a commercial solar farm.

Microinverter — A small inverter attached to a single panel, improving performance when part of an array is shaded.

Module (panel) — Multiple solar cells wired and sealed together into one weatherproof unit.

Molten-salt storage — Hot salt used to store concentrating-solar heat so a plant can generate power after sunset.

Monopile — A single large steel tube driven into the seabed as a fixed foundation for an offshore turbine.

MPPT (maximum power point tracking) — An inverter function that continuously finds the voltage yielding peak power as light and temperature change.

Nacelle — The housing atop a turbine tower containing the gearbox, generator, and controls.

Nameplate capacity — See Capacity (nameplate).

NdFeB magnet — A powerful neodymium-iron-boron permanent magnet used in direct-drive generators and electric-vehicle motors.

Net metering — A billing scheme that credits rooftop-solar owners for surplus power they send to the grid.

Offshore wind — Wind farms built in the sea, where winds are stronger and turbines can be larger.

Onshore wind — Wind farms built on land, the cheapest source of new electricity by levelized cost as of 2024.

Oversupply (glut) — When manufacturing capacity exceeds demand and crashes prices, as with solar modules in 2023 and 2024.

PERC (passivated emitter and rear cell) — A once-dominant efficiency-boosting silicon-cell design now giving way to TOPCon and HJT.

Perovskite — A class of crystal materials that make cheap, efficient thin solar cells; promising for tandem cells but still maturing on durability.

Permanent-magnet synchronous generator (**PMSG**) — A gearless generator using strong magnets, common offshore and in direct-drive turbines.

Photon — A particle, or quantum, of light; each carries energy that can free an electron in a solar cell.

Photovoltaic (**PV**) — The direct conversion of light into electricity using semiconductors.

Photovoltaic effect — The generation of voltage or current in a material exposed to light, discovered by Becquerel in 1839.

p–n junction — The boundary between positively and negatively doped semiconductor regions; the built-in electric field there drives a solar cell's current.

Polysilicon — Highly purified silicon, the feedstock for most solar wafers and cells.

Power coefficient — See Cp.

Power purchase agreement (**PPA**) — A long-term contract to buy a project's electricity at an agreed price.

Power tower — A concentrating-solar design with a field of heliostats focusing sunlight on a receiver atop a central tower.

Production Tax Credit (**PTC**) — A US per-kilowatt-hour tax credit for wind and other generation.

Pumped-storage hydro — Storing energy by pumping water uphill and releasing it through turbines later, the largest form of grid storage.

Rare earth elements (**REE**) — Metals such as neodymium and dysprosium used in permanent magnets; supply is concentrated in China.

Rated speed (rated power) — The wind speed at which a turbine reaches its maximum, or nameplate, output.

Reactive power — A component of alternating-current power that inverters and generators supply to help stabilize grid voltage.

Renewable Portfolio Standard (RPS) — A mandate requiring utilities to source a set percentage of power from renewables.

Repowering — Replacing older turbines with fewer, larger, more efficient modern ones on an existing site.

Rotor — A turbine's rotating assembly of blades and hub that captures the wind.

Semiconductor — A material such as silicon whose conductivity lies between that of a conductor and an insulator; the heart of a solar cell.

Shockley–Queisser limit — The maximum theoretical efficiency, about 33.7 percent, of a single-junction solar cell under one sun.

Silicon (crystalline) — The semiconductor used in roughly 95 percent of solar cells, made as either monocrystalline or polycrystalline material.

Solar constant — The power of sunlight reaching the top of the atmosphere, about 1,361 watts per square metre.

Solar thermal — Using sunlight as heat, for hot water, process heat, or concentrating-solar power, rather than making electricity directly.

Solar tracker — A mount that tilts panels to follow the sun across the sky, boosting output.

String (PV) — A series-wired chain of solar modules feeding an inverter.

String inverter — An inverter serving a whole string of panels, the most common utility and commercial type.

Swanson's law — The observation that solar module prices fall about twenty percent for every doubling of cumulative shipments.

Tandem (multi-junction) cell — A cell stacking layers of different band gaps to capture more of the spectrum and beat single-junction limits.

Terawatt (TW) — One trillion watts. Global solar capacity passed roughly 1.6 terawatts in 2023.

Thermalization — Energy lost as heat when a photon carries more energy than the band gap needs.

Thin-film solar — Cells made from thin layers such as cadmium telluride or amorphous silicon on a substrate, using less material than wafers.

TOPCon (tunnel oxide passivated contact) — A leading high-efficiency silicon-cell technology succeeding PERC.

Tower (wind) — The tall structure supporting a turbine's nacelle and rotor.

Tracker — See Solar tracker.

Transmission — The high-voltage lines that move power over long distances, a key bottleneck for connecting remote wind and solar.

Utility-scale — Large, power-plant-sized solar or wind projects feeding the grid, as opposed to rooftop or distributed systems.

Value-adjusted LCOE (VALCOE) — A cost metric that accounts for when and how usefully power is produced, not just its average cost.

Variable renewable energy (VRE) — Weather-dependent sources such as wind and solar whose output varies.

Vertical-axis wind turbine (VAWT) — A turbine spinning around a vertical shaft, used in niche and small-scale applications.

Virtual power plant (VPP) — A network of distributed solar, batteries, and flexible loads coordinated to act like one power plant.

Wafer — A thin slice of silicon cut from an ingot, the basis of a solar cell.

Watt (W) — The basic unit of power; solar panels are rated in watts.

Watt-peak (Wp) — A solar panel's rated output under standard test conditions.

Wright's Law — The general principle that costs fall a fixed percentage for every doubling of cumulative production; Swanson's law is the solar case.

Yaw — Rotating a turbine's nacelle horizontally so the rotor faces into the wind.

Appendix A — The Physics Toolkit & By the Numbers

Everything in this book rests on a handful of physical ideas and a handful of numbers. This appendix collects them in one place, in plain language, so you can reason about any solar or wind claim you meet in the wild. None of it requires mathematics beyond arithmetic.

Energy versus Power

The single most common confusion in energy talk is the difference between power and energy. **Power is a rate; energy is an amount.** Power is how fast energy flows right now, and energy is how much has flowed over some stretch of time. A kitchen tap gives you a useful analogy: the flow rate from the tap, litres per second, is like power, while the total water that ends up in the bath is like energy.

Power is measured in watts. Energy is measured in watt-hours, or more usefully kilowatt-hours: a device drawing one kilowatt of power for one hour uses one kilowatt-hour of energy. A solar panel rated at 400 watts is making a claim about power, its best instantaneous output in full sun. How much energy it delivers over a year depends on how many hours of good sun the panel actually sees, which is why capacity factor (below) matters so much.

The Watt Ladder

Power spans an enormous range, from a single LED to the entire human enterprise, so it helps to have everyday anchors for each rung.

- **Watt (W):** the base unit. An LED bulb draws around 8 to 10 watts; a human at rest radiates heat at roughly 100 watts.
- **Kilowatt (kW), one thousand watts:** an electric kettle or a hair dryer draws about 2 kilowatts; a typical home rooftop solar system is 5 to 10 kilowatts.
- **Megawatt (MW), one million watts:** a single large modern wind turbine is rated at 3 to 15 megawatts; a megawatt is roughly the peak draw of several hundred to a thousand homes.
- **Gigawatt (GW), one billion watts:** the scale of a big power station. A large nuclear reactor or a major coal plant is around one gigawatt.
- **Terawatt (TW), one trillion watts:** the scale of whole civilizations. Total human power demand across all uses averages on the order of 18 to 20 terawatts; global installed solar passed roughly 1.6 terawatts of capacity in 2023.

The Kilowatt-Hour

The kilowatt-hour is the unit on your electricity bill and the most useful everyday measure of energy. One kilowatt-hour will run a 10-watt LED for 100 hours, boil water for many cups of tea, or power a typical refrigerator for most of a day. An average home in a wealthy country uses somewhere between 8 and 30 kilowatt-hours a day, depending on climate, size, and habits. Keeping the kilowatt-hour in mind lets you sanity-check almost any energy figure: a "100 megawatt-hour battery," for instance, could supply a hundred average homes for a full day.

Sunlight: The Solar Constant and "One Sun"

Above the atmosphere, sunlight arrives at about **1,361 watts per square metre**, a figure called the solar constant. By the time it has passed through air, cloud, and dust, clear-sky sunlight at the ground around midday is close to **1,000 watts per square metre**. That round figure is called "one sun," and it is the standard brightness at which panels are tested and rated.

The headline fact that follows from these numbers is worth memorizing: **the sunlight striking Earth in a single hour carries more energy than all of humanity uses in a year.** The challenge of solar power has never been the size of the resource; it has always been the cost of catching it.

Wind: The Cube Law

The power carried by moving air follows a simple and consequential rule. The power in the wind is proportional to the air density, to the area the rotor sweeps, and to **the cube of the wind speed**. Two consequences dominate the whole of wind engineering:

- **Doubling the wind speed gives eight times the power** (two cubed). This is why turbine sites are chosen for wind quality above almost all else, and why a modest increase in average wind is worth a great deal.
- **Doubling the blade length quadruples the swept area** (area grows with the square of the radius), which multiplies the power caught. This is the relentless

logic behind ever-larger rotors.

Betz's Limit

No turbine can capture all the wind's energy, because if it did the air would have to stop dead behind the rotor and pile up. In 1919 Albert Betz proved that the most any rotor can extract is **16/27, about 59.3 percent**, of the kinetic energy in the air passing through it. Real modern turbines reach a power coefficient of roughly 0.35 to 0.50, which is to say about 75 to 85 percent of the Betz limit under good conditions. The limit is not an engineering weakness to be overcome; it is a law of physics.

The Shockley–Queisser Limit

Solar cells have their own ceiling. A single-junction cell, like a standard silicon panel, can convert at most about **33.7 percent** of the sunlight falling on it into electricity, a bound derived by William Shockley and Hans Queisser in 1961. Two unavoidable losses set the limit: photons with too little energy pass straight through, and photons with too much energy waste the surplus as heat. Stacking cells of different band gaps in a **tandem** cell sidesteps the single-junction limit; laboratory perovskite-on-silicon tandems have already passed 34 percent, and their theoretical ceiling reaches into the mid-forties.

Capacity Factor

A power plant almost never runs flat out all the time. The **capacity factor** is the energy it actually produces divided by the energy it would produce running at full nameplate power every hour of the year. Typical ranges:

- **Solar PV: roughly 15 to 25 percent**, depending on latitude and climate.
- **Onshore wind: roughly 30 to 40 percent.**
- **Offshore wind: roughly 40 to 55 percent**, thanks to stronger, steadier sea winds.

Capacity factor is why a gigawatt of solar and a gigawatt of gas are not the same thing over a year, and why comparing nameplate capacities alone can mislead.

Typical Sizes

- **Solar cell:** a wafer roughly 15 to 18 centimetres across, producing about half a volt.
- **Solar panel (module):** commonly 350 to 600 or more watts, around 20 to 23 percent efficient, roughly one by two metres.
- **Home rooftop system:** typically 5 to 10 kilowatts, perhaps 15 to 30 panels.
- **Wind turbine:** onshore machines are commonly 3 to 6 megawatts with rotors 130 to 170 metres across; the largest offshore units reach 15 megawatts and beyond, with rotors past 300 metres.

Key 2024 Figures (as of 2024, estimates vary)

These headline numbers move every quarter; treat them as a snapshot and refresh from the sources in Appendix C.

- **Global installed solar capacity:** roughly 1.6 terawatts (end of 2023, rising fast).
- **Global installed wind capacity:** roughly 1.13 terawatts (about 1,052 gigawatts onshore plus 83 gigawatts offshore, end of 2024).
- **Cost of new solar (levelized):** about 4.3 US cents per kilowatt-hour, down roughly 90 percent since 2010.

- **Cost of new onshore wind (levelized):** about 3.4 US cents per kilowatt-hour, the cheapest of any source.
- **Share of world electricity:** solar 6.9 percent, wind 8.1 percent, together close to 15 percent, with all clean sources past 40.9 percent.

Appendix B — How to Read a Panel, a Turbine, and an Energy Bill

Spec sheets are written to impress, not to inform, and the single number in the headline is rarely the one that matters. This appendix teaches you to read the three documents you are most likely to encounter: a solar panel's datasheet, a wind turbine's fact sheet, and your own electricity bill. At the end there is a worked example that estimates what a real roof might produce.

Reading a Solar Panel

A panel's datasheet is a page of numbers under standard test conditions, meaning one sun of brightness (1,000 watts per square metre), a cell temperature of 25 degrees Celsius, and a standard sky spectrum. Look for four things.

Rated power (watts-peak, Wp). The big number, perhaps 400 or 450 watts. It is the panel's best output under those ideal test conditions, which almost never occur in the field.

A hot roof at midday will sit well above 25 degrees, so real output is routinely lower. Treat rated power as a ceiling, not a promise.

Efficiency. The percentage of sunlight the panel converts, typically 20 to 23 percent for a good modern module. Higher efficiency mainly matters when roof space is tight, because it packs more watts into each square metre. If you have plenty of roof, a slightly cheaper, slightly less efficient panel may be the better buy.

Temperature coefficient. A small negative number, often around minus 0.3 percent per degree Celsius. It tells you how much power the panel loses for every degree above 25. This is why panels underperform their rating on the hottest, sunniest afternoons, and why the same panel yields more in a cool, bright climate than in a scorching one.

Warranty. Two warranties actually matter: the product warranty against defects (commonly 12 to 25 years) and the performance warranty, which guarantees the panel will still produce, say, 85 to 90 percent of its rated power after 25 years. Panels degrade slowly, typically a few tenths of a percent per year, so a 25-year-old panel still works; it just works a little less.

Reading a Wind Turbine

Turbine fact sheets are built around four numbers.

Nameplate capacity. The maximum power, in megawatts, that the turbine produces once the wind reaches its rated speed. A modern onshore machine might be 4 to 6 megawatts; the largest offshore units exceed 15. As with panels, this is a peak, not an average.

Rotor diameter. The width of the circle the blades sweep, often larger than a football pitch is long. Because the power caught grows with the square of the rotor diameter and the cube of the wind speed, rotor size is one of the two levers, along with hub height, that has driven turbines ever upward.

Hub height. How high the rotor's centre sits above the ground. Wind is stronger and steadier higher up, so taller towers reach a better resource; hub heights of 100 to 150 metres are now routine onshore.

Capacity factor. Often quoted for the intended site, this is the fraction of nameplate power the turbine is expected to deliver averaged over a year, typically 30 to 40 percent onshore and 40 to 55 percent offshore. It is the number that actually tells you how much energy the machine will make. A 5-megawatt turbine at a 40 percent capacity factor produces the same annual energy as a 2-megawatt turbine running flat out.

Reading Your Electricity Bill

Your bill is denominated in **kilowatt-hours**, the same unit solar output is measured in, which makes it the bridge between the two. Find three things on it.

Total consumption. The kilowatt-hours you used this billing period, and ideally over a full year, since a single month can mislead. Annual consumption is the number to size a system against.

The price per kilowatt-hour. What you pay to import a unit from the grid. This is the figure rooftop solar competes against, and it is usually well above the wholesale cost of solar power, which is why generating your own can pay even when large solar farms sell more cheaply still.

Any time-of-use or export rate. Increasingly, bills charge more at peak evening hours and pay less for power you export than the price you pay to import. This gap is the entire economic case for pairing rooftop solar with a home battery: store your cheap midday surplus and use it in the expensive evening rather than selling it back at a discount.

Estimating Rooftop Solar: A Worked Example

Here is the back-of-envelope method that professional estimators start from. It needs only your system size, your local sun resource, and a loss factor.

Annual production, in kilowatt-hours, is roughly the system size in kilowatts, multiplied by the local "peak sun hours" per day, multiplied by 365, multiplied by a performance ratio of about 0.8 to account for real-world losses (heat, wiring, dust, inverter, and shading).

Peak sun hours is a tidy way to fold a whole day's varying sunlight into one number: it is the equivalent hours of full one-sun brightness the site receives per day, typically 3 in a cloudy northern climate, 4 to 5 in a temperate one, and 6 or more in a desert.

Suppose you install a **6-kilowatt** system on a well-oriented roof in a place with **4.5 peak sun hours** a day. The estimate is:

6 kilowatts, times 4.5 hours, times 365 days, times 0.8, which comes to roughly **7,900 kilowatt-hours a year**.

If your household uses about 8,000 kilowatt-hours a year, that system would cover close to your whole annual consumption, though not hour by hour: you would export a midday surplus and import in the evenings and winter unless you add storage. Shade from a single chimney or tree, a roof facing away from the equator, or a steep unfavourable tilt can each knock 10 to 30 percent off the figure, which is why a real quote uses site-specific modelling. But this simple calculation will get you within striking distance and let you judge whether a sales quote is honest.

Appendix C — Key Reports, Documents & Data Sources

The statistics in this book move quarterly, and any honest account depends on returning to the primary sources. This appendix is a guide to the handful of institutions whose data forms the living backbone of the field: what each one publishes, why you can trust it, and how to find the latest numbers. When two of them disagree, it is almost always because they count differently, grid-connected versus shipped capacity, calendar year versus fiscal year, and the honest response is to cite the source and note that estimates vary.

The Institutions

IEA — International Energy Agency. The intergovernmental energy body based in Paris. Its annual *Renewables* report and its flagship *World Energy Outlook* are the standard references for global capacity, generation, and forward projections, alongside specialist studies such as *Solar PV Global Supply Chains*. The IEA is authoritative and cautious; historically it underestimated solar's growth, a bias worth remembering when reading its forecasts. Find the latest at iea.org.

IRENA — International Renewable Energy Agency. The intergovernmental agency dedicated specifically to renewables. Its *Renewable Power Generation Costs* series is the definitive record of how levelized costs have fallen, and its *Renewable Capacity Statistics* give the most-cited installed-capacity figures. When you see the claim that solar cost has fallen roughly 90 percent since 2010, it traces to IRENA. Find it at irena.org.

GWEC — Global Wind Energy Council. The wind industry's global association, whose annual *Global Wind Report* is the standard source for wind installations, onshore and offshore, by country and region. Because it is an industry body, cross-check its figures against IRENA and national statistics, but its granularity on turbines and markets is unmatched. Find it at gwec.net.

Ember. An independent energy think tank whose annual *Global Electricity Review* is the go-to source for the electricity *mix*, the share of generation from wind, solar, and every other source. Ember is the origin of widely quoted figures such as wind at 8.1 percent and solar at 6.9 percent of world electricity in 2024. Independent and transparent, with open data. Find it at ember-energy.org.

NREL — National Renewable Energy Laboratory. The US Department of Energy's renewables laboratory, whose famous Best Research-Cell Efficiency chart is the authoritative record of every certified solar-cell efficiency record since the 1970s. NREL also publishes deep technical and cost analyses. When a perovskite tandem sets a new record, NREL certifies it. Find it at nrel.gov.

Lawrence Berkeley National Laboratory (LBNL). A US national lab whose annual market reports, including the wind and utility-scale solar "Land-Based Wind Market Report" and companion studies, are the most detailed public account of US project costs, prices, and performance. Rigorous and independent. Find it at emp.lbl.gov.

US EIA — Energy Information Administration. The statistical arm of the US Department of Energy, publishing detailed, neutral US generation, capacity, and price data through its *Electric Power Monthly*, *Annual Energy Outlook*, and open data browser. The reference for anything US-specific. Find it at eia.gov.

Our World in Data. A University of Oxford-affiliated project that gathers energy and electricity datasets from the sources above into clear, free, downloadable charts and long-run series. It is not a primary source but an excellent, transparent gateway to them, ideal for seeing trends over decades at a glance. Find it at ourworldindata.org.

Also Worth Knowing

For solar manufacturing and supply-chain shares, the industry associations and analysts, notably China's CPIA and firms such as Wood Mackenzie, together with policy analysis from bodies like CSIS, provide the granular figures on where polysilicon, wafers, cells, and modules are made. For prices of modules and cells, the trade press, pv-magazine, PV Tech, and BloombergNEF's regular surveys, tracks the market week to week.

How to Find the Latest Numbers

Every figure in this book was current at the time of writing but will have moved by the time you read it. The reliable habit is this: for capacity and additions, check the most recent IRENA statistics and the IEA *Renewables* report; for costs, the latest IRENA *Renewable Power Generation Costs*; for the electricity mix, the newest Ember *Global Electricity Review*; for wind specifics, the current GWEC *Global Wind Report*; and for cell-efficiency records, NREL's chart. Search each institution's name plus the current year, favour the primary publication over a news summary of it, and note the year and the counting method whenever you quote a number.

Appendix D — Myths, Corrected

Few technologies attract as much confident misinformation as solar and wind, some of it honest confusion, some of it motivated. This appendix gathers the book's corrections in one place. The aim is not boosterism: where a criticism holds a grain of truth, the correction says so. Good nonfiction acknowledges the real problems without surrendering to the false ones.

Myth: "A solar panel never generates as much energy as it took to make." False, and by a wide margin. The energy used to manufacture a modern panel is repaid, on average, within one to two years of operation, against a working life of 25 to 30 years or more. Over its lifetime a panel produces on the order of ten to thirty times the energy embodied in it. The figure has improved steadily as manufacturing has grown more efficient.

Myth: "Wind turbines kill vast numbers of birds." Overstated. Turbines do kill birds, and poorly sited early wind farms such as Altamont Pass were genuinely harmful to raptors. But in absolute terms turbines account for a tiny fraction of human-caused bird deaths compared with domestic cats, collisions with buildings and windows, and vehicles, each of which kills orders of magnitude more. Careful siting, radar-triggered shutdowns, and better design continue to reduce the toll, and the fossil-fuel alternative is far worse for birds through habitat loss and climate change.

Myth: "A grid full of renewables needs an equal amount of fossil backup, so nothing is really saved." False as stated. High-renewable grids are managed today through a portfolio of tools, geographic spread, accurate forecasting, transmission, demand response, and storage, so that the amount of firm backup needed is far less than the renewable capacity installed. The old idea of one-for-one backup treats every plant as if it fails at the same moment, which weather and geography make vanishingly unlikely across a large enough system.

Myth: "Solar panels are toxic waste at the end of their life." Misleading. Most of a panel by weight is glass and aluminium, both readily recyclable, with silicon and small amounts of silver; some thin-film types contain trace regulated materials that are handled and recovered accordingly. Panels last decades, so the waste stream is only now beginning to grow, and recycling capacity is expanding to meet it. Panels are not hazardous waste in the way the phrase implies.

Myth: "Renewables only survive because of subsidies." No longer true, whatever its history. In most markets new onshore wind and utility-scale solar are now the cheapest sources of new electricity on an unsubsidized basis, which is precisely why they are being built at record scale even where subsidies have lapsed. Subsidies were decisive in the early decades, driving the deployment that drove the cost down, and every energy source, fossil fuels emphatically included, has been subsidized. But the technology now stands on its own economics.

Myth: "Rare earth elements are rare, so we will run out." Doubly misleading. Rare earths are not geologically rare; they are moderately abundant in the Earth's crust. The real issue is that their mining and especially their processing are heavily concentrated in China, which is a supply-chain and geopolitical concern, not a scarcity one. Moreover, most solar power uses no rare earths at all, and geared wind turbines use far fewer than direct-drive designs, so the dependence can be, and is being, reduced.

Myth: "Solar and wind only work in sunny or windy places." Overstated. Germany, one of the world's solar leaders, is cloudier than Alaska; Denmark generates a large share of its power from wind on a flat, temperate coast. Panels produce usefully in diffuse light, and wind resources are widespread. Resource quality certainly affects economics, the best deserts and coastlines are cheaper, but "it isn't sunny enough here" is rarely the real constraint in practice.

Myth: "It's nuclear versus renewables, and you have to pick a side." A false choice. Both are low-carbon sources with different strengths: renewables are cheap and fast to build but variable, while nuclear is firm and dispatchable but slow and costly to build. Most credible analyses of deep decarbonization find a role for both, alongside storage and flexible demand. Framing them as enemies mostly serves the incumbents that both would displace. (See the essay "Nuclear and Renewables, Fairly.")

Myth: "Making a wind turbine emits more carbon than it ever saves." False. A turbine repays its manufacturing carbon in a matter of months of operation, and over a twenty-year life its lifecycle emissions per kilowatt-hour are a small fraction of those from coal or gas. The same holds for solar.

Myth: "Wind turbines cause cancer, or 'wind turbine syndrome.'" No credible evidence supports either claim. Large reviews of the health literature have found no established link between turbines and disease; reported symptoms track more closely with attitudes and media coverage than with proximity to turbines.

Myth: "Panels stop working in winter or the cold." Backwards, in fact. Solar cells are slightly more efficient in cold temperatures; what falls in winter is the number of daylight hours and the sun's angle, not the panel's ability. Snow can temporarily cover panels, but it also reflects light and usually slides off tilted modules quickly. Cold, bright climates can be excellent for solar.

Myth: "There isn't enough lithium, cobalt, or other material to build all the batteries and turbines we need." A real concern that is routinely exaggerated. Demand for critical minerals will multiply, and bottlenecks in mining and processing are genuine near-term risks. But global reserves are large, new sources and chemistries (such as cobalt-free lithium-iron-phosphate batteries) keep emerging, and recycling will eventually supply a growing share. The problem is one of investment, geography, and lead times, not ultimate scarcity.

Myth: "Solar and wind can't provide power at night or when the wind drops, so they can never run a grid." The honest kernel here, that both are variable, is exactly why storage, transmission, and complementary sources matter. But the leap to "never" is unwarranted. Several grids already run for long stretches on very high shares of wind and solar, and the cost of the batteries that cover the gaps has itself collapsed. The transition is an engineering and investment challenge, not an impossibility. (See the essay "The Grid Is the Real Machine.")

Appendix E — Where You Meet It Today

Solar and wind are no longer futuristic. They are ordinary infrastructure, woven so thoroughly into daily life that most people stop noticing them. This appendix is a field guide to spotting the sun-and-wind age around you, from the calculator in a drawer to the record-breaking turbines you can name, and to the pioneer sites where it all began.

On and Around Your Home

Rooftop panels. The most visible face of the revolution, now on tens of millions of roofs. Look for the dark rectangular modules tilted toward the equator, and for the inverter, usually a box on an outside wall or in a garage, that turns their direct current into the alternating current your house uses. A smart meter or app often shows the flow in real time.

The numbers on your own bill. You meet solar and wind every month whether or not you own a panel, because a growing share of the grid electricity you buy comes from them. Your bill's kilowatt-hours, its price per unit, and any time-of-use or export rates are where the economics of the transition reach into your household budget. (Appendix B shows how to read them.)

The Small Solar You Already Own

Pocket calculators. The oldest mass-market photovoltaic product. That thin strip above the display is an amorphous-silicon cell, and calculators like it have run on light since the late 1970s, quietly making solar the most widespread semi-conductor power source on Earth long before rooftops.

Solar road signs, speed displays, and parking meters. Along highways and in car parks, countless small stan-dalone systems, a panel, a battery, and a controller, run flashing signs, radar speed displays, emergency phones, and ticket machines where running a grid cable would cost more than the device. These are the direct descendants of the remote-power markets that first made terrestrial solar viable in the 1970s.

Garden lights, chargers, and gadgets of every kind. Solar garden lamps, backpack chargers, and camping panels put photovoltaics literally in your hands.

The Big Machines You Can Name

Onshore wind farms. Ridgelines and plains across the world now carry rows of white three-blade turbines. If you can watch one closely, notice the slow majestic turn of the rotor, roughly ten to twenty times a minute, and the way the whole nacelle swivels, yaws, to keep the blades facing the wind.

Offshore arrays. Out at sea sit some of the largest machines humans have ever built. Britain's Dogger Bank and Hornsea developments in the North Sea, among the world's biggest, and China's rapidly growing offshore fleet, use turbines whose single blades are longer than a football pitch. China's

Dongfang unveiled a 26-megawatt unit in 2024 with a rotor roughly 310 metres across.

Grid-scale batteries. Increasingly common beside substations and solar farms, these container-sized lithium-iron-phosphate installations store midday solar and discharge it into the evening peak. Australia's Hornsdale battery in South Australia became famous as an early proof that storage could stabilize a grid; thousands of such projects now follow.

Utility-scale solar farms. Fields of panels stretching to the horizon, sometimes on trackers that pivot to follow the sun, sometimes shading grazing sheep or crops in an agrivoltaic layout. The largest now rank among the biggest power plants of any kind.

Pioneer Sites and Museums

Altamont Pass, California. The birthplace of the American wind industry, filled during the 1980s "wind rush." Its early machines were hard on birds, and many have since been replaced by fewer, larger, better-sited turbines, making it a living lesson in how the technology matured.

Vindeby, Denmark. The site of the world's first offshore wind farm, built in 1991 and decommissioned in 2017, its eleven modest turbines the ancestors of the giants now dotting the North Sea.

Maadi, Egypt. Where Frank Shuman built the world's first solar-thermal power station in 1912–13. Nothing of the plant survives, but the place remains a pilgrimage in the imagination for anyone who knows the story.

Museums and preserved machines. Denmark's Poul la Cour Museum at Askov honours the father of wind electricity on the site of his laboratory. Science and technology museums around the world, from the Smithsonian, which holds a Bell Labs 1954 solar cell, to Germany's Deutsches

Museum, display the artefacts of both technologies' origins. Many working wind farms and solar plants now run visitor centres of their own.

The Wider View

Once you learn to see it, the sun-and-wind age is everywhere: in the satellites overhead, every one solar-powered since Vanguard 1 in 1958; in the traffic signals and telecom towers on remote roads; in the solar-lit villages of the developing world that skipped the grid entirely; and in the ordinary wall socket, into which, on a sunny and breezy afternoon, a growing share of the current now flows from nothing but light and moving air.

Appendix F — Frequently Asked Questions

The questions readers actually ask, answered honestly and briefly. Where the true answer is "it depends," this appendix says so rather than pretending otherwise.

Do solar panels work on cloudy days? Yes, just less. Panels respond to diffuse light as well as direct sun, so on an overcast day they typically produce somewhere between 10 and 30 percent of their clear-sky output. This is why cloudy Germany became a solar leader: enough usable light falls even under grey skies to make the numbers work.

What happens at night? A panel produces nothing in the dark. At the level of a single home, evening and overnight demand is met by the grid or by a battery charged during the day. At the level of a whole system, night-time is covered by wind (which often blows harder after dark), by stored energy, by hydro, and by other dispatchable sources. Solar and storage are increasingly built together precisely to bridge the night.

Are solar panels worth it for my home? Usually yes in sunny regions with high electricity prices, and often yes even in cloudier ones, but it depends on your roof, your local rates, the up-front cost, and any incentives. A well-oriented, unshaded roof in a place with expensive grid power and a good export or net-metering arrangement can pay back the investment in well under a decade and then generate nearly free power for two decades more. Appendix B shows how to estimate your own case before trusting any sales quote.

Do turbines and panels ever pay back the energy and carbon used to build them? Emphatically yes. A solar panel repays its manufacturing energy in one to two years and its carbon in a similar span, against a life of 25 years or more. A wind turbine repays its energy in a matter of months. Over their lifetimes both produce many times more energy than they cost to make, at a small fraction of the emissions of fossil generation.

Can the world really run on renewables? A growing body of serious analysis says a very high share is achievable, and several grids already run for long stretches on wind and solar alone. The honest caveats are that it requires major investment in storage and transmission, that the last stubborn percentage points are the hardest, and that firm low-carbon sources (hydro, nuclear, perhaps hydrogen) likely help fill the gaps. "Mostly" is well supported; "100 percent everywhere, easily" is not yet proven.

What about nuclear? Nuclear is a low-carbon, firm, dispatchable source, a genuine complement to variable renewables rather than a rival. Its drawbacks are high cost and slow construction; its strength is steady output regardless of weather. Most credible decarbonization studies find room for both. Treating it as an either-or choice tends to help only the fossil incumbents that both would replace. (See the essay "Nuclear and Renewables, Fairly.")

Why is China so dominant in solar and batteries? Sustained industrial strategy. Over two decades China poured investment into every step of the supply chain, quartz to polysilicon to wafers to cells to modules, achieving enormous scale and driving costs down faster than anyone else.

As of 2024 it made the large majority of the world's polysilicon, wafers, cells, and modules, and it leads in battery cells too. This has made clean energy cheap for everyone while concentrating supply-chain risk in one country, a genuine double-edged outcome.

Do panels work in winter and the cold? Yes, and cold actually helps. Solar cells are slightly more efficient at low temperatures. What shrinks winter output is fewer daylight hours and a lower sun angle, not the cold itself. Snow can briefly cover panels but usually slides off tilted modules and reflects extra light onto them in the meantime.

Is there enough lithium and other materials? For the foreseeable build-out, the constraints are mining capacity, processing, and lead times rather than ultimate scarcity. Reserves are large and growing as exploration continues, chemistries are shifting toward more abundant materials (lithium-iron-phosphate uses no cobalt; sodium-ion uses no lithium), and recycling will eventually recover much of what is already in circulation. Bottlenecks are real but solvable with investment.

Don't wind turbines kill huge numbers of birds? They kill some, and bad early siting hurt raptors, but the numbers are small beside the birds killed by cats, buildings, and cars, each of which takes orders of magnitude more. Better siting and technology keep reducing the toll, and unchecked climate change would harm far more birds than turbines ever could.

How long do panels and turbines last? Solar panels are typically warrantied for performance over 25 years and often keep working, at slightly reduced output, well beyond that. Modern turbines are designed for roughly 20 to 30 years, after which sites are frequently "repowered" with fewer, larger, more efficient machines.

What happens to them at the end of life? Panels are largely glass and aluminium and are increasingly recycled; the waste stream is only now growing as the first big wave of installations ages. Wind blades are harder, being tough composite materials that have often been landfilled, though recyclable resins and dedicated recycling processes are emerging. Neither is a solved problem, but neither is the crisis it is sometimes portrayed as.

Why does my rooftop solar cost more per unit than a giant solar farm? Scale. A utility solar farm buys panels by the shipload, mounts them cheaply on open land, and connects at wholesale, so its cost per kilowatt-hour is very low. A rooftop system carries the fixed costs of a small custom installation, permits, labour, and a home-sized inverter, spread over far fewer panels. It still often pays, because it competes against the higher retail price you would otherwise pay the grid.

What is the "duck curve" I keep hearing about? It is the shape of daily electricity demand once a grid has a lot of solar. Midday solar output pushes net demand down into a belly, then as the sun sets and people come home, demand ramps up steeply into the evening, forming a curve that looks like a duck. Managing that steep evening ramp, largely with batteries and flexible demand, is one of the central operational puzzles of a solar-heavy grid.

Are offshore turbines harming whales? The available scientific evidence does not support claims that offshore wind surveys or construction are killing whales; marine-mammal deaths along affected coasts have other documented causes. Offshore wind does raise legitimate questions about marine life, fisheries, and noise that are studied and regulated, but the specific whale-mortality claim is not backed by the science.

Do I need a battery with my solar? Not necessarily. Without one, you export surplus midday power to the grid and buy it back in the evening; whether that is a good deal depends on your export and import rates. A battery makes most sense where the gap between those rates is large, where the grid is unreliable, or where you want backup during outages. It adds cost, so it is an economic choice, not a technical requirement.

Is solar or wind better? Neither; they are complementary. Solar peaks at midday and in summer; wind often peaks at night and in winter and storms. A grid drawing on both, spread over a wide area and backed by storage, is far steadier than one leaning on either alone. The book's title pairs them for exactly this reason.

Why did renewables sometimes get more expensive around 2022 to 2024 if the technology keeps improving? Because renewables are capital-intensive, most of their lifetime cost is the up-front build, financed by borrowing. When interest rates rose sharply after 2021, financing costs rose with them, nudging levelized costs up even as hardware kept getting cheaper. Offshore wind, the most capital-heavy, was hit hardest. It is a reminder that the economics depend on finance as much as on physics.

Will my house go dark during a blackout if I have solar? Ordinarily yes, unless you have a battery and the right setup. For safety, grid-tied inverters automatically shut down during an outage so they cannot energize lines that workers may be repairing, a feature called anti-islanding. To keep power flowing when the grid is down you need a battery and an inverter designed to "island" your home safely off the grid.

Can renewables power heavy industry, shipping, and aviation? Directly, not easily, which is the frontier. Electricity from wind and solar can decarbonize much of transport (electric vehicles) and heating (heat pumps), but the hardest sectors, steel, cement, long-haul shipping, and aviation, need very high heat or dense fuels. The leading candidate is to use cheap renewable electricity to make green hydrogen and derived fuels, still costly today but a major focus of research and investment.

Is the transition actually happening fast enough? That is the honest open question. Deployment is breaking records every year and the technology is cheaper than the alternatives, which is unprecedented. Whether it is fast enough to meet climate targets depends on grids, permitting, supply chains, and politics, not on the cost of panels and turbines, which is no longer the obstacle. Optimism about the machines and realism about the system are both warranted.

Thematic Essays

Six short essays that step back from the narrative to look at the ideas beneath it. Each can be read on its own.

The Sun as the Source of Almost All Energy

It is a curious fact, easy to state and startling to absorb, that almost every form of energy humanity has ever used is sunlight in disguise. The book you are holding treats solar and wind as two technologies, but at a deep enough level they are one, and so is nearly everything else on the energy ledger. Trace any power source back far enough and, with only a handful of exceptions, you arrive at the sun.

Begin with the obvious case. A solar panel converts sunlight directly, photon to electron, with no intermediary. There is no disguise at all; the relationship is naked. Wind is only slightly less direct. The sun heats the Earth unevenly, more at the equator than the poles, more over land than sea, more by day than by night. That uneven heating creates differences in air pressure, and air flowing from high pressure to low is wind. Roughly one to two percent of the solar energy reaching the Earth is converted into the motion of the atmosphere. A wind turbine, then, is a solar machine at one remove: it harvests sunlight that has first been turned into moving air.

Hydropower is sunlight at two removes. The sun evaporates water from the oceans, lifting it as vapour; that vapour falls as rain and snow on high ground; and the water running back downhill, carrying the gravitational energy the sun invested in raising it, spins a turbine. A hydroelectric dam is a solar collector whose absorbing surface is the entire ocean and whose storage medium is the mountains. Biomass, the wood and dung and crop waste that fuelled humanity for most of its history, and the biofuels that supplement transport today, is sunlight captured by photosynthesis and stored as chemical bonds in living tissue. Burning a log releases sunlight that a tree gathered over decades.

Then come the fossil fuels, and here the disguise becomes a magician's trick of time. Coal, oil, and natural gas are ancient sunlight, photosynthesis from the deep past, buried and pressure-cooked over hundreds of millions of years. The plankton that became oil and the swamp forests that became coal were solar collectors of the Carboniferous and later ages. When we burn petrol, we are spending sunlight that fell on a Jurassic sea. The whole fossil-fuelled edifice of industrial civilization is, viewed from a great enough height, a two-century binge on a solar inheritance that took geological ages to accumulate. That framing reveals why the inheritance is finite and why its combustion returns long-buried carbon to the sky.

Only a few energy sources escape the solar lineage. Nuclear power draws on the binding energy of atomic nuclei forged in dying stars other than our sun, a genuinely different inheritance. Geothermal heat rises from the Earth's hot interior, warmed by primordial heat and radioactive decay in the rock. Tidal power comes chiefly from the gravitational pull of the moon. These are the true exceptions, and they are modest in the global picture. Everything else, the overwhelming majority of the energy that lights our cities and moves our vehicles and smelts our metals, is solar in origin, differing only in how the sunlight was stored and how long ago.

Seeing the unity has a practical payoff, not merely a poetic one. It reframes the energy transition as a change not in the ultimate source, which has always been the sun, but in the delay between capture and use. Fossil fuels tap sunlight stored for aeons; renewables tap sunlight arriving now or stored for hours, days, or a season. The advantage of the ancient store was its density and its convenience: millions of years of accumulation packed into a barrel you could pour. The disadvantage is that the store is finite and its combustion loads the atmosphere with carbon. Wind and solar trade that convenient density for a resource that is effectively infinite on human timescales and clean to use, at the cost of having to catch it as it comes rather than digging it up on demand.

The sun's generosity is almost absurd. It radiates about 3.8 times ten to the twenty-sixth watts in every direction, of which the Earth intercepts a sliver, and yet that sliver, the sunlight striking the planet in a single hour, carries more energy than all of humanity uses in a year. The problem of energy has never been supply. The sun pours down thousands of times more than we could ever need. The problem has always been capture: building the collectors, cheaply enough and at sufficient scale, to catch a useful fraction of what arrives free every morning. That is the whole story of this book, and it is why the title names the two most direct ways we have yet found to catch today's sunlight rather than yesterday's.

The Learning Curve as an Idea

The most important idea in the modern history of solar and wind is not a piece of physics but a piece of economics, and it is one that experts repeatedly failed to believe even as it happened in front of them. The idea is the learning curve: the observation that the cost of making a thing falls by a predictable percentage every time the cumulative quantity ever produced doubles. It sounds almost too simple to matter. It has reshaped the world.

The principle is old. In 1936 an aeronautical engineer named Theodore Wright noticed that each time the total number of aeroplanes built doubled, the labour required to build the next one fell by a consistent fraction, around twenty percent. The regularity was uncanny: not a vague tendency but a straight line on the right kind of graph, holding across many doublings. What became known as Wright's law turned out to describe far more than aircraft. Transistors, hard drives, LED bulbs, and, most consequentially, solar modules all follow learning curves, each with its own characteristic rate.

For solar the rate is roughly twenty percent per doubling of cumulative shipments, a relationship the semiconductor pioneer Richard Swanson made explicit and that *The Economist* christened "Swanson's law" in 2012. The number seems modest until you appreciate the power of compounding across many doublings. Solar shipments have doubled again and again and again, dozens of times over the decades, and each doubling shaved a fifth off the price. Multiply a fifth-off by itself enough times and the cumulative effect is staggering: solar module prices have fallen by roughly ninety-nine point nine percent since the 1970s. What cost a fortune per watt in the space age costs a few coins per watt today. Wind follows a shallower but real learning curve of its own, and lithium batteries a steep one that is now doing to storage what Swanson's law did to panels.

Why should cost track cumulative production so faithfully? Because a learning curve is a summary of ten thousand small improvements that only happen when you actually build things. Workers get faster. Factories get bigger and reap economies of scale. Engineers find ways to use less silicon, less silver, less labour per watt. Suppliers compete and specialize. Defect rates fall. None of these gains arrive from thinking about the problem in the abstract; they arrive from the grinding, accumulating experience of manufacturing at scale. This is the deep reason the learning curve is driven by *deployment*, not by research alone. You cannot laboratory your way down the curve. You have to build.

That single insight rewrites the logic of energy policy. For decades the intuitive approach to an expensive clean technology was to wait: fund research, and deploy only once the technology became cheap. The learning curve says this gets the causality backwards. The technology becomes cheap *because* it is deployed. Every panel installed, even an overpriced one, buys down the cost of the next. Germany's feed-in tariff of 2000 is the great worked example. Germany is not especially sunny, and critics mocked it for paying lavish subsidies to install solar under grey northern skies. But the guaranteed prices created enormous, reliable demand, that demand pulled a global manufacturing industry into existence, and the scale of that industry drove costs down the learning curve for everyone on Earth. German ratepayers, in effect, paid the tuition for the whole world's solar education. The panels going onto roofs in Lagos and Lima today are cheap partly because Bavarians overpaid two decades ago.

The learning curve also explains why experts kept being wrong, always in the same direction. Official forecasts from respected institutions repeatedly projected that solar costs would flatten out, and repeatedly the costs kept falling as the curve predicted. The forecasters were reasoning from the physics and the current price; the curve was reasoning from cumulative volume, and the curve kept winning. It is a humbling lesson in which kind of model to trust. When a technology is early on a steep learning curve, betting against further cost declines is betting against one of the most robust empirical regularities in all of economics.

There is a caution to hold alongside the enthusiasm. Learning curves are descriptions, not laws of nature; they can bend if a raw material becomes scarce or a supply chain seizes up, as pandemic and interest-rate shocks briefly reminded everyone in the early 2020s. And not every technology has a steep curve; some, like large civil-engineering projects, barely learn at all. But for the modular, mass-manufactured devices at the heart of the sun-and-wind age, the curve has held with remarkable fidelity for half a century, and understanding it is the single best tool a reader has for judging where the energy transition is headed next.

The Grid Is the Real Machine

It is natural to think that the solar and wind revolution is a story about panels and turbines. They are the photogenic parts, the objects you can point at on a hillside or a rooftop. But the harder and more interesting engineering, the part that decides whether the revolution succeeds, happens in something almost nobody looks at: the electricity grid. The grid is the real machine, and solar and wind are only components plugged into it.

To grasp why, you have to appreciate what the grid does every second of every day. Electricity, at grid scale, is almost impossible to store in bulk cheaply, which means that supply and demand must be balanced continuously and exactly, instant by instant. The moment a nation switches on its kettles at the end of a popular broadcast, generators somewhere must ramp up to meet the surge, or the grid's frequency sags and, in the worst case, the whole system trips into blackout. For a century this balancing act was performed by dispatchable power plants, coal, gas, hydro, that could be turned up and down on command like the throttle of an engine. The grid was a machine you drove.

Wind and solar break that comfortable arrangement, because they cannot be dispatched. They produce when the wind blows and the sun shines, following the weather rather than demand. A grid operator cannot ask a cloud to move or a calm to lift. This is the property called variability, and it is the central technical challenge of the transition, more fundamental than the cost of any panel. Integrating a large share of variable generation means rebuilding the grid's whole logic, from a system that adjusted supply to match demand into one that also adjusts demand to match supply, and that leans on new kinds of flexibility.

The tools for doing this turn out to be several, and no single one is a silver bullet. The first is geographic spread: it is rarely calm everywhere or cloudy everywhere at once, so a grid drawing on wind and sun across a wide enough area, stitched together by long transmission lines, sees a much smoother aggregate output than any one site. This is why transmission, unglamorous and endlessly delayed by permitting fights, is arguably the most important infrastructure of the transition. The second tool is forecasting: modern weather models let operators predict wind and solar output hours and days ahead, so the system can prepare rather than react. The third is storage, above all the lithium batteries whose cost has lately collapsed, which soak up the midday solar surplus and release it into the evening. The fourth is flexible demand, the growing ability to shift electricity use, charging cars, running heat pumps and factories, toward the hours when power is abundant and cheap. And the fifth, for now, is a shrinking backbone of dispatchable plants, increasingly gas and hydro, that fill the remaining gaps.

The "duck curve" is the emblem of this new world. Chart the electricity a solar-heavy grid must supply from its other sources across a day, and you get a shape with a deep belly at midday, when solar floods in and net demand collapses, and a steep neck in the early evening, when the sun sets just as people come home and switch everything on. That evening ramp, the need to bring vast amounts of power online in a couple of hours, is one of the defining operational puzzles of a solar grid, and batteries charged during the midday glut are the leading answer. Curtailment, the deliberate throttling of wind and solar when there is more than the grid can use or carry, is another face of the same challenge, wasteful-looking but often cheaper than the alternatives, and reduced by exactly the same tools of storage and transmission.

The deeper point is that the old vocabulary no longer fits. Words like "baseload," which described the always-on plants that met a grid's minimum demand, were coined for a system built around inflexible coal and nuclear that ran best when never switched off. In a grid built around cheap, variable renewables and fast, flexible storage, "baseload" describes a solution to a problem that is disappearing. The valuable quality is no longer steadiness but flexibility, the ability to move power across space with transmission and across time with storage. Even the way we measure cost has to evolve: a plant's levelized cost of electricity, its average price per unit, matters less at high penetration than the *value* of its output, whether it produces when power is needed or when it is already in surplus. Newer metrics that weight power by when it arrives are the tools of this more sophisticated accounting.

None of this is science fiction. Denmark, South Australia, and parts of Germany, Spain, and Texas already run for long stretches on very high instantaneous shares of wind and solar, holding their grids stable with the toolkit above. They are the working prototypes of the machine the rest of the world is now building. The lesson they teach is that the barrier to a renewable future was never really the panels and turbines, which have become almost absurdly cheap. It is the grid, the vast, invisible, continent-spanning machine into which they plug, and rebuilding that machine is the true work of the decades ahead.

Nuclear and Renewables, Fairly

Few debates in energy generate more heat and less light than the argument between champions of nuclear power and champions of renewables. Each camp can produce a devastating case against the other, and each is partly right, which is precisely why the framing as a contest is so misleading. Both are low-carbon ways to make electricity; both would displace the fossil fuels that are the actual problem; and a fair reading of the evidence suggests the sensible question is not which to choose but how much of each, and where.

Start with the honest case for nuclear. A nuclear plant is firm and dispatchable: it produces a steady, controllable stream of electricity regardless of whether the wind blows or the sun shines. It does this with an extraordinarily small land footprint and with lifecycle carbon emissions as low as wind's. It has an enviable safety record measured in deaths per unit of energy, better than fossil fuels by an enormous margin and comparable to renewables, despite the outsized fear that a handful of dramatic accidents have inspired. In a decarbonized grid, that steady firm output is genuinely valuable, filling exactly the gaps that variable renewables leave. A country that already has a nuclear fleet running well is, from a climate standpoint, fortunate, and shutting reactors early, as some nations have done, tends to mean burning more coal or gas in the interim.

Now the honest case against, or rather the honest account of nuclear's difficulties. New nuclear plants in most Western countries have become extraordinarily expensive and slow to build, running years late and far over budget, for reasons that are as much about lost industrial know-how, one-off designs, and financing as about the technology itself. Where a solar farm or wind project can be built in months to a couple of years and its cost is known with fair confidence, a nuclear plant is a decade-long, multi-billion commitment with a real risk of overrun. That combination of high cost and long lead time is fatal in a world where renewables keep getting cheaper while you build. It is why, on pure economics, new-build nuclear struggles to compete with new-build wind and solar for bulk energy.

The honest case for renewables is by now the burden of this whole book: they are the cheapest new sources of electricity in most of the world, they can be deployed astonishingly fast, and they scale down as easily as up, from a single rooftop to a gigawatt farm. Their honest weakness is equally clear: they are variable, and a grid leaning heavily on them needs storage, transmission, and flexibility to stay reliable, all of which add cost and complexity that a plant-level price tag hides. The last stubborn stretch, the deep decarbonization of the final fraction of the grid through long, calm, cloudy spells, is where variable renewables strain hardest and where firm sources look most attractive.

Set the two honest accounts side by side and the complementarity almost draws itself. Renewables are cheap and fast but variable; nuclear is firm and steady but costly and slow. The weaknesses of each are the strengths of the other. A grid that pairs a large, cheap bulk of wind and solar with a firm backbone, whether nuclear, hydro, geothermal, or eventually clean hydrogen, is more robust and often cheaper than a grid that bets everything on either alone. This is not a fudge or a diplomatic evasion; it is what a great many detailed decarbonization studies actually find when they optimize a whole system rather than cheer for a favourite.

Why, then, the venom? Partly it is tribal, energy having become a marker of political identity. Partly it is a genuine disagreement about a scarce resource: the money and political attention available for the transition are limited, and every advocate fears the other technology will crowd theirs out. And partly it is a failure to distinguish two very different questions, whether to keep existing nuclear running (almost always yes, on climate grounds) and whether to build vast amounts of new nuclear (a much harder economic call that depends on local costs and the alternatives). Conflating them produces arguments in which both sides talk past each other.

The most useful posture for a reader is to notice who benefits from the fight. Every hour spent arguing whether nuclear or renewables is the true saviour is an hour not spent displacing the coal and gas that both would replace. The fossil incumbents are the quiet winners of the nuclear-versus-renewables war, because a low-carbon movement divided against itself builds less of everything. The fair conclusion is neither the renewables purist's dismissal of nuclear nor the nuclear advocate's contempt for "unreliable" wind and solar. It is that the atom and the elements, fission and sunlight and wind, are allies against a common problem, and that a serious climate strategy uses whatever clean tools work, chosen by cost and circumstance rather than by tribe.

The Developing-World Leapfrog

There is a pattern in the history of technology that recurs often enough to deserve a name, and it is usually called leapfrogging. It happens when a place that lacks an old, expensive, infrastructure-heavy technology skips it entirely and jumps straight to a newer one that needs less infrastructure. The classic case is the telephone. Across much of Africa and South Asia, most people never got a landline, because stringing copper wire to every village was ruinously expensive and slow. Then mobile phones arrived, needing only scattered towers rather than a wire to every home, and within a couple of decades hundreds of millions of people who had never owned a telephone were carrying one in their pocket. They leapt over the landline generation altogether.

Something very similar is now unfolding in electricity, and it may prove even more consequential. The old model of electrification, the one the rich world built over the twentieth century, is a giant centralized grid: enormous power stations connected by a spreading web of transmission and distribution lines that must physically reach every home before that home has power. It is a magnificent machine, but it is also expensive, slow to build, and financially brutal in places with low population density and low incomes, because you must sink the cost of the wires long before you earn much revenue from the households at their ends. For the roughly seven hundred million people who still lack reliable electricity, most of them in sub-Saharan Africa and parts of South Asia, waiting for that grid to arrive has meant waiting, in some cases, forever.

Solar power, especially solar paired with cheap batteries, offers a way to skip the wait. Because a solar panel generates power wherever the sun falls, and because it scales smoothly from a single small module upward, it does not require a wire stretching back to a distant power station. A household can buy a small panel, a battery, and a couple of LED lights and phone chargers, and have electricity tonight, with no grid in sight. This is the solar-home-system model, and it has spread across rural Africa and Asia with the same viral speed the mobile phone did, often sold on the very same pay-as-you-go financial rails, a small mobile payment unlocking the system for another week. One leapfrog riding on the back of another.

Above the single household sits a second tier: the minigrid, a village-scale solar array with batteries serving a cluster of homes and small businesses through a local network. Minigrids can power not just lights and phones but the productive uses that actually lift incomes, refrigeration for a clinic's vaccines, a mill, a workshop, an irrigation pump, an internet connection. They can be built in weeks rather than the years a grid extension would take, and they can later connect to the national grid if and when it arrives, rather than blocking it. The falling cost of both panels and batteries, the very learning-curve story told elsewhere in this book, is what has made this tier viable where a decade ago it was a charity project rather than a business.

The parallel with mobile phones is worth pressing, because it illuminates both the promise and the limits. Mobile telephony did not merely replace the landline; it enabled things landlines never could, mobile banking above all, which reached people no bank branch had ever served. In the same way, distributed solar is not simply a cheaper path to the same electricity the grid would have delivered; it can enable a different development pattern, decentralized, incremental, owned closer to the user, and resilient against the failures of a distant central system. A village with its own solar minigrid does not go dark when a transmission line fails two hundred kilometres away.

But the limits are real, and honesty requires naming them. A solar-home system with a small battery can light a house and charge a phone; it cannot yet cheaply run the heavy machinery of industrialization, the electric furnaces and mass manufacturing that turned poor countries rich in the twentieth century. Leapfrogging lit the mobile revolution but did not, on its own, build heavy industry, and the same caveat applies here. The most ambitious futures for developing economies still involve serious grids and serious firm power, whether that grid is reached by extension or grown outward from interconnected minigrids. Leapfrogging is a way to deliver electricity's first, transformative increment, light, communication, refrigeration, pumping, fast and cheap to people the old model failed. It is a beginning, an extraordinary one, not a complete substitute for the whole apparatus of a modern power system.

Still, the significance is hard to overstate. For the first time, the cheapest way to bring first electricity to a remote poor household is also the cleanest, a decentralized panel rather than a diesel generator or a coal-fired grid. The rich world electrified by burning its way through fossil fuels because nothing else was available or affordable. The parts of the world electrifying now have a genuine choice their predecessors never did, and increasingly the clean option is simply the cheap one. That is the quiet revolution inside the loud one, and it may touch more human lives, more directly, than anything else in this book.

What "Energy Independence" Really Means in a Solar and Wind World

"Energy independence" is one of the most durable phrases in political speech, invoked by leaders across a century and across the ideological spectrum, and for most of that time it meant something fairly specific: producing enough oil, gas, and coal within your own borders that you need not depend on foreign suppliers who might raise prices, cut you off, or drag you into their quarrels. The oil shocks of the 1970s, when a cartel of exporters could bring Western economies to their knees by turning a valve, burned the ideal into a generation's politics. In a world remade by solar and wind, the phrase survives, but its meaning quietly and profoundly changes.

The first change is the most hopeful. Sunlight and wind are not traded commodities that anyone can embargo. No country can shut off another country's sun; no cartel can corner the wind. A nation that meets its electricity needs from domestic panels and turbines has, in the most literal sense, an energy supply that cannot be cut off by a foreign government, because the fuel is delivered free every morning by the sky. This is a genuinely new kind of security, and it explains why energy-importing nations from Europe to India to China have embraced renewables partly on national-security grounds that have nothing to do with climate. For a country that has spent its history anxiously dependent on imported oil, the appeal of a fuel that falls out of the air over its own territory is easy to understand.

But the second change complicates the first, and any honest account must sit with the tension. The energy of a solar or wind system may be free and domestic, but the hardware is not. The panels, the turbines, the batteries, and above all the raw and refined materials that go into them, polysilicon, lithium, the rare-earth metals in the magnets, are made and processed in a global supply chain that is, as of the 2020s, extraordinarily concentrated. China dominates nearly every step of solar manufacturing, leads in battery cells, and controls most of the world's rare-earth processing. So the old dependence on a handful of oil exporters has not vanished; it has migrated, from a dependence on those who sell the fuel to a dependence on those who make the machines that catch the free fuel. Independence in energy has become, to a degree, dependence in industry.

These two dependencies are, however, importantly different in character, and the difference is the crux. Dependence on imported fuel is perpetual and recurring: you must buy the next tanker, and the next, forever, and your supplier holds a valve he can close at any moment. Dependence on imported hardware is front-loaded and finite: once a wind turbine or a solar farm is built and standing on your soil, it generates for twenty or thirty years using no imports at all. A foreign supplier of panels can decline to sell you the *next* batch, which is a real and serious vulnerability for future growth, but he cannot switch off the panels already on your roofs. The leverage of a fuel exporter is a hand on the tap; the leverage of a hardware exporter is more like a hand on the factory door. One controls your ongoing survival; the other controls your future expansion. Both matter, but they are not the same kind of grip, and the renewable kind is the looser.

This reframing is what drives the industrial policies now reshaping the world, from the American Inflation Reduction Act to the European Union's critical-minerals initiatives to India's manufacturing incentives. Their logic is that true energy independence in a solar and wind age is not only about generating your own power but about being able to *build* the means of generating it, or at least about not depending on a single supplier for the machines. Hence the scramble to reshore polysilicon and battery manufacturing, to diversify the mining of critical minerals across friendly nations, and to build recycling that turns yesterday's panels and batteries into tomorrow's raw materials, closing the loop so that a mature renewable system eventually feeds itself. Recycling, unglamorous as it sounds, is a strategy of independence: the metals in a retired turbine or battery are a domestic mine that needs no foreign supplier at all.

There is a further, subtler sense in which the meaning shifts. Because sunlight and wind are diffuse and available almost everywhere, at least to some degree, renewables tend to distribute energy production rather than concentrate it. The old fossil order made a few geologically lucky regions into energy superpowers and left everyone else dependent on them. A solar and wind order is more egalitarian in its raw resource: nearly every country has some sun or wind or both, and can generate at least a meaningful share of its own power. The result need not be perfect self-sufficiency, which is rarely even desirable, but a world of less lopsided dependence, where energy leverage is spread more evenly and the ability to strangle a rival by cutting off fuel largely disappears.

The honest conclusion, then, is neither the triumphalist claim that renewables deliver perfect energy independence nor the cynical claim that they merely swap one master for another. It is that they change the *nature* of dependence, from a recurring, coercive reliance on foreign fuel to a front-loaded, contestable reliance on foreign hardware, and that this is, on balance, a far better kind of dependence to have, and one that domestic manufacturing, diversified supply, and recycling can steadily loosen. Energy independence in a solar and wind world is less a destination you arrive at than a quality you build, panel by panel, factory by factory, and increasingly it is within reach of countries that could never have hoped for it in the age of oil.

Sources & Further Reading

A guided path through the literature, arranged so you can enter at whatever depth suits you. "Start here" is for the curious general reader; "Go deeper" is for those who want the fuller history, science, and argument; "The data" is where to find the living numbers when this book's figures have aged. A formal, alphabetized list follows in the Bibliography.

Start Here

Let It Shine: The 6,000-Year Story of Solar Energy — John Perlin. The most readable long history of solar power, from ancient sun-worship and Roman bathhouses through to modern photovoltaics. The natural companion to this book's solar chapters.

Powering the Dream: The History and Promise of Green Technology — Alexis Madrigal. A lively, story-driven tour of America's forgotten clean-energy experiments, early wind chargers, solar engines, and the visionaries behind them. Delightful and humane.

Sustainable Energy — Without the Hot Air — David J. C. MacKay. A physicist's calm, numbers-first reckoning with what renewables can realistically deliver, written to cut through slogans. Available free online, and still the best cure for muddled energy thinking, even where its 2008 cost assumptions have been overtaken by events.

Electrify: An Optimist's Playbook for Our Clean Energy Future — Saul Griffith. A brisk, practical case for electrifying everything with clean power, strong on how the pieces fit together in a modern life. Read it for momentum and a sense of the whole.

Go Deeper

Taming the Sun: Innovations to Harness Solar Energy and Power the Planet — Varun Sivaram. A leading technologist's argument about the innovation, finance, and grid integration still needed to scale solar to its potential. The best single book on solar's near future.

Energy and Civilization: A History and **Power Density** — Vaclav Smil. Sweeping, data-dense, deliberately skeptical histories of how humanity's energy systems actually change, and how slowly. Smil is the essential counterweight to optimism; reading him keeps the transition honest.

The Grid: The Fraying Wires Between Americans and Our Energy Future — Gretchen Bakke. A cultural and technical history of the electricity grid and why integrating renewables into it is so hard. The perfect complement to this book's essay on the grid as the real machine.

Wind Energy in America: A History — Robert Righter. The definitive scholarly history of American wind power, from Brush's Cleveland dynamo through the California wind rush. Rigorous and readable.

Let It Shine and **The Boom** aside, for the industrial and political story try **Switching to Solar** — Bob Johnstone, on how German and Japanese feed-in tariffs built the industry, and **Short Circuiting Policy** — Leah Stokes, on why US clean-energy policy so often stalled against utility opposition.

Reinventing Fire — Amory Lovins and Rocky Mountain Institute, for a systems-level roadmap to a fossil-free economy, and **Superpower** — Russell Gold, for a narrative of the fight to build the transmission the transition needs.

For the underlying engineering, the standard texts are **Wind Energy Explained** — Manwell, McGowan, and Rogers, **Renewable Energy: Power for a Sustainable Future** — Godfrey Boyle (editor), and **Solar Cells** — Martin Green. These carry the mathematics this book deliberately leaves out.

The Data

When the figures in this book have aged, and they will, go to the primary sources rather than to news summaries of them. Appendix C describes each in detail; in brief:

- **IEA** (iea.org) — the annual *Renewables* report and the *World Energy Outlook* for global capacity, generation, and forecasts.
- **IRENA** (irena.org) — *Renewable Power Generation Costs* for the definitive cost history, and *Renewable Capacity Statistics* for installed capacity.
- **GWEC** (gwec.net) — the annual *Global Wind Report* for wind installations, onshore and offshore.
- **Ember** (ember-energy.org) — the *Global Electricity Review* for the electricity mix and the share of wind and solar.
- **NREL** (nrel.gov) — the Best Research-Cell Efficiency chart for every certified solar record.
- **Our World in Data** (ourworldindata.org) — free, clear charts and long-run series gathering all of the above in one place.

The reliable habit is to search each institution's name plus the current year, prefer the primary publication, and always note the year and counting method when you quote a figure.

Bibliography

Books, major reports, and key sources consulted, alphabetized by author or issuing organization. A note on the fast-moving figures follows the list.

Books

Bakke, Gretchen. *The Grid: The Fraying Wires Between Americans and Our Energy Future.* New York: Bloomsbury, 2016.

Boyle, Godfrey (editor). *Renewable Energy: Power for a Sustainable Future.* Oxford: Oxford University Press / Open University, 3rd–4th editions.

Gold, Russell. *Superpower: One Man's Quest to Transform American Energy and Seize the Future.* New York: Simon & Schuster, 2019.

Green, Martin A. *Solar Cells: Operating Principles, Technology, and System Applications.* Englewood Cliffs, NJ: Prentice-Hall, 1982.

Griffith, Saul. *Electrify: An Optimist's Playbook for Our Clean Energy Future.* Cambridge, MA: MIT Press, 2021.

Johnstone, Bob. *Switching to Solar: What We Can Learn from Germany's Success in Harnessing Clean Energy.* Amherst, NY: Prometheus Books, 2011.

Lovins, Amory B., and Rocky Mountain Institute. *Reinventing Fire: Bold Business Solutions for the New Energy Era.* White River Junction, VT: Chelsea Green, 2011.

MacKay, David J. C. *Sustainable Energy — Without the Hot Air.* Cambridge: UIT Cambridge, 2008. (Also available free online.)

Madrigal, Alexis. *Powering the Dream: The History and Promise of Green Technology.* Boston: Da Capo Press, 2011.

Manwell, James F., Jon G. McGowan, and Anthony L. Rogers. *Wind Energy Explained: Theory, Design and Application.* 2nd edition. Chichester: Wiley, 2009.

Perlin, John. *Let It Shine: The 6,000-Year Story of Solar Energy.* Novato, CA: New World Library, 2013.

Righter, Robert W. *Wind Energy in America: A History.* Norman: University of Oklahoma Press, 1996.

Righter, Robert W. *Windfall: Wind Energy in America Today.* Norman: University of Oklahoma Press, 2014.

Sivaram, Varun. *Taming the Sun: Innovations to Harness Solar Energy and Power the Planet.* Cambridge, MA: MIT Press, 2018.

Smil, Vaclav. *Energy and Civilization: A History.* Cambridge, MA: MIT Press, 2017.

Smil, Vaclav. *Power Density: A Key to Understanding Energy Sources and Uses.* Cambridge, MA: MIT Press, 2015.

Stokes, Leah Cardamore. *Short Circuiting Policy: Interest Groups and the Battle Over Clean Energy and Climate Policy in the American States.* New York: Oxford University Press, 2020.

Major Reports and Data Series

Ember. *Global Electricity Review 2025.* Ember, 2025. ember-energy.org.

Global Wind Energy Council (GWEC). *Global Wind Report 2025.* Brussels: GWEC, 2025. gwec.net.

International Energy Agency (IEA). *Renewables 2024.* Paris: IEA, 2024. iea.org.

International Energy Agency (IEA). *World Energy Outlook 2024.* Paris: IEA, 2024.

International Energy Agency (IEA). *Solar PV Global Supply Chains.* Paris: IEA, 2022.

International Energy Agency (IEA). *Global Hydrogen Review 2024.* Paris: IEA, 2024.

International Renewable Energy Agency (IRENA). *Renewable Power Generation Costs in 2024.* Abu Dhabi: IRENA, July 2025. irena.org.

International Renewable Energy Agency (IRENA). *Renewable Capacity Statistics 2025.* Abu Dhabi: IRENA, 2025.

International Renewable Energy Agency (IRENA). *Critical Materials for the Energy Transition: Rare Earth Elements.* Abu Dhabi: IRENA, 2022.

Lawrence Berkeley National Laboratory (LBNL). *Land-Based Wind Market Report* and *Utility-Scale Solar* editions. Berkeley, CA: LBNL, annual. emp.lbl.gov.

National Renewable Energy Laboratory (NREL). *Best Research-Cell Efficiency Chart.* Golden, CO: NREL, updated continuously. nrel.gov.

Our World in Data. *Energy and Electricity datasets.* University of Oxford / Global Change Data Lab, ongoing. ourworldindata.org.

US Energy Information Administration (EIA). *Electric Power Monthly* and *Annual Energy Outlook*. Washington, DC: US Department of Energy, ongoing. eia.gov.

Reference and Background Sources

American Physical Society (APS). "This Month in Physics History: April 25, 1954, Bell Labs Demonstrates the First Practical Silicon Solar Cell." *APS News*.

Center for Strategic and International Studies (CSIS) and China Photovoltaic Industry Association (CPIA). Analyses of China's solar manufacturing supply-chain shares, 2024–2025.

PVEducation.org. Reference articles on the photovoltaic effect, cell physics, and device history.

Wikipedia. Reference articles cross-checked against primary sources, including "Betz's law," "Charles F. Brush," "Poul la Cour," "Frank Shuman," "Augustin Mouchot," "Smith–Putnam wind turbine," "Renewable Energy Sources Act (Germany)," and "Swanson's law."

A Note on the Figures

The fast-moving statistics in this book, installed capacity, annual additions, levelized costs, record auction prices, manufacturing shares, and battery deployment, were drawn from the latest available releases of the IEA, IRENA, GWEC, Ember, and NREL as of 2024–2025, and are dated inline throughout. Because these numbers move every quarter and because sources sometimes count differently (grid-connected versus shipped capacity, calendar versus fiscal year), readers seeking current figures should consult the newest editions of the reports listed above; Appendix C explains how to find and compare them.

www.ingramcontent.com/pod-product-compliance
Lightning Source LLC
LaVergne TN
LVHW100132061126
851857LV00010B/54
* 9 7 9 8 2 3 5 4 1 3 5 3 5 *